MAKING AMMO

A Beginner's Guide to Handloading

2ND AMENDMENT MEDIA®

Whitman Publishing, LLC
PUBLISHING SINCE 1934
www.whitman.com

© 2015 Whitman Publishing, LLC
3101 Clairmont Road • Suite G • Atlanta, GA 30329

ISBN: 0794842577 EAN: 9780794842574
Printed in the United States of America.

FOREWORD

Small arms technology took a mighty leap when metal-cased ammunition replaced paper "cartridges" more than 150 years ago. Between the late 1850s and the early 1870s when the first patents expired for metal-cased ammunition, firearms innovators and inventors raced to create the first central-fired firearms and the ammo to feed them. A decade later, more advancements took place, and by 1886, the French issued the 8mm Label firing the first military-issued smokeless powder cartridge.

Viewed from many perspectives, today's firearms and modern ammunitions have not changed much in 125 years. The basics of a brass case, primer, powder, and bullet are still essentially the same. The various components have, of course, been improved over time, but the essentials are still the same.

In today's nomenclature, "reload" bears little resemblance to a battlefield command given to troops who needed to be reminded to make their weapons ready for a follow-up volley. Now, the act of reloading encompasses taking a spent cartridge case and reforming it so that it will once again fit its parent firearm, replacing a spent primer with a new one, charging the case with an appropriate amount and type of gun powder, and finally seating a new projectile.

The reloading, or handloading, process is pretty straightforward if all you want to do is hurl a new chunk of copper and lead down range. If you have performance standards, then a myriad of nuances can quickly complicate this simple process. Safety is another issue that can't be ignored if you value your eyesight and digits. Add a few layers of funky politics and whacky socioeconomics, and you might find yourself on the wrong end of an ammunition shortage. That last factor is one of the biggest reasons that handloading has seen a sharp increase over the past few months. In those few months, the top-selling firearms books have been handloading manuals that offer thousands of ammunition recipes of powder, bullets, and primers. What's been left behind in the ammunition shortage/handloading surge are the basics of how to put these ammunition recipes together. The book in your hands offers an easy-to-understand format that's heavy on the "how-to" of the process.

Going beyond the basics of how to load center-fire rifle and handgun ammunition, *Making Ammo: A Beginner's Guide to Handloading* will delve into many of the processes that will give you the ability to craft ammunition that will perform with accuracy and reliability. It doesn't matter if you simply like to punch paper or climb lofty peaks in pursuit of fleet-footed critters, reliable and accurate ammunition matters. If you stuck around long enough to read the last sentence, we'll let you in on a tightly-held secret: If you learn to handload, you can shoot more for less money, and you won't stand in line at the sporting goods counter at Monster Mart and wonder if the guy in front of you will get the last box of ammo again.

Lock and Load...
The Editors

>> Contents

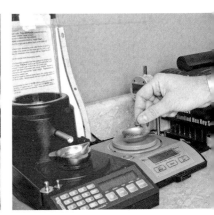

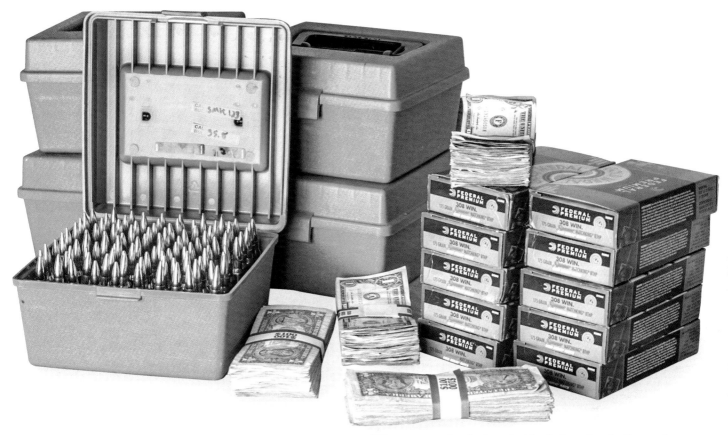

When comparing factory fodder to handloads, the cost is about a third when you do it yourself.

To Reduce Costs

The answer to the question "Why should I start reloading?" is the same for many shooters: "Money." Ammunition has been more expensive and less available over the last year than at any time in modern history. As I write this paragraph, ammunition is just starting to return to dealer's shelves and websites in small quantities here and there following the panic caused by the latest gun-control push from the political left and their counterparts across the political spectrum. Any ammo we do find available for purchase is often grossly overpriced (compared to what we consider the "normal" pricing of typical market conditions), or disappears from those shelves and websites with dizzying speed. Gun owners have emptied the industry of product from raw material to retail shelf, and even through a highest-bidder resale market. The trade in rifles, pistols, magazines, ammunition, and reloading components has resembled shortage-driven commodity markets in communist countries at times. Lines of people stretching several city blocks seeking to gain entry into gun stores, gun shows, and sporting goods shops rumored to have received "a shipment" can be likened to similar lines of people waiting in hopes of having the opportunity to buy toilet paper, blue jeans, or shoes in the cold-war era Soviet Union. Scarcity of supply coupled with booming demand equals bare shelves and higher prices, one way or another.

While you may have noticed reloading components on the list of shortage items above, the justification of reducing costs is still valid, even in times like these. Consider the following example based on today's component and ammunition prices at a well-known online retailer (shipping and hazardous materials charges not included). We'll add together the prices of the components needed to load our own ammo: cartridge case, primer, powder, and bullet. We'll then compare that total to factory ammunition and show how we save more money by reusing our cases for subsequent loadings.

.308 Winchester Cartridge Case	$0.38 each
175gr Sierra Match King Bullet	$0.33 each
Federal 210m Primer	$0.04 each
41.7gr IMR4064 Powder	$0.14 each
Total to load first time:	$0.89 per round
Total to load second time and beyond (reusing the same cartridge case each time):	$0.51 per round
Per-round average cost, assuming five loads before brass failure or loss:	$0.59
Per-round cost of Federal Gold Medal Match factory-loaded .308 ammunition:	$1.56
Savings from reloading:	$0.97 per round

Think about that: Every time you pull the trigger, you spend almost a dollar less than you would have spent if you were shooting factory ammo. In other words, you can shoot 2,644 rounds of reloaded ammunition for the same price as 1,000 rounds of factory ammunition. It is easy to see the advantage the reloader has over the non-reloader!

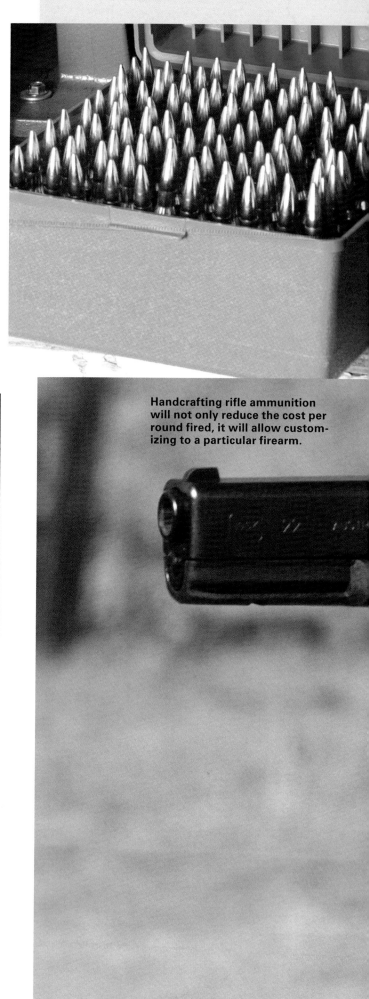

Handcrafting rifle ammunition will not only reduce the cost per round fired, it will allow customizing to a particular firearm.

Let's look at an example of the cost reduction a reloader can achieve when loading the 9mm Luger handgun cartridge. Consider the following example based on today's component and ammunition prices at the same well-known online retailer we used for the aforementioned .308 Winchester comparison (shipping and hazardous materials charges are again not included):

9mm Luger Cartridge Case	$0.20 each
115gr Full-Metal Jacket (FMJ) Bullet	$0.11 each
Small Pistol Primer	$0.03 each
5.7gr Winchester Super Field Powder	$0.02 each
Total to load first time:	$0.36 per round

Total to load second time and beyond (reusing the same cartridge case each time):	$0.16 per round
Per-round average cost, assuming five loads before brass failure or loss:	$0.20
Per-round cost of Winchester 115gr FMJ 9mm Luger factory-loaded ammunition:	$0.31
Savings from reloading:	$0.11 per round

I'm sure you've noticed that in the 9mm Luger comparison above, the cost of our first round of reloaded ammunition is actually $0.05 per round higher than our factory-loaded ammo. That's an indication of where we are in the panic, boom, bust

cycle that I mentioned earlier in the chapter. Ammunition supplies go down and prices go up during a panic before reloading component supplies and prices follow. Once the worst of the panic has passed, ammunition supplies return and prices fall a bit sooner than component supplies and prices do. I can't put hard scientific numbers on paper to quantify that effect. I speak only from my own experience and only anecdotally. This assessment is solely based on my study and observation of a number of these panic-induced cycles as a shooter (consumer) and as an industry insider over the last many years. In "normal" times, if we ever see them again, reloaded ammo is almost always slightly cheaper than factory ammo the first time we load it and significantly cheaper on subsequent loadings.

That being said, the reloader can still save money compared to buying factory ammunition when component prices are high. Ammo prices and component prices will generally rise and fall together, with ammo as the leader i.e., ammo prices are first to climb and first to fall. The cost advantage of reloading our brass is undeniable.

Setting up a reloading shop to load several different cartridges is a big investment but worth it.

To Improve Accuracy And Precision

One of the biggest benefits of reloading is being able to tailor your ammunition to a specific rifle or pistol. This allows us to improve accuracy, precision, and even reliability in our guns. Modern ammunition is better than it has ever been and, in some cases, is an incredible display of performance from a mass-produced product. Despite this fact, manufacturers can never know the specific characteristics of your personal firearms.

Producing accurate loads for most cartridges, like the .458 SOCOM, requires consistent case length.

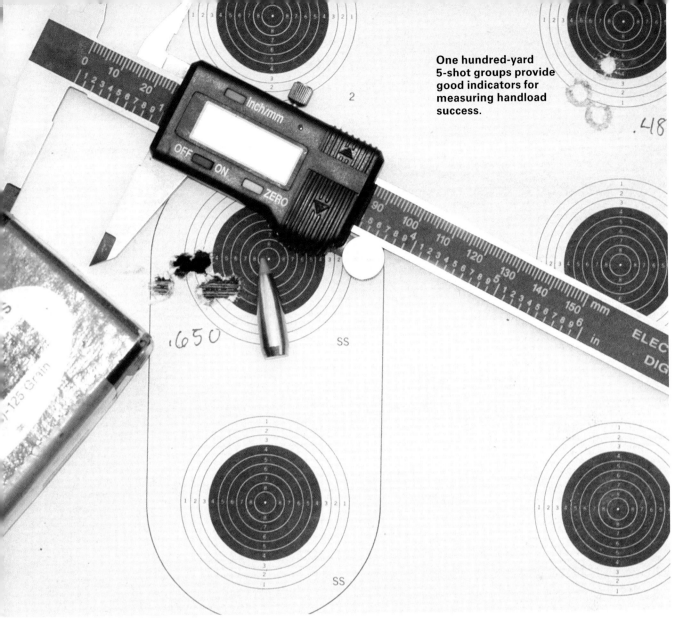

One hundred-yard 5-shot groups provide good indicators for measuring handload success.

Mass-produced ammo must be made to an industry standard that may or may not produce acceptable results in your unique firearm. The reloader is able to manage the size of his cartridge cases relative to his rifle's chamber, the positioning of the bullet relative to the rifling in the barrel, the amount and type of gun powder in each case, the type and intensity of primer that ignites the powder charge, and the amount of neck tension and crimp (if any) applied to each cartridge. The shooter can control these variables to improve the performance of his individual firearm.

To Create Ammunition For Use In Rare, Obsolete, Or "Wildcat" Firearms

As new cartridges are created, others disappear. The .296 Whiz-Bang might be the hottest topic in every gun magazine this year, but it could fade into near obscurity by the time our kids are building rifles and loading ammunition. In some cases, it doesn't take long (do some research on the WSSM cartridges). Another scenario repeated far too often in the gun industry happens when the factory "parent" of a new cartridge fails miserably at marketing their new baby. It is just as likely that the whole thing was a bad idea to start with. How many medium-sized .30 caliber cartridges do we really need? It is important to note that "obscurity" and "obsolescence" are not the same and can be relative terms.

One case of this from our not-too-distant past is the .260 Remington (.260). Although the .260 neared obscurity not many years after its introduction, it is not obsolete. The cartridge has neither performance problems nor problems with its physical properties. In this case, at least, seri-

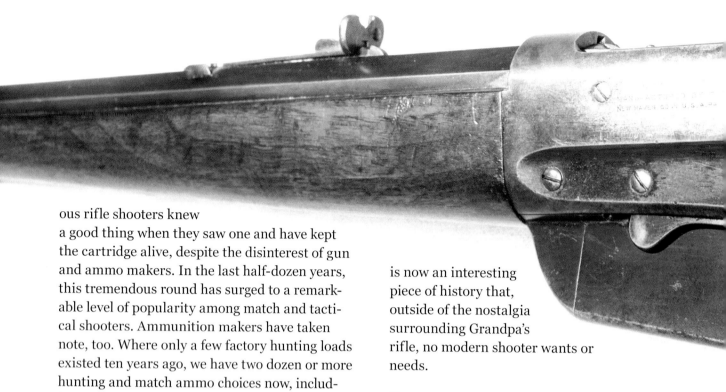

ous rifle shooters knew a good thing when they saw one and have kept the cartridge alive, despite the disinterest of gun and ammo makers. In the last half-dozen years, this tremendous round has surged to a remarkable level of popularity among match and tactical shooters. Ammunition makers have taken note, too. Where only a few factory hunting loads existed ten years ago, we have two dozen or more hunting and match ammo choices now, including Federal Gold Medal Match which is widely regarded as the finest and most accurate mass-produced factory ammunition ever made. Since it has the external ballistics of a .300 Win Mag and the recoil of a .243 Winchester, we can count ourselves lucky with the revival of the .260 Remington. In the interest of full disclosure, the .260 is one of my two favorite cartridges. I'm not blind to the attributes and liabilities of any cartridge; however, I found almost all of what I personally want in the .260.

A contrary example can be found in the 30-40 Krag. According to *Nosler Reloading Guide 7*, in the early 1890s, the Krag was a state-of-the-art military cartridge created in the infancy of smokeless powder. The Krag soon faded into true obsolescence and obscurity, killed by advancing technology in weapons and ammunition performance. The Krag's eventual successor, the 30-06 Springfield, is considered obsolete *as a military cartridge* but is still one of the most popular centerfire rifle cartridges in America after 100+ years due to wide commercial acceptance and industry support. A quick search of manufacturers' catalogues tells me that nearly every bolt-action hunting rifle made is offered in 30-06. The Krag is now an interesting piece of history that, outside of the nostalgia surrounding Grandpa's rifle, no modern shooter wants or needs.

For Projectile Selection

I've already bragged on the fine quality of today's factory ammo, but it's simply impossible for every manufacturer to load every available bullet into ammunition for every available cartridge. If I'm a sophisticated rifleman who appreciates the ballistic performance of the 7mm Winchester Short Magnum, for example, it's difficult to find factory ammunition loaded with a 145gr Barnes LRX bullet. While this is a fantastic combination capable of effectively taking any big-game animal in the lower 48, it just isn't popular enough to justify the expense of running a full lot of factory ammo, which is often counted in the tens of thousands of rounds or more. This is a simple problem for the handloader to solve, however. We jump online, order our chosen combination of components, wait for the big brown truck to arrive, then load and shoot. We can create any combination we want (within some limits of safety if we intend to keep our fingers and eyeballs).

Take a look at the 6.5x47 Lapua (6.5). It's a great cartridge, very similar to the .260 Remington and 6.5 Creedmoor (Creedmoor). According to the Lapua manufacturer's website, the 6.5 was created as a competition cartridge, unlike the .260.

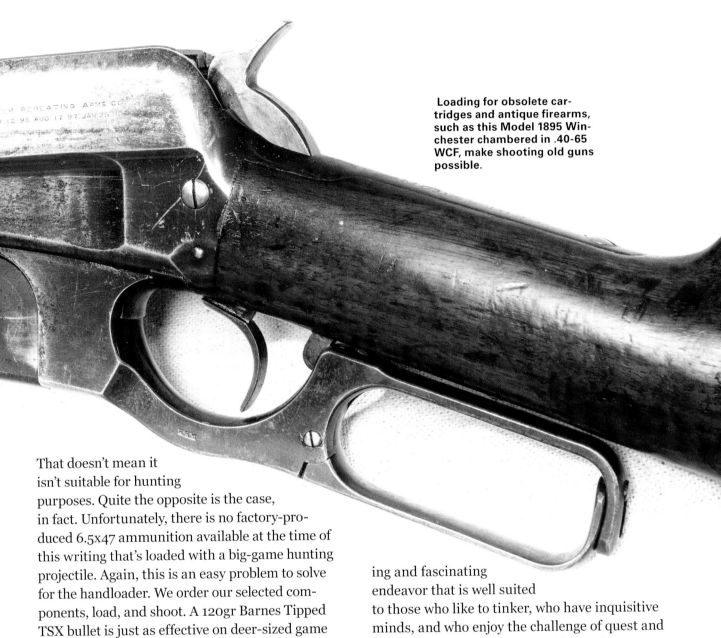

Loading for obsolete cartridges and antique firearms, such as this Model 1895 Winchester chambered in .40-65 WCF, make shooting old guns possible.

That doesn't mean it isn't suitable for hunting purposes. Quite the opposite is the case, in fact. Unfortunately, there is no factory-produced 6.5x47 ammunition available at the time of this writing that's loaded with a big-game hunting projectile. Again, this is an easy problem to solve for the handloader. We order our selected components, load, and shoot. A 120gr Barnes Tipped TSX bullet is just as effective on deer-sized game with a 6.5x47 Lapua as it is with a .260 Remington, 6.5 Creedmoor, .264 Winchester Magnum, or 6.5 Remington Magnum (Speaking of obscure cartridges...). Handloading gives us vastly more flexibility than if we relied solely on factory-produced ammunition.

For Hobby/Recreation

Finally, we choose to handload for recreational purposes or as another hobby related to our interest in the shooting sports. Just as often, I believe, our recreational interest in handloading develops over time without any real intent to create a new hobby. Handloading is a challeng-ing and fascinating endeavor that is well suited to those who like to tinker, who have inquisitive minds, and who enjoy the challenge of quest and mastery. Remember when I said that the new handloader's intent to "reduce costs" is often misstated as one to "save money?" If you're like me and love to learn, be challenged, overcome obstacles, and celebrate breakthroughs, I can absolutely guarantee you that you will not "Save Money" by handloading! You will spend more money than ever, but you will also get more enjoyment from your sport than ever before. Still, if we look at the cost of the components per round of ammunition loaded, we realize significant cost reductions by loading our own ammunition. Lucky for us, that means we can shoot even more!

The Author doing what he loves: engaging targets in the field. There are no bench rests out here.

Maintain Good Working Order

Safely loading and shooting ammunition requires the use of properly maintained and working firearms in serviceable condition with secure fasteners, proper sights, and parts made and installed according to the manufacturer's specifications. I can't tell you the number of times I've seen shooters experience problems with their guns that are directly, solely, and completely related to improper assembly, the installation of ill-designed aftermarket parts, loose fasteners, or lack of basic maintenance. In fact, despite my best efforts, decades of experience, and a career and lifestyle based on firearms and their advanced application, it happened to me less than one week ago!

I noticed that the accuracy of one of my favorite custom bolt-action rifles wasn't as good as it had been in the past. The rifle hasn't had enough rounds fired through it yet for a worn barrel to be the issue. The bore had been properly cleaned only a couple dozen rounds prior to my last range visit so fouling wasn't a concern based on the rifle's history. Perplexed, I unloaded the rifle and began to inspect components and fasteners from one end to the other. I actually started with the ammunition thinking I'd erred in a measurement or adjustment of my loading dies. All seemed well with the remaining ammo I had with me. The action fit well in the stock with no detectable movement and the bolt locked up properly and snugly on a loaded cartridge. (A small amount of movement between the bolt and receiver may be present when the bolt is closed on an empty chamber, even in a fully custom and precision trued bolt-action rifle.) No cracks were visible in the stock, the barrel was floating freely without contacting the stock at any point, the trigger broke cleanly and at approximately the correct weight of pull, magazines inserted and fell away freely with no binding, the safety operated on and off

Be sure that your firearm is in good working order before the first round goes in the chamber.

smoothly and without hesitation, and the scope ring caps were torqued exactly to the manufacturer's recommended specification. I grabbed the scope and briskly pushed and pulled side-to-side and top-to-bottom, checking for looseness of the base-to-receiver fit. Nothing was out of place. Everything seemed just fine.

The next suspect on the list was the rifle's scope. Perhaps an internal mechanism within the scope had failed? That's when I noticed it! Though the scope seemed tight in its mounts and didn't budge an inch when vigorously pulled and prodded during my inspection, I could see the faintest marks on the tube of the scope where the rings had slipped under recoil. Inertia on display! Though the rings were tightened to spec, the scope was able to slip very slightly within. Perhaps the inside diameter of the rings is a bit larger than intended or the scope tube is a bit smaller, maybe both. Who knows if the slippage occurred every time the rifle fired, every other time, every fifth shot, etc. The point is even though I was shooting an expensive custom rifle made with personally selected top-shelf components by one of the most respected gunsmiths in the country and I'd meticulously installed a proven high-quality optic according to the manufacturer's recommendation, a few screws that were slightly under torqued caused serious problems with the accuracy of my rifle system. Once I moved the scope back the 1/4 inch or so that slipped and re-torqued the scope ring caps, this time to 5 inch-pounds MORE than the manufacturer's recommended setting, the rifle once again shot into tiny little groups that would make any rifleman proud.

Take my word for it; the little details we miss can cause big problems. If you don't have the knowledge and experience to thoroughly evaluate a firearm for safety, proper assembly, and function, get some help before proceeding. If you encounter a problem or even *suspect* there is a problem, stop and get help from someone with the experience and training to diagnose and resolve the issue.

If you're working with a rifle you acquired second hand, or maybe third, fourth, or fifth hand,

The first rule in firearms safety when handling a weapon is to visually check the chamber to be sure it is unloaded.

it is doubly important that you properly inspect your firearm before firing. It is not uncommon to see older rifles re-bored, rebarreled, "wildcatted" (modified to fire a non-standard cartridge), or otherwise altered or in some cases, downright bastardized by someone's crazy uncle Larry in his garage "gunsmithing" shop. I've seen examples of this in my life that have so disturbed me that I can barely stand to think of them. One in particular involved an early 1960s-era Belgian-made Browning shotgun with perfect deep bluing and the most incredible blonde wood you can imagine. The gun was alongside a claw hammer and a flathead screwdriver in the rear cargo rack of an ATV parked outside a strip mall. I tried to rescue the beauty, but Bubba wasn't interested. He was determined to turn a stunningly beautiful and classic firearm, a manufactured work of art from an era of craftsmanship long gone, into a camo-clad plastic-stocked turkey gun for his poor, doomed seven-year-old son instead of buying him the brand new turkey gun of his choice with the money I offered! Lord, have mercy...

Suffice it to say that a few dollars spent to ensure the safety and proper function of a used gun will not be wasted. You just don't know what indignity the gun suffered along the way. At a minimum, make sure the bore is free of obstructions, the safety mechanism functions properly, all fasteners are secure, and you have the correct ammunition for the gun's current chambering.

Something as simple as proper scope mounting can play havoc with your efforts to shoot small groups. Sweating the details makes a difference in the reliability of your ammo as well as your weapon's performance.

Shoot Factory Ammo For A Baseline

Isn't this a book on reloading? Why am I suggesting that you go spend (more) money on factory ammo when the whole point is to get away from expensive factory ammo? That is a perfectly valid question. The answer is to establish a baseline of performance. If your new carry pistol malfunctions two or three times out of ten with high-quality factory ammunition, even with multiple types, you will know that your re-loaded ammunition isn't to blame if the problem continues.

Conversely, if you've fired 3,000 rounds of factory ammunition through your favorite AR15 without ever experiencing a single malfunction, you can be reasonably certain that the rifle will work properly when you assemble reloaded ammo to the correct dimensions and powder charge. If your trusty long gun begins malfunctioning every other time you pull the trigger on

your reloaded ammo, the rifle may not be the problem.

If Dad's 30-06 deer rifle (AKA "Ol' Buck Slayer") groups factory ammo into 1/2 inch five-shot groups at 100 yards, you know what level of accuracy you should be able to replicate with handloads. You should also email me immediately so I can write you a check for the purchase of Ol' Buck Slayer. If your handloads group into two inches at the same distance, you know that with continued work and experimentation you can improve the accuracy of your handloads because you've already established your baseline of performance as higher. Similarly, if "Ol' Buck

Slayer" has yet to print a group under four inches at 100 yards with factory ammo, you can feel pretty darn good when your handloads group into two inches!

A couple of notes about the process of establishing a baseline with factory ammo... Always shoot more than one group with each type of ammunition. One really tight group means absolutely nothing. Verify that you can repeat that level of accuracy. Fire on targets made from 8.5 x 11 inch sheets of paper. There are a number of sites online where you can download and print targets in hundreds of varieties. I refined my preferences over the years and posted a few of my favorite sample targets,

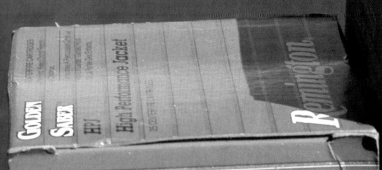

Hornady AMMUNITION

216 CARTRIDGES

CUSTOM™
Hornady®
45 AUTO
#9112   200 gr XTP®

Muzzle   50 yds 100 yds
900 fps   856   817

GOLDEN SABER   HPJ   High Performance Jacket

Remington

Remington. GOLDEN SABER™

.45 Auto   MADE IN THE U.S.A.
230 GRAIN
BRASS JACKETED HOLLOW PT.   GS45APB

Remington. GOLDEN SABER™

.45 ACP (+P)   MADE IN THE U.S.A.
185 GRAIN
BRASS JACKETED HOLLOW PT.   GS45APC

Remington®

EXPRESS® **PISTOL & REVOLVER**

45 AUTOMATIC
230 GR. MC R45AP4

Black Hills Ammunition
.45 ACP
230 Gr Jacketed Hollow Point

Remington®

EXPRESS® **PISTOL & REVOLVER**

45 AUTOMATIC
185 GR. JHP R45AP2

AMERICAN EAGLE®
.45 AUTOMATIC PISTOL
230 GRAIN FULL METAL JACKET
AE45A

Factory ammunition can serve to provide the handloader with a base line
for performance. Once you know
what factory loads do on paper, you
can work to match or improve
performance.

along with other great resources for the shooter and reloader, on the Resources page of my website: (*www.gamescoutusa.com*). Keep these target sheets! Make copious notes! Record more info than you think you will possibly ever need directly on these targets. Buy an inexpensive three-ring-binder and a pack of clear plastic sheet protectors. Place your annotated targets inside these sheet protectors and keep them for future reference. You will learn much more about the data requirements and solutions that will make you a more effective loader and shooter later in the book.

Take Measure

If you own a chronograph, using it while firing factory ammo lets you know what velocities are reasonable from your pistol or rifle with a given bullet weight. Since we usually don't know what type of gunpowder the ammo maker used, we can't expect to exactly duplicate the velocity of factory ammo in every case. We may end up finding a good reliable and accurate load at a lower velocity or even slightly higher.

What we don't want is for the velocities of our handloads to be significantly higher than factory ammo. In the 9mm Luger, for example, most factory ammo firing a 115 grain jacketed bullet will average around 1,150 feet per second (fps). We can find "+P" (plus pressure) loads that deliver velocities in the 1,250 fps range. If our handloads fly through the chronograph at between 1,100 and 1,200 fps, we have reached an acceptable level of velocity and, generally speaking, pressure. If our bullets sing through the chronograph at 1,300-1,400 fps, we have a problem! We know we are FAR outside the range of what is considered normal and should stop shooting immediately. Disassemble any remaining ammunition loaded similarly and recycle into loads at a lower powder charge. Firing high-quality factory ammo helps us know what to strive for and expect from our reloaded ammunition.

Chronograph Basics

The primary source of accuracy is uniformity. Uniformity includes a well built gun, a shooter's hold, etc., as well as ammunition where each round fired is just like the previous.

If your ammo puts all shots into one hole, then it's pretty obvious that your load is uniform. If you are dissatisfied with group size, the best place to start is to make sure your load's velocity is uniform. With a chronograph, you can measure velocity and uniformity, and that the average velocity meets objectives safely.

There's no guarantee that ammo with uniform velocities will always shoot the smallest groups, but it's a good place to start. If you know the uniformity and the average velocity of each tested load, you can find the best load in less time.

Standard deviation is a widely used measurement of variability or diversity used in statistics and probability theory. It shows how much variation or "dispersion" there is from the "average" (mean, or expected/budgeted value). A low standard deviation indicates that the data points tend to be very close to the mean, whereas high standard deviation indicates that the data are spread out over a large range of values.

Several variables affect chronograph performance and velocity tabulations. Temperature is the biggest factor, be it ambient temperature or barrel temp. Barrel temperature influences velocities a couple of ways. First, the metal's expansion can change velocities. Secondly, a hot chamber can heat up ammo, raising its working pressure and velocity, too. I've seen the evidence of this several times when running rapid strings during testing.

The next variable is light and its direction in relation to the chronograph. That's why diffusers help even out the light that reaches the sensors. Muzzle blast impacts a chronograph, as it commonly "sees" the blast wave even if a bullet does not pass through the instrument. The spacing between sensors, as we'll later discuss, can have a detrimental impact on results. Even the power source, be it batteries or AC current, can change the readings, too.

Setup Tips

It is important that the chronograph be set up at a consistent distance from the muzzle to gather statistically comparable data. Ten feet is a common distance and can be determined with a tape measure. Some chronographs use a rail for mounting the screens. Not placing the screens in the exact position every time can have a tremendous impact on results. For instance, a rail-specified 24 inch spacing between screens can produce results that are in error 4.17% if the screens are accidentally spaced 23 inches apart. Readings would be higher than actual. Spacing the screens too far apart will have equally detrimental results. In these scenarios, faulty data can lead to dangerous overloads.

Be sure to use a steady mounting surface, such as a table or tripod. In windy conditions, extra weight should be added to the tripod to prevent it from blowing over and possibly damaging the unit.

Shooters follow different philosophies when shooting through their chronographs. Some hold that targets should not be shot at the same time, owing to the risk of accidentally putting a bullet through the device and destroying it. Many believe that the weapon must be held firmly and the same way each time to get viable velocity results.

I typically shoot targets when chronographing for rifles, but don't shoot targets when testing handguns. Though I have never accidentally hit a chronograph with a bullet, I have had a muzzleloader sabot strike one of the side panels of an Oehler 35P's Sky-screen with disastrous results. The energy of the sabot, which had separated from the bullet, was sufficient to shatter the Sky-Screen and the sensor housing.

One solution that I have seen is to horizontally hang a strip of masking tape two-thirds up from the lens between the legs supporting the diffuser and marking a black dot in the middle. Firing a rifle with the scope an inch above bore will put each shot below the marked tape.

Calibrating a measuring device lets the user know that the end results are derived from a constant basis. When handloading, it's a common practice to calibrate a scale so you have confidence that the measured weight is the same weight each time it's used. Since chronographs don't have a calibration feature, a good technique is to shoot a string of 10 shots from the same .22 rimfire rifle shooting ammo from the same lot. Record the average velocity and the ambient temperature in a log book and keep it with the chronograph. That way, you will have a "constant" to compare and help you evaluate when examining the final results of a day's shooting.

Learning the velocity by chronographing factory loads with a given bullet weight tells you approximate velocities to achieve with handloads.

Powder is most safely stored in a wooden container near the floor.

Most of us have a general idea about what it takes to be safe on the range with our firearms. Keep your finger off the trigger until you're ready to shoot, never let your muzzle point at anything you're not willing to destroy, be aware of your target and what's around and beyond it, and always treat every gun as if it were loaded. We will also need to be extremely vigilant and conscientious of safety practices while reloading.

Safety on the bench covers several areas: powder handling and storage, primer handling and storage, chemicals and compounds, and sparks and fire.

Powder Handling

Smokeless gunpowder (powder) is the fuel that propels all modern small-arms cartridge ammunition. Today, reloaders have hundreds of powder options to choose from, ranging from rela-

tively fast-burning pistol powders to relatively slow-burning magnum rifle powders. Take note of the word "relatively" in the preceding sentences. None of these powders burn slowly. We're talking fractions of a second here. Still, powders do have a range of burning rates. Those relative burn rates are one factor that we will use (or that the publishers of our reloading manuals have already used) to select powders for our particular application. A quick Internet search for "powder burn rate chart" will turn up a number of options. Each powder manufacturer offers a burn rate chart as well. From a safety standpoint, it is important to recognize that these powders are EXTREMELY dangerous when exposed to heat, flame, or spark. You must remain vigilant and establish routines and procedures for handling gunpowder. I will review some of these in the next few paragraphs. You must maintain cleanliness on your bench and treat powder spills as

a serious safety hazard. A few kernels of powder may not cause the house to burn down today, but the accumulation of small spills here and there over years of reloading can create a dangerous situation.

Avoid any type of spark-producing situation around powder and primers. This is a big "no, no."

Powder should always be kept in its original container. Original labels must remain intact. If you transfer powder into powder measures or charging devices, you must label those devices too. When you accidentally mix different types of powder in a container or measure, THROW IT AWAY! The price of a jug of powder is far smaller than the value of your firearm, your eyes, your fingers, or the safety and well-being of friends, family members, or others at the firing range. While some mix-ups may not result in the immediate destruction of your gun or injury to yourself or another person, other

powder mix-ups are deadly. Fill a rifle case with a fast-burning pistol powder and you can essentially have a grenade in your hands. Pull the trigger on that cartridge and your rifle can literally burst into dozens of shards of jagged metal that can travel many feet at high speed. Anyone nearby has a chance of injury or death. As much as it hurts your wallet and your pride (Ask my father in-law about the H4895-tainted 8 lb. jug of 2000-MR he just dumped out.), just get over it and move on.

Let's talk about how to avoid that situation in the first place. My personal powder handling routine consists of about a dozen steps that are designed to nearly eliminate the possibility of dangerous mistakes. Follow this process step-by-step, and you will reduce your chances of a mix-up or mishap.

1) Ensure there are no obvious dangers nearby, such as cigarette butts smouldering in an ashtray across the shop or maybe some burning scented candles that your wife uses to cover up the

It's easy to get distracted and return powder to the wrong container. **THROW IT AWAY** to avoid a catastrophe by inadvertently using the blended powder later and potentially wrecking a gun ... or yourself.

manly smells of your reloading room. Fall Spice, Apple Dumpling, Clean Laundry – you know the kind I'm talking about. Get rid of all of that. Also be aware of any electrical hazards that may result in an arc or a spark.

2) Place your 3 x 5-inch loading card for the load you are about to assemble on the bench next to your loading tray or powder measure.

3) Ensure that the bench is free of powder and powder containers used during previous reloading sessions. If any powder remains in measures, charging devices or other reservoirs, clear the labels on each device of the powder name, then return that powder to its proper container before doing anything else.

4) Remove all powder containers from the bench and empty all measures and reservoirs, then select the powder you intend to use for the next reloading session according to your 3 x 5 inch loading card.

5) Label the empty powder measures and/or charging devices with the name of the new powder.

6) Pour required amount of powder into measure and into any related tools like a powder trickler, for example. Place the lid or cap on the powder measure to reduce the likelihood of spills.

7) Replace and tighten the lid on the original powder container, even if empty. Always do this for consistency of process.

8) Place the original container of the powder you are currently using on the bench next to your powder measure. This will be the only container on the bench while you are loading.

You are now ready to load.

9) After completing your loading session, remove powder in the reverse order of the above process, making certain that you remove the labels from any powder measures or charging devices before placing the original powder container back in storage.

10) Clean the bench and all surrounding areas of any spilled powder, primers, case lube, tumbling media, and anything else that is out of place.

One final note: powder manufacturers recommend that you do not mix containers of the same powder when those containers come from different manufacturing lots. Treat containers of the same powder that are from different manufacturing lots as though they are different powders altogether.

Powder Storage

You now know the importance of keeping powder in original containers. One other storage

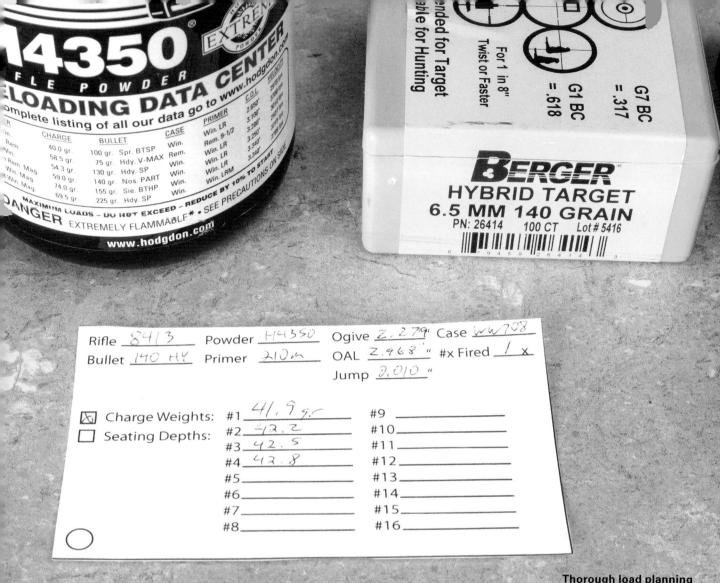

Thorough load planning is step one and should be completed before primer and powders are dispensed.

consideration that many don't think about is the outer container, box, bin, or other place we keep our original containers of powder. In addition to safety concerns, this may be important to us for legal and insurance reasons in some places. For example, a fire-department acquaintance tells me that smokeless powder in quantities over 20 pounds must be kept in a specific type of wooden cabinet according to the fire codes in some residential areas. Now, I'm not aware of any instances of the reloading police (tongue-in-cheek) doing random safety checks of a reloader's home (yet), but the aftermath of a house fire almost anywhere in the developed world will include some sort of investigation. If your personal reloading components helped to contribute to the fire because you did not store them properly according to local regulations, you may diminish or destroy your claims for insurance reimbursement. The standards suggested by the National Fire Prevention Association (NFPA) can be found at (*www.saami.org*), the website of the Sporting Arms and Ammunition Manufacturers' Institute, Inc.

In general, an almost airtight, non-metallic, and non-pressure sealed container constructed of a material that will offer some insulation (albeit temporary) from heat is ideal. Common sheetrock is another option reloaders can use to create an enclosure or barrier inside a closet to offer temporary protection. Heat rises, so keep powder and other combustibles stored as low to

the floor as possible. This will give you a bit of a time cushion before your components burn in the event of a fire. This is for your information and consideration only. Use common sense and good judgement and always err on the side of caution.

Hazardous Materials Handling

When dealing with the types of hazardous materials we use in reloading, it is imperative that you maintain a clean work area. As mentioned before, a few kernels of powder here and there over time can add up to create a real fire hazard. Drop a couple of primers, and you have an instant fire emergency lurking under your feet. Take these spills very seriously. Stop what you

A neat and clean powder charging area reduces the likelihood of a mistake and improves safety.

are doing immediately and clean up any spills. Keep a plastic drinking cup containing a bit of water or water-dampened paper towels nearby. Permanently label the cup POISON, CAUTION, or TOXIC and keep it out of reach of children. Use this as a trash bin to dampen any spilled powder, reducing the fire hazard. Empty this cup frequently and ensure the discarded residue is disposed of properly. Always keep a serviceable fire extinguisher nearby, just in case.

I encourage you to avoid reloading over a carpeted floor where powder, primers, and residue is often much harder to clean up after a spill. If the only spot you have is over carpet, get yourself

Cast lead bullets and the process to create them can lead to lead poisoning if not handled properly.

a linoleum or plastic flooring remnant from a local home center or flooring showroom, and put that down in front of your bench. Take the remnant outside for a vigorous shaking to remove small spills. Discard and replace remnants after a big spill or when it gets soiled over a period of time. Do NOT attempt to vacuum spilled powder, primers, or other debris from your reloading bench area. Just imagine a spark followed by a slowly smouldering fire INSIDE your vacuum cleaner while it sits in the hall closet.

Even spent primers may contain small amounts of potentially dangerous residue that can accumulate if a proper cleaning regimen isn't in place. At a minimum, there is a small amount of lead residue in most primers. Avoid handling and breathing the dust from these. Keep any discarded or damaged primers in your water-filled cup for several days before discarding.

Dispose of waste components and materials properly.

Make sure that you remove any loose, damaged, or spent primers from your bench at the end of each reloading session. It is easy to mix primer types. Using the wrong primer in the wrong load may result in equipment damage or personal injury.

Cartridge cases are another potential source of lead exposure. When handling these, avoid eating, dipping snuff, picking your nose, and anything else that requires your lead-covered fingers to enter your mouth or nose. When I have a big pile of spent brass to work through I'll throw on a pair of the cheap cotton gardening gloves. Latex medical gloves are even better.

Dirty tumbling media is full of lead contaminants which can end up in your lungs while you are emptying the tumbler, separating brass from media and such. Try not to breathe the dust and keep this in mind when selecting the location of your case cleaning area. This dust will get everywhere. Set up case cleaning outdoors or in a garage or shed where you will not mind the dust and media spillages that will happen. Pick

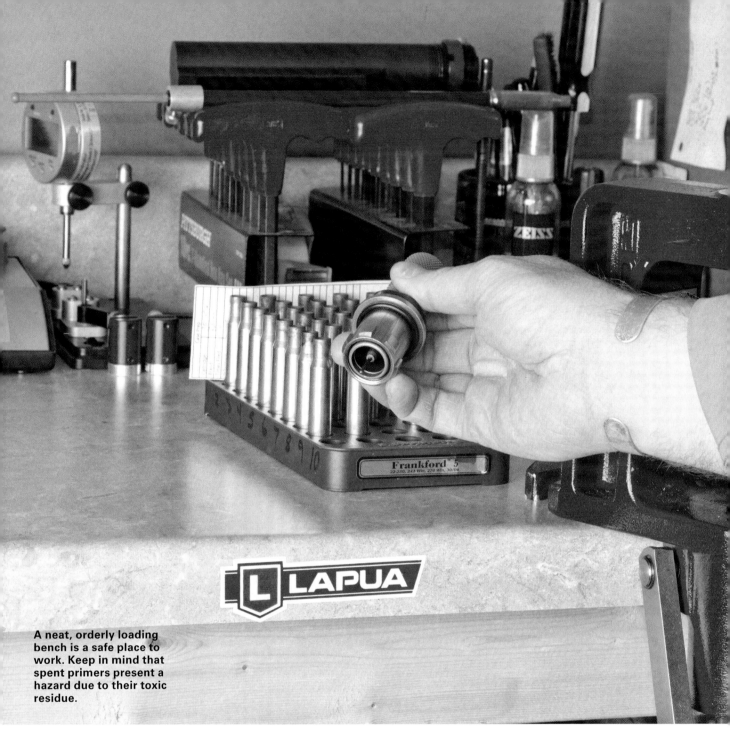

A neat, orderly loading bench is a safe place to work. Keep in mind that spent primers present a hazard due to their toxic residue.

an area where children are not exposed to these contaminated materials. At the very least, wash your hands thoroughly after handling fired cases and dirty guns. While the risk from handling lead-contaminated components is low, it is in our power to eliminate the risk almost completely. The levels of lead and other substances we are talking about may not be a significant threat to a 220-pound man, but they can affect a developing 22-pound toddler.

We will use a number of chemicals during our reloading efforts. While some are optional, others are required. A quick glance around my bench reveals aerosol case lube, acetone, gun oil, brake cleaner, and lithium grease. In the garage near my case-cleaning area, we'd find mineral spirits and liquid case cleaning compound in addition to all the normal garage chemicals we all accumulate. Handle these carefully, maintain a clean work area, and be hyper-vigilant about spark and flame hazards. We're not dealing with nuclear waste, obviously, but use your brain, pay attention, and keep the cumulative effects of these substances in mind over the long term.

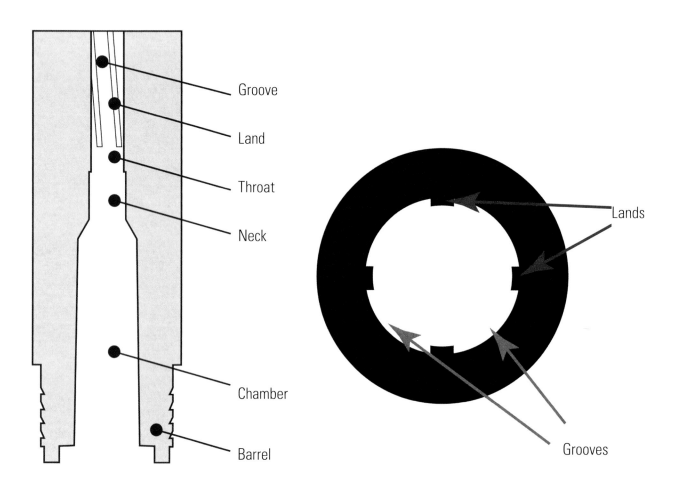

Sectioned-diagram of a rifle barrel
Nomenclature for the reader to know
includes: chamber, neck, throat, land,
and groove.

Labels: Groove, Land, Throat, Neck, Chamber, Barrel

Cross-section of a rifled gun barrel
The grooves define the bore diameter. The
lands form the actual "rifling" that imparts
spin stabilization to the bullet.

Labels: Lands, Grooves

Ammunition Nomenclature

Almost all modern, centerfire small-arms ammunition consists of four separate components: cartridge case, primer, smokeless powder, and projectile. Some ammunition loaded with non-jacketed lead bullets will have a small copper disk called a "gas check" placed between the bullet and the powder. In a rimfire cartridge (the .22 long rifle is the most obvious example), there is no separate primer. Instead, the chemical priming compound is inserted directly inside the base of the cartridge case. The case is then spun rapidly on its long axis, causing the priming compound to be distributed under the case rim by centrifugal force. Rimfire ammunition is not generally reloadable.

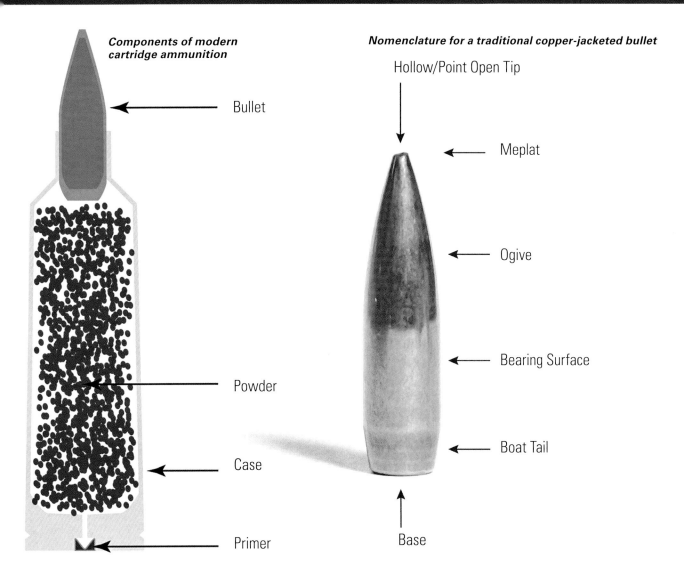

Components of modern cartridge ammunition

Bullet

Powder

Case

Primer

Nomenclature for a traditional copper-jacketed bullet

Hollow/Point Open Tip

Meplat

Ogive

Bearing Surface

Boat Tail

Base

Projectile (Bullet)

The bullet is what actually moves down the gun's barrel, flies through the air, and hits the target (or whatever it was pointed at when the gun fired). While it may seem unnecessary to explain this concept to many who are reading this, I guarantee someone somewhere is under the impression that the entire cartridge flies through the air. After all, this depiction has appeared on TV, in music videos, in news graphics, and other media for years. Once we get through this chapter, the role of each component and their interaction with each other will be clear.

Bullets have several named parts that help describe their shape, purpose, or design. The tip or point of the bullet is called the meplat. From the meplat moving rearward to the base we find the ogive, or the (usually) curved portion of the bullet that increases in diameter until full diameter is reached at the bearing surface. The bearing surface is the full-diameter portion of the bullet that engages the rifling inside the barrel. This is sometimes called the "shank" as well. Behind the bearing surface is a flat base or a boat tail. Some jacketed bullets will have a circumferential reduced-diameter band that interrupts their bearing surface near the ogive. This band is called a cannelure and allows the case mouth to be crimped into its smaller diameter to control and maintain bullet position during feeding and under recoil.

Modern Rifle And Pistol Bullets

Modern rifle and pistol bullets are constructed primarily from two general types of material. Most have a lead-alloy core with a copper or copper-alloy jacket. A couple of years ago, I was able to spend the day at the Hornady ammunition manufacturing facility in Nebraska. This was a fascinating experience where I witnessed large blocks of lead being turned into bullet cores and then finished bullets ready to load into ammunition. I saw flat copper being made into jackets for those lead-core bullets, brass cups being punched into cartridge cases, and ammunition being assembled and packaged for shipment. I cannot discuss everything I learned that day due to confidentiality agreements, but descriptions of the general process are available in articles and videos online. You can get a good idea of how the process works by watching these videos and reading the articles that turn up with a simple Internet search.

Lead core and copper jacket

Lead Core Bullets

The process of making lead-core bullets begins with huge lead ingots that are melted and formed into large shapes that are then forced through a tool that produces a thick lead "wire" which gets rolled onto a spool. This wire is cut to the required length to become a bullet core. This piece of lead is inserted into a jacket cup (more on this later), and the two are forced into the appropriate size and shape together by a series of dies. As you can imagine, the forces used to shape these materials into bullets are significant, measured in tons and tens of tons. If the manufactured bullet requires a plastic tip, the tip will be inserted before the final step.

Copper Jacket

Jackets start out as thin sheets of rolled copper that are punched into disks (think of a cookie cutter), smashed into roughly the shape of a cup, and paired with a lead core. Once the two

Cartridge loaded with a jacketed-lead style bullet
Note the exposed lead tip above the copper jacket.

Jacketed lead-core bullets fill just about every need a handloader has.

are together, the forming process continues as described above.

The result is a bullet constructed of two different materials, physically formed into one shape. Jacketed lead-core bullets have been the standard for a century and offer fantastic performance in a number of roles. The most accurate and effective match bullets made are jacketed lead-core bullets. A huge percentage of the total number of rounds of small-arms ammunition fired by militaries worldwide is a jacketed lead-core, or partially lead-core, bullet. Hunters and target shooters are satisfied with the results achieved with jacketed lead-core bullets.

Monolithic Bullets

Monolithic bullets are increasingly popular and offer a number of advantages and disadvantages. The term "monolithic" can actually describe any bullet that is made from just one material: lead, copper, brass, etc. The most common use of the term is to describe bullets made from a solid copper alloy or pure copper with no lead core. Part of the impetus behind the development of monolithic bullets is based on legal and environmental requirements in some areas. Since the ban on lead shot in wetland areas for waterfowl or upland hunting decades ago, interest in "non-toxic" projectiles has grown out of necessity. It didn't take long for the environmentalists to draw dubious connections between lead-core bullets used in big-game hunting and some negative effect to other animals that got lead-core projectiles banned in areas frequented by Condors, Big Bird, the Loch Ness Monster, and who knows what else.

Part of the drive behind these copper-based monolithic bullets is pure performance, however. I've shot my last dozen or so deer with Barnes or Nosler monolithic copper-based bullets. I am so impressed by the performance of these monos that I switched to them entirely for all of my big-game hunting. There are a couple of reasons for this, but the primary factor in this decision is reliable penetration. Lead-core bullets deform, fragment, and disintegrate as they travel through tissue at high speed. While this can increase the wounding effectiveness of the bullet, it also means the bullet weighs less and less as it sheds its lead core. Less weight (in this case, and in others we won't get into) means less penetration. It is common for lead-core bullets to retain only 40 to 60 percent of their original weight after striking and traveling through tissue. Monolithic bullets will commonly retain 95 to 100 percent of their weight after traveling through tissue. Note that the lead-core bullets I'm referring to are *designed* to shed and retain certain percentages of their original weight. That's how they work. It isn't bad; it's just different than the deep-penetrating monolithic bullets. Since the monos keep their weight

Monolithic copper bullets from Nosler (l) and Barnes (r)

throughout the penetration process, we can start out with a lighter weight bullet. Lighter bullets have the advantage of being pushed at higher velocities than heavier bullets in a given cartridge, so we can achieve some minor short to mid-range trajectory advantages with monolithic hunting-style bullets compared to lead-core hunting bullets. This benefit is generally small, if it exists at all, because of the (typically) lower ballistic coefficients of the monolithic bullets. Most of the monolithic rifle bullets have circumferential grooves around their bearing surfaces that reduce their ballistic coefficients. These grooves reduce copper fouling and pressure while the bullet travels through the gun's barrel. Each type of projectile has plusses and minuses.

Monolithic pistol bullets offer a viable alternative to traditional lead-core pistol bullets. These haven't grabbed the market share from leaded bullets like monolithic rifle bullets have for a number of reasons, but they deserve serious consideration when terminal performance is an issue. Since monolithic pistol bullets do not have a lead core that separates

Monolithic bullets provide textbook expansion in tissue.

from a copper jacket, they retain their mass just like their rifle counterparts. Expansion occurs quickly, and diameters can more than double in tissue. A 9mm (0.358") diameter bullet may expand to three-quarters (0.75") of an inch, for example. Lighter weight monolithic projectiles can penetrate like their heavier leaded cousins. Advancements in bonded bullets (more to come) have blunted some of this advantage. If you're a handgun hunter or you intend to load defensive ammo for zombie invasions, or whatever your favorite TEOTWAWKI scenario may be, look at monolithic pistol bullets when making your buying decisions. Check the bullet manufacturer's recommendations, as there may be some differences in load data or loading technique.

Bonded Bullets

Bonded bullets are similar in construction to traditional lead-core jacketed bullets, but their copper jackets and lead cores are "bonded" together chemically, electro-chemically, or thermally to significantly reduce the possibility that a core-jacket separation will occur. The desired result is increased penetration of tissue, even if the bullet must first pass through other substances. Much of the drive to improve bonding technology derives from the needs of the law enforcement community. Faced with heavy clothing, auto glass, sheet metal, wallboard and other barriers, police bullets are often defeated or made much less effective before ever reaching their intended target. When a traditional lead-core bullet engages such a barrier, it is common for the jacket to be completely stripped away from the lead core. With the soft lead core exposed, the remainder of the bullet is easily deformed and penetrates less effectively than desired. This can clearly be a dangerous situation for a police officer. Bonding is one way to improve the durability and penetration of the bullet.

Big-Game hunters also want maximum penetration through heavy hide, bone, and muscle, and they are leaders in the advancement of bullet bonding, especially in rifle bullets. A great example of this technology includes the Accubond product line from Nosler and the Fusion line from Federal. Bonding is just one technique used to create bullets that will stay intact and pen-

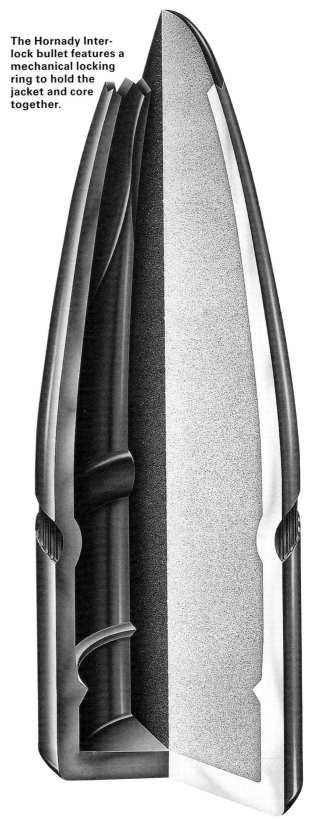

The Hornady Inter-lock bullet features a mechanical locking ring to hold the jacket and core together.

etrate through tissue and intervening barriers. Based on the huge popularity of bonded bullets, I think we will see them for decades to come. If you are a hunter or defensive shooter, plan some of your load-development with bonded bullets.

Lathe-turned projectile

Specialty Bullets

Some specialty bullets are turned on a lathe from brass, copper, or other suitable or alloyed material. These specialty bullets seek to offer features and construction that are not accomplished with traditional forged jacketed bullets. Jacketed bullets will have a small range of variation in the alignment of their cores with their jackets. These variations may only measure in the ten-thousandths of an inch. Still, any misalignment between the lead core and copper jacket of a projectile that flies through the air at multiples of the speed of sound while spinning on its long axis thousands of times per second can lead to accuracy-robbing imbalance that strips the non-concentric bullet of its efficiency and consistency. Think about spinning a top (a child's toy) where the center of gravity is not in the center of the toy (one side is heavier than the other). Very quickly the spinning toy would wobble, tilt, and veer off in an unpredictable direction as it loses gyro-scopic stability. In the long-range shooting game, consistency is king. With bullets turned on high-precision computer numerically controlled (CNC)

lathes from solid metal, no misalignment can exist between jacket and core. Since these bullets are not forced into a die under high pressure, they can be manufactured without the same constraints that manufacturers of traditional jacketed bullets face. Turned bullets have their own set of issues as well, of course. A primary constraint is the cost of these bullets. Turned bullets may cost two to ten times as much as jacketed bullets. Many turned bullets require non-standard twist rates for adequate stabilization so your standard rifle may not be able to fire these effectively. Lathe-turned solid bullets are a specialty item. As a beginning reloader, I suggest you put these in the back of your mind until you're more than comfortable with traditional bullets.

Materials And Processes

In addition to the varying materials and processes used to make bullets today, reloaders have a wide range of different bullet types intended for different purposes. It's easier to show you many of these than to describe them so take a look at the pictures that accompany each description.

9mm Luger Cartridge loaded with 115gr Full Metal Jacket (FMJ) bullet

Examples of "Hollow-point" bullets loaded in rifle (l) and pistol (r) cartridges

The Full Metal Jacket

The full metal jacket (FMJ) is one of, if not the most commonly handloaded bullets. FMJs are available for rifles and pistols and are one of the most affordable types of bullets for general use. The name Full Metal Jacket is almost literal and descriptive of the construction of the bullet. FMJs are lead-core copper-jacketed bullets where the jacket completely covers the nose and outside surfaces of the bullet. When I said the name was almost literal, I was allowing for some FMJs that have a small area of exposed lead at their base. This is not visible when the bullet is loaded into a case and is only a remnant of the manufacturing process where the jacket is formed around the bullet from the nose end. The jacket wraps around the base of the bullet without completely meeting in the middle.

Hollow-Point Bullets

Hollow-point bullets are the most common type of defensive handgun projectiles used by civilian and law enforcement shooters in the U.S. Hollow-point bullets are popular in some rifle applications, especially big-game hunting and varmint bullets. Intended for defensive use, hollow-point bullets expand in a controlled manner upon striking soft tissue. As tissue (fluid) enters the hollow cavity in the nose of the bullet, the walls of the cavity collapse outward forming an enlarged frontal area. This creates a larger wound path than the bullet's original diameter would create.

Hollow-point bullets designed for varmint hunters using centerfire rifles often expand rapidly and violently to create maximum fragmentation and tissue disruption in light game animals and varmints. This occurs even though the size of the hollow cavity is significantly smaller than the cavity found on hollow-point bullets used in defensive pistols. The thin copper jacket and high velocity of rifle cartridges help these small hollow-points open, fragment, and/or deform upon impact. Hollow-point rifle bullets made for big-game hunting offer more controlled expansion suitable for penetration of heavy tissue and bone. The popular TSX all copper bullets from Barnes are an example of a hollow-point bullet intended for big-game hunters. An identical bullet is marketed to tactical shooters under the Tac-X name.

Open-Tipped Match

Open-Tipped Match or OTM Bullets are essentially hollow-points but renamed for various reasons (Search "Open-Tip Match" at *www. dtic.mil* for an interesting review.). I list OTMs and hollow-points separately because I believe that one of those various reasons is significant enough to be of importance to handloaders and anyone using such bullets. The justification for the name change is revealed in the name itself and has more to do with the purpose of the bullet than anything. OTMs differ from hollow-points in that the purpose of the OTM design is to achieve the utmost in accuracy and precision while the purpose of the hollow-point is to expand upon impact with tissue and achieve maximum wounding effect. An OTM bullet is designed without regard to achieving expansion in tissue. For this reason alone, it is worth making the distinction between the two bullet designs.

"Hollow-point" or "Open-Tipped Match" bullet? You're correct either way.

The new .300 AAC Blackout likes Sierra Match King bullets up to 230 grains in weight.

Jacketed soft-point bullets like the Hornady Varmint are designed to expand rapidly.

How does the design of the OTM enhance accuracy? The most important aspect of the OTM design (in contrast with the hollow-point) is the consistency of the bullet's tail and/or base. Think about it this way; what is the last point of contact between the rifle and the bullet? What is the final opportunity we have to ensure that the bullet is guided along a stable flight path towards the target? Answer: the base of the bullet. For this reason, the design and consistent manufacturing of the base of an OTM bullet is crucial. If the bullet doesn't get a smooth and even departure from the rifle's muzzle, it can be ever so slightly rocked off axis by the jet of expanding gasses that escapes around the bullet upon exit. Design and manufacturing focus must remain on the *front* of the bullet to make a good hollow-point. Design and manufacturing focus must remain on the *rear* of the bullet to make a good OTM. This does not suggest that bullet designers neglect other aspects of bullet design and manufacture, but if a compromise is necessary in regards to the design or manufacture of a hollow-point bullet, that compromise will happen somewhere other than in the *front* (hollow-point) of a bullet designed to expand in tissue. If a compromise is necessary in regards to the design or manufacture of an OTM bullet, that compromise will happen somewhere other than in the *rear* (base/boattail) of a bullet where maximum consistency leads to accuracy.

It is important to understand that common use of the name "OTM" is relatively new (within the last couple of decades). Many bullets on the commercial market have hollow-point labels but actually fit the definition of an OTM. The Sierra Match King (SMK) is a perfect example. The SMK is commercially sold as a "Hollow-Point Boat Tail" but loaded into military ammunition as an OTM. Some manufacturers have adopted the OTM moniker for their match-grade bullets (Berger), and some have not. Neither is "wrong," but it is helpful to categorize the *intended use* of a bullet by differentiating between OTM and hollow-point.

Jacketed Soft-Point Bullets

Jacketed soft-point bullets (soft-points) are a long-time standard for hunters pursuing deer and other medium game. Soft-points are copper-jacketed lead-core bullets where the lead core is exposed at the nose of the bullet. This allows the bullet to deform and expand upon striking tissue, creating a larger wound path. Soft-points are relatively affordable and effective when used within the appropriate velocity range. I think it's safe to say that more deer have been killed in the U.S. with soft-points than any other design. When pushed at excessive velocities, soft-points will often exhibit core separations, fragmentation, and limited penetration. However, in some cases, such as coyote or varmint hunting, limited penetration is actually desirable. When using soft-points on larger game, ensure that impact velocities will be within the manufacturer's intended velocity range.

There have been numerous soft-point handgun projectiles over the years, but the design is not generally well suited to the slow velocity levels of defensive handgun cartridges where rapid expansion and maximum wounding is desired. Exceptions to this include large-bore handguns used as hunting or defensive arms in bear country or pursuing heavy game or hogs with a handgun. Classic examples of this are the 240-grain jacketed soft-point used in the .44 Magnum or the 300-grain jacketed soft-point used in the massive .454 Casull. In this role, the limited expansion and deep penetration of the heavy soft-point moving at moderate speed is an ideal combination.

Jacketed soft-point bullets have been taking game for a century and still work just fine for many uses.

Lead Bullets

Lead bullets, also called "cast" bullets, are solid lead or lead-alloy projectiles without a copper jacket. Lead bullets are usually cast from molten lead that's been hardened by adding tin and antimony. Handloaders can cast their own bullets at home with the proper equipment. Raw lead or pre-alloyed lead can be purchased in ingots from cast bullet manufacturers or reloading suppliers. Handloaders will frequently collect old wheel weights from tire shops or scrap lead from various sources to be melted and cast. A shooting mentor of mine who has since left us for the Lord's Range scored a barn full of scrap lead from telephone crews replacing old lines many years ago. We cast fishing lures and duck decoy weights for years and I don't think the pile ever got smaller.... If you have the materials source to keep you supplied, cast bullets can be darn near free for the home caster.

Once molded, the cast bullets are then pushed through a sizing die to bring them to the correct final diameter and a waxy lube is applied to the outside of the bullet. Lead bullets are the least expensive of the more common bullet types and are suitable for general plinking and competition use. You can find lead bullets in a few primary shapes or styles. Some examples of lead bullets include the round-nose bullet, flat-point bullet, the semi-wadcutter bullet, and the wadcutter. These bullets have circumferential rings around the bearing surface of the bullet that hold lube and reduce the amount of lead in contact with the bore. This reduces the amount of lead fouling or "leading" that occurs. There are some high-end lead bullets used primarily as hunting bullets for heavy game animals. Higher quality lead bullets have an increased hardness as a result of higher levels of tin and antimony, both of which are more expensive than lead. These harder lead bullets allow higher velocities which is especially important to rifle shooters firing cast bullets.

The hardness of your lead bullet should match the velocity of your load. There is no need to use a hardened high-end cast bullet in a mild load that's only traveling at 800 feet per second. In fact, using a hardened high-end bullet at lower velocity can actually cause *more* leading than you will see with a softer bullet. As always, follow the recommendations of your bullet manufacturer when loading cast bullets.

The Wadcutter is a cast lead-alloy bullet that cuts nice round holes in paper targets for easy scoring during competition use. With only a few exceptions, wadcutters are for use in revolvers or single-shot handguns only.

Gas Checks

Gas checks are thin copper discs that you can add to the base of some lead bullets to provide a protective barrier between the lead bullet and the hot gasses of the combusting gunpowder during firing. Hot expanding gasses can cut or erode the soft lead at the base of the bullet as it travels down the bore with loads developing high pressure and velocity. As the lead erodes and displaces, it is left in the gun's barrel as lead fouling. You can add gas checks to the base of the bullet as a barrier of harder copper to mitigate this issue.

Casting bullets at home is a subject all on its own and could very easily be its own book. It's mentioned here for your familiarity, but I don't recommend that a new reloader jump into casting just yet. After you have successfully loaded and fired a few thousand rounds of ammo for a couple different cartridges and guns, it may be worth your time to look into casting bullets at home. The Los Angeles Silhouette Club has a great website with loads of technical information on bullet casting and safety.

With enough magnification, we can see the unevenness and deformation of the meplat on this OTM bullet. Meplat deformation can result in slightly diminished accuracy and efficiency in flight.

Polymer-Tipped Bullets

Polymer-tipped bullets offer similar or superior performance with several advantages compared to traditional jacketed hollow-points or OTMs. In some cases, polymer-tipped bullets offer slightly higher and more consistent ballistic coefficients and enhanced expansion compared to their open-tipped match or hollow-point counterparts. Since the meplat of an open-tipped match bullet is a result of a manufacturing process that forces the open end of the copper jacket together into a point, there will always be some small amount of variability in the finished shape, consistency, and length of the finished bullet. The polymer tip insert used to create polymer-tipped bullets has a more consistent point (Meplat) than some open-tipped match bullets. The finer point and more consistent shape of this polymer insert gives the polymer-tipped bullet its potential advantage.

Polymer tips can also be used to initiate the expansion of a bullet designed for varmint or big-game hunting. Hollow-point bullets rely on hydraulic expansion as a result of soft tissue (water) entering the open nose cavity of the bullet and forcing the walls of the bullet's nose outwards. If the open nose cavity becomes plugged with clothing, hide, wallboard, or other media, expansion may be reduced or eliminated as tissue is prevented from entering the nose cavity. Polymer tips on expanding bullets act as a mechanical punch that is forced into the center of the bullet's nose cavity upon impact with the target. Since the cause of expansion is a mechanical force rather than hydraulic force, reliable expansion occurs without the risk of clogging the nose cavity with clothing, hide, wallboard, or other barrier.

The polymer tip on this bullet is nearly perfect and can be made to a consistent shape every time.

Military Surplus Bullets

When they are available, military surplus bullets can offer a great way to save money on plinking and practice rounds. Surplus bullets are usually bullets that were pulled (removed) from loaded ammunition and deemed unserviceable, improperly stored, out of spec, or obsolete. Though some "pull marks" may be present, the bullets are typically perfectly fine for casual use, blasting zombies, or whatever type of recreational shooting you do on the weekends. If the pull-down process proves to be too harsh on the bullets, resizing is sometimes an option. If this has been done, it will often be included in the bullet description provided by the seller. After using a boatload of these pulled surplus bullets over the years, I found them to be satisfactory for my intended purpose of training, blasting, and practicing. I wouldn't enter a long-range rifle match with ammo loaded with pulled bullets, but I would sure practice positional shooting for that match with pulled bullets.

The most common type of surplus bullet you will see is the FMJ. You may also come across tracers or armor-piercing projectiles. Armor-Piercing (AP) projectiles are made to have penetration characteristics that are greater than standard lead-core or mild-steel core projectiles. Just because you are shooting surplus AP projectiles, do not expect to be able to punch through the engine block of a locomotive. Armor-piercing capabilities are relative to the armor. Think along the lines of half-inch thick sheets of cold-rolled steel, and you will understand the type of penetration we are talking about. If I had to take out an oncoming Suburban full of brain-hungry zombies, I wouldn't mind having the additional penetration of some surplus "black-tip" .30 caliber projectiles to defeat the sheet metal, auto glass, and other components of the vehicle. The commonly used "green tip" 5.56x45 NATO M855 ammunition uses a 62-grain "penetrator" projectile with a mild-steel core. This is NOT the same as AP ammo and can be used without much additional concern. Some ranges do not allow shooters to engage steel targets with M855 ammo or the SS109 projectile it contains due to concern over accelerated target wear, so keep that in mind.

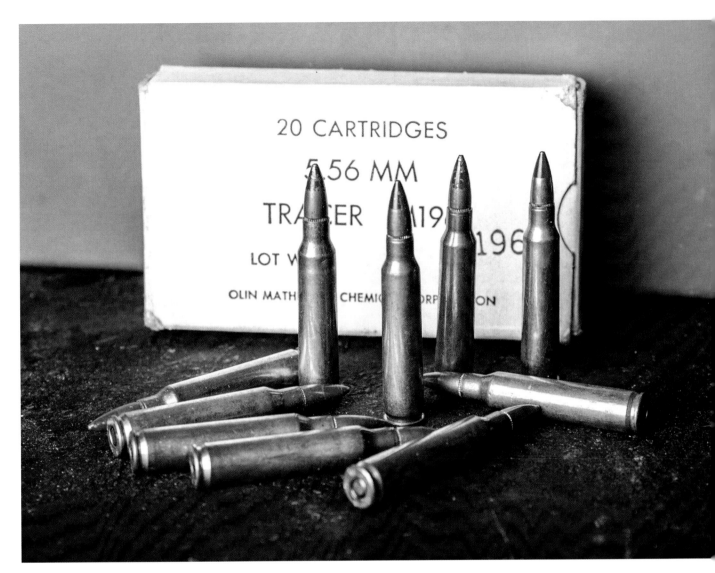

Tracers

Outside of military use, tracers are novel and fun to shoot. Each tracer round has a chemical compound(s) inserted into its base that will be ignited by the burning gunpowder upon firing. The result is an extremely bright, but short-lived chemical fire in the base of your bullet that creates a very visible glowing "trace" as your bullet arcs toward the target. Tracers are also intended for marking purposes where one shooter sees the target, but other shooters do not. Announce, "On my trace!" before firing the shot and everyone knows to look where you're shooting to locate the target. Again, we are talking about invasions, zombies, or whatever fantasy scenario you want to envision where these might have some utility. Military shooters use tracers for a number of purposes, but for most of us, they are simply fun to shoot.

Tracers do have one inherent and unavoidable risk as a result of their design. The burning chemical compound responsible for the trace WILL start a fire at every opportunity. The tracer can ignite dry grass, leaves, paper, or anything else that can burn once it comes to rest. I can't even begin to count the number of times I witnessed shooters on military ranges cease firing in order to stamp out the range fires started by tracers. Under no circumstances should you fire a tracer in an indoor range. Many indoor ranges have backstops filled with ground tires or other recycled rubber products that help diminish the chance of splatter, bounce back, etc., from fired projectiles. Launching a burning tracer into this combustible mess of petroleum-borne solids can cause an incredible fire.

While tracers and AP bullets are not inherently unsafe, I would generally recommend against using these for your initial load workups. As a beginning handloader, you already have enough to concentrate on without starting fires or punching AP bullets through the backstop at the range.

Primers

Primers provide the spark that ignites the gunpowder inside our ammunition. The spark is created by the percussion from the impact of the firing pin or striker on the face of the primer cup. When the firing pin or striker impacts the primer cup, the primer cup deforms under the force of this impact. The geometry and placement of the anvil against the primer pocket allow it to remain relatively immobile during this process. As the cup deforms, the priming mixture inside the cup is smashed between the cup and the anvil. The speed and pressure of this impact causes the explosive priming mixture to detonate, sending a shower of sparks, flame, and heat through the flash hole into the waiting column of gunpowder.

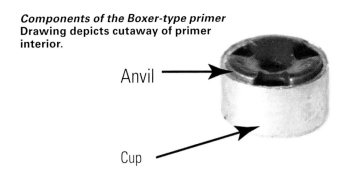

Components of the Boxer-type primer
Drawing depicts cutaway of primer interior.

Anvil

Cup

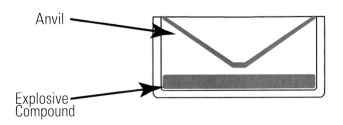

Anvil

Explosive Compound

Primer Types

Centerfire rifle and pistol ammunition will be loaded with one of two primer types. Practically all American-made ammunition and most commercially available ammunition in the U.S. uses a boxer-type primer. A good portion of the imported foreign military surplus ammunition that's available in the U.S. uses a Berdan-type primer. Though not strictly true, cartridge cases using Berdan-type primers are generally considered to be non-reloadable. Almost all of the reloading equipment available in the U.S. is intended to be used with boxer-primed cases. In order to tell what priming system your cartridge cases are designed to use, look through the case neck into the inside of the cartridge case where you will observe the flash hole or holes in the case head. You may need to use a flashlight to inspect this area. If you see one flash hole in the center of the case head, then that brass is boxer primed. If you see two or more flash holes, that case is Berdan primed. In those two examples we're talking about fired brass that has not yet

been deprimed. This generally represents spent brass from an unknown source at the range. If you're inspecting brand new brass that has not been primed you can simply look at the primer pocket from the outside. There are a couple more technical differences between Berdan and boxer priming systems, but none matter to us for practical purposes. All discussion of primers and reloading in this book focuses on boxer-primed cases and boxer-type primers.

Primer Sizes And Strengths

In addition to the two priming systems discussed above, there are also different sizes of primers and primers of different strengths. Primers will come in either small or large sizes. You must know what size primer your brass requires, as these are not interchangeable. Small primer pockets will measure approximately 0.170" in diameter. Large primer pockets will measure approximately 0.210" in diameter. Note the use of the word "approximately" as different brands

Large (l) and small (r) primer sizes are used in both rifle and pistol cartridges.

Choose primers based on pocket size and intended use.

of brass will have slightly varying dimensions. These dimensions will also change somewhat as the cases are fired and reloaded. It is not uncommon for one cartridge to have brass from various manufacturers that use different primer sizes. For example, the .45 ACP may be found loaded with small primers and large primers. The same can be said for the 6.8 SPC, .308 Winchester, and others, I'm sure. Remington 6.8 SPC brass uses a large primer while Hornady (and most other manufacturers) 6.8 SPC brass uses a small primer. Thoroughly inspect your brass, especially if using range discards or secondhand brass to ensure that you are using primers of the appropriate size. Attempting to seat a large primer in a small primer pocket can result in a primer detonation on the press. Not only is this a loud and startling event, but the potential for fire is significant. This is one of many reasons why I outlined the strict powder-handling process in the safety portion of this book.

Picking Primers

Not only should we use primers of the appropriate size, but we must also ensure that we are using rifle primers for rifle cartridges and pistol primers for pistol cartridges. To further complicate matters, primers are available in standard and magnum versions in both large and small rifle and pistol types. If you think we've covered all the options, there's more. Primers are also available in military and "match" grades. Military primers will generally have slightly harder cups or priming mixtures designed to ensure reliability in all conditions, especially in semi-automatic or fully automatic military-type guns. Match-grade primers are generally the same composition as standard primers, but will have added quality control processes or oversight during their manufacture. I can tell you match-grade primers are supposed to provide more consistent results, better accuracy, and so forth, but you can only confirm what works best in your guns by trying the different grades for yourself. I've seen standard (less expensive) primers provide more consistency than "benchrest" match-grade primers in some cases, and I've seen the exact opposite in others.

Primer Comparison

In order to examine the variables of primer selection, let's look at the popular .308 Winchester/7.62 NATO (.308) cartridge for loading into our long-range competition rifle. .308 cases are found in both boxer and Berdan primed variants and can be had with small or large primer pockets. In this example, we've inspected our cases and found them to have primer pockets that measure approximately 0.210" in diameter. This tells us our brass has large primer pockets.

Next, we'll select *rifle* primers since the .308 is a rifle cartridge (even if we use it in a gun that is legally classified as a pistol). The .308 has a powder capacity of around 42-47 grains, depending on the bullet we're loading. This is well under the level of a magnum cartridge where we may use 75 grains of powder or more so we'll use standard (not magnum) primers. Finally, we'll select match-grade primers since we're loading for accuracy in our competition rifle. We may achieve the results we desire with the first primer we choose, but I always try at least two different brands when nailing down the final details of load development. With large rifle match primers, I'll usually start with the Federal 210m and try the CCI BR2 when I've ironed out the major details of the load. There is no reason for that sequence other than it just being the way I've found success over the years with the components I usually have on hand. With large rifle magnum primers, I'll start with CCIs and then try Federals.

A single flash hole located in the center of the primer pocket is evidence of the boxer-style priming system.

I will point out that simply changing primers at the end of the load development cycle often makes no difference. It HAS made a noticeable difference for me in some cases, though. One .308 load using a 168 AMAX bullet and Varget powder became noticeably more consistent when I switched to CCI primers. A .260 Remington load using H4350 became noticeably more consistent when I switched to Federal primers. As I said before, you need to try these combinations in your own firearms. There are no hard and fast rules to primer brand selection.

Straight-walled case

Bottleneck cartridge case

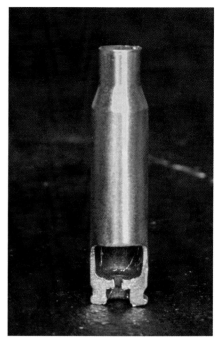

Cutaway view of the case head

Cases

The cartridge case is the protective shell, the delivery capsule, and the safety system of your loaded ammo. The case holds the primer in place and the powder inside, and it holds and positions the bullet in place for the shot. When the powder inside ignites, the case expands to conform to and hold on to the walls of the chamber, creating a tight seal to keep all those hot gasses and pressure moving in the right direction—out of the *front* of the barrel (muzzle)!

Each case has several features or parts that you should know and understand. Pistol and rifle cartridges are very similar but have a few differences that will be readily apparent in the photos to follow. Most modern rifle cartridges have a bottleneck design and most pistol cartridges are straight-walled. Of course, there are always examples that blur lines and cross boundaries. Plenty of those exist when it comes to cartridge cases.

Cartridge Head

The cartridge head is at the base of the case. The head is the thickest and hardest part of the case. Here we'll find the extraction rim or groove, depending on the cartridge design, primer pocket, flash hole, and head-stamp.

Examples of rimmed (r and c) and rimless (l) handgun cartridges.

The Extractor Rim

The extractor rim or groove provides a place for the gun's extractor to hook on to the case in order to pull it out of the chamber after firing, when unloading, or when clearing a malfunction. Generally speaking, rimmed cases are older designs or are intended for use in older-style guns, including lever-action rifles and revolvers. These work just fine, obviously, but aren't suitable for use in modern semi-auto or double-stack magazine-fed guns. "Rimless" cases will have a groove cut around the circumference of the case near the base of the head. This groove provides a place for the extractor to grip the case for removal from the chamber. Since there is no protruding rim, rimless cases can stack easily inside a magazine. Without a protruding rim, these can't be used in revolvers without a special device to position them in the chambers and allow extraction when necessary.

The Primer Pocket

The primer pocket is the recess in the case head that receives and holds the primer. Primers are held in by an interference or friction fit in handloaded ammo. Military and some commercial tactical ammo adds a crimp to the primer pocket to ensure that primers don't pop loose from their pockets during rough handling, transport, feeding, firing, extraction, and ejection. Loose primers floating around inside your gun can lodge within critical mechanisms (like the trigger group) and completely disable the gun.

The Flash Hole

The flash hole is found in the center of the primer pocket. The flash hole provides an orifice through which the exploding primer compound sends sparks and flames into the powder column inside the case. When using new brass for the first time, check to be sure that the flash hole exists and is free from debris. Avid shooters chasing minute accuracy gains will deburr the flash hole for uniformity to ensure a consistent flow of sparks into the powder column. I have found this to be totally unnecessary for all purposes outside high-level precision shooting competitions where shooters are achieving accuracy of approximately one-quarter minute of angle or better already. If you're that good, you probably aren't reading this basic handloading book. Save yourself some time, money, and effort and don't worry about your primer pockets being uniform at this point.

The long case body taper on the .375 H&H (left) aids in extraction.

Case Body

Moving up the case towards the mouth, we next find the case body. The body provides the largest portion of the "gripping" area the case uses to seal the chamber and form a protective barrier against hot expanding gasses during firing. Each case design will have a carefully selected amount of case body taper that matches the type of firearm, pressure level, and intended use of the cartridge. Greater taper to the case body generally eases extraction. Increased taper results in a corresponding reduction in the powder charge. Trade-offs must be made according to the intended role of the cartridge.

Shoulder

The case shoulder is the bridge from the body diameter to the case neck diameter. Typical shoulder angles range from 20-35 degrees, but more and less angled shoulders are found on some cartridges. A more shallow shoulder angle eases feeding while a more significant shoulder angle reduces the growth of your brass from firing (less case trimming!) and increases case capacity. Trade-offs again.... We can also use a case comparator to measure from the base of our fired cases to the datum on the case shoulder. This isn't a true headspace measurement since our cartridge case will have pulled back a couple thousandths from its fully expanded size

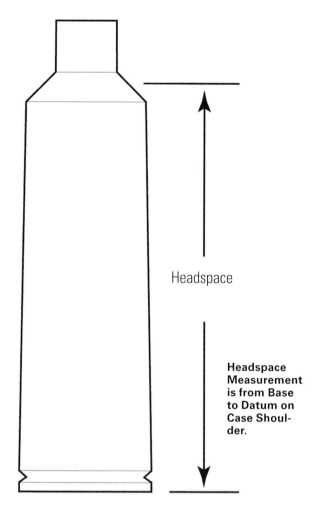

Headspace

Headspace Measurement is from Base to Datum on Case Shoulder.

during firing, but if our chamber has excessive headspace we can determine this by measuring the length of our expanded (fired, but not yet resized) cases. If our measurement provides a reading that is longer than the recommended max length it's a good idea to have the rifle, checked for excessive headspace. Ideally, we'll resize our cases down 0.002"-0.003" smaller than our actual chamber dimensions to ensure fit and function without excessive crush during locking.

Case Neck

The case neck is what actually holds the bullet in position during transport, feeding, and chambering. The neck holds the case through an interference or friction fit. In some applications, a crimp will also be applied to the case neck near the mouth open-

The case neck ends at the case mouth, the opening in the neck where we insert the powder and projectile during the loading process.

ing to apply additional holding pressure. This can be especially important in some semi-auto rifles and most pistol applications. While the neck of a bottleneck-style case is pretty obvious, straight-walled cases have less distinct necks. The case neck ends at the case mouth, the opening in the neck where we insert the powder and projectile during the loading process.

Case Materials

Cartridge cases are made from several materials. U.S. military and most U.S. commercial ammunitions use a copper/brass-alloy cartridge case. Aluminium and steel cases are also used in ammunition that is intended to sell at a lower price point and are not intended to be reloaded. A good portion of the foreign military ammunition made around the world uses a steel case that has been coated with a metallic or lacquer-based protective layer to prevent corrosion. These steel cases are not intended to be reloaded, but it may be possible to reuse them up to a couple times if no other option exists.

Plastic cases are in development but are not a viable or proven technology at this point. If this technology can be perfected, the benefits to the shooter would be significant. Significant weight reductions could be achieved through the use of polymer-cased ammunition. Polymer-cased ammunition is also said to transfer less

Minor dents are inevitable and not a worry for the reloader.

heat to the gun. This can result in cooler operating temperatures during extended firing.

Case Inspection

It's important to inspect your cases before loading each time. In addition to checking for length (trimming when necessary), look for cracks, corrosion, stretching, rim damage, dents, or enlarged primer pockets. If you find significant examples of any of these issues you should discard that case immediately. Even minor cracks in any part of the case justify culling. Cracks or splits in your brass are usually indicative of overuse (work hardening), metallurgical imperfections, or stretching, which is indicative of a headspace problem with your gun. In any instance, brass with cracks or splits must be discarded.

In today's world of semi-automatic firearms and concrete firing ranges, it is inevitable that our cases will occasionally receive minor dents. Our resizing dies will take care of most of these dents. Any small dents that remain after resizing are generally a nonissue and will disappear when that case is fired again. Large dents that can't be eliminated through the sizing process may prevent chambering and justify the elimination of that cartridge case.

Steel (l) and brass (r) cases are commonly used in the manufacture of commercial and surplus ammo. Generally speaking, steel cases are not reloadable, but boxer-primed brass cases may be reloaded many times over.

Corroded brass cases this far gone should be discarded.

Corrosion

Mild to moderate corrosion can usually be polished out during dry tumbling. The removal of corrosion is one area where the stainless steel wet-tumbling process excels. Stainless media tumbling can make corroded brass that would be otherwise lost to the scrap bucket look like brand new again. Make sure the corrosion is only surface deep. Any cases that exhibit deep or significant corrosion should be discarded. Be especially wary using heavily corroded bottleneck rifle cases with pitting or erosion inside the neck. This can cause inconsistent neck tension and reduce the accuracy of your ammunition. Mild tarnish and corrosion can be reduced or eliminated with a standard vibratory tumbler.

Primer Pockets

Primer pockets should remain tight for a dozen or more loadings when pressures are kept to appropriate and safe levels. Handloaders who insist on running loads up to the red line will find their primer pockets become loose and unable to retain a primer after only a few firings. If you notice that your primer pockets are loose or require significantly less force for proper primer seating after being loaded a couple times, that's likely a sign of excessive pressure in your hand-loads. Some brass is softer than others, and if that's the case, the primer pockets may get loose even though the pressure may be completely reasonable in your handloads. This is one area where a hand-priming tool really helps. With a good hand-priming tool you can really develop a feel for how your primers fit in your primer pockets. When you find a case with a pocket that has noticeably loosened, pitch that case in the trash and replace it. Leaking primer pockets can cause a number of problems with the function, performance, and longevity of your firearms.

Rim Damage

Some semi-auto firearms, primarily rifles, can cause damage to the rims of your cases during extraction. I've experienced this with factory .308 Winchester ammo. The problem in that instance turned out to be a new chamber that was cut too tight (leading to increased pressure). In many cases, though, rim damage caused by the extractor is a sign of excessive pressure or timing and/or cycling problems with the gun. In other cases, you may simply be experiencing the incompatibility of your chosen components with that particular gun. Select a different brand of brass, a different type of powder, a different bullet and the problem may disappear. A batch of military-grade .50 BMG

The spent primer is now free from the case and can be discarded. Primer residue is visible inside the primer pocket. This can be removed, if desired.

ammo I worked with several years ago caused very noticeable damage to the cartridge rims of just about every round fired through several .50 BMG rifles. By simply switching to a different brand of ammo (different case, different powder, different bullet) the problem disappeared, and the guns ran just fine with no extraction issues. This wasn't a case of "bad" ammo, but simply a misalignment between the component mix and loading specifications of that ammo and the characteristics of the rifles we were using. Another example where we may see rim damage is when using a powder that is of the incorrect burn rate for our semi-auto firearm. The classic M1 Garand rifle is a prime example. The Garand must be used with powders in a specific range of burn rates to avoid significant damage to its operating rod. There is plenty of information on Garand loading available online and in reloading manuals. Do your homework before selecting your components. If you haven't loaded your ammo in accordance with guidelines developed specifically for the Garand and you see extraction marks on your cartridge rims, it may be too late to avoid damage to your rifle.

Head Separation

Head separation is a very serious condition that indicates the possibility that your firearm has major problems and is absolutely unsafe to fire. Keep an eye out for the formation of a whitish ring forming on the lower case body near the case head. This is a sign of the brass in that area stretching beyond its limit and nearing a rupture. Case head separation is typically a sign of excessive headspace. As the cartridge is fired, the cartridge case expands to fill the chamber. When the dimensions of the chamber are excessive, the brass must stretch beyond its capabilities. Head separation may occur upon the first firing in a too-long chamber, or it may take several firings for the problem to occur. When the case fails, a rapid and dangerous venting of gasses can be sent into the action of the gun or into the shooter's face and hands. If you see what you think may be signs of head separation, cut a case in half lengthwise to look inside. That's where you'll see a significant thinning of the case wall if this problem exists. If you suspect there is a headspace issue stop shooting that gun immediately and take it to a qualified gunsmith.

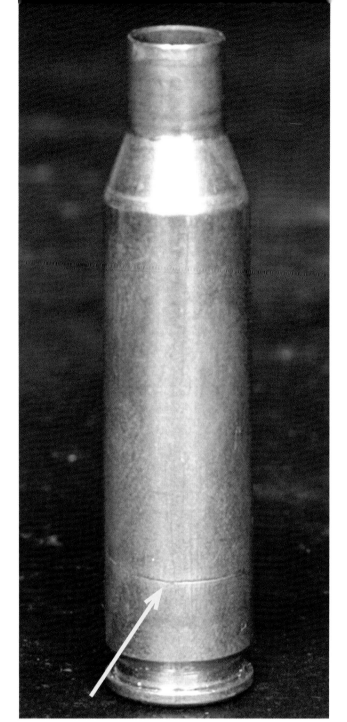

Powder

Smokeless gunpowder is used in all modern cartridge loads. Smokeless powder (powder) is available in a dizzying variety of choices today. We're blessed with so many good options as handloaders that we can achieve outstanding results with several powders in almost any cartridge. It is important to understand that powder must be selected with extreme care and understanding, though. Always use the EXACT powder specified in loading references provided by industrial sources. Accurate 4350 is close to, but not the same as IMR 4350 which is close to, but not the same as

Modern smokeless powders are most commonly found in spherical, flake, or extruded styles.

Hodgdon 4350. It's easy to see the potential for confusion, right? Each powder company has load recommendations available free of charge, as do most bullet companies. When in doubt, call the manufacturer of your chosen powder for guidance. We only use rifle powder in rifle cartridges. We only use pistol powder in pistol cartridges. You'll eventually find data to support some crossover in very limited and specialized circumstances, but keep the rifle/pistol powder rules firmly n your mind until you are an experienced reloader working under the specific guidance of the powder manufacturer. As with primers, shooting a rifle cartridge in a gun that is legally considered a pistol doesn't change the fact that you're firing a rifle cartridge. Always load it as such!

Powder Compositions

Modern smokeless powder comes in two compositions: single-based and double-based. Single-based powders contain Nitrocellulose. Double-based powders contain Nitrocellulose and Nitro-glycerine. There are advantages and disadvantages to each. I primarily rely on single-based powders for use in rifles because of their tendency to display somewhat reduced sensitivity to temperature variations. This is important for a hunter that may zero his rifle in the fall when temperatures are in the 60s or 70s but hunt on top of a mountain months later with temperatures near zero degrees. There are excellent powders available in both single and double-based configurations. Until you're significantly advanced from the basic level, I wouldn't even worry about this detail. As long as you follow the guidelines provided by the powder manufacturer there is no practical difference that will impact your results.

Powder

Powder is available in several basic physical shapes. The most common of these are extruded (think of a cylinder shape like a beer can), flake (think of a flat disk like a Frisbee or dinner plate), and spherical (think of a baseball that's been slightly flattened). Once the powder is inside your cartridge case, there is little practical difference between the granule shapes. Where the shape of your selected powder does make a difference is when you're charging those cases with powder. The smaller spherical powders will flow through a powder measure MUCH more freely and consistently than the extruded powders will. In fact, the kernels of the largest extruded powders are not-so-lovingly referred to as "Lincoln Logs" because of their incredibly long (relatively speaking, of course) shape and the difficulty with which they move through most powder measures. Some measures work better than others with these long extruded powders, of course. Still, inconsistent throws, bridging (the long kernels stop flowing through the measure when they interlock and form a "bridge" across the measure's discharge orifice or measuring chamber), and shearing or cutting of kernels will always be more common with these large-kernel powders. Some manufacturers of extruded powders have successfully managed to develop so-called "Short-Cut" versions of their longer extruded powders. These short-cut versions show much improved performance in manual and automatic powder measures and I choose them whenever I can over their longer counterparts.

Flake Powders

Flake powders are almost always used in shotgun or handgun-loading applications. Flake powders may be single-based, although most are double-based. Though it's primarily a shotgun powder, I have used Unique, a double-based powder in my 9mm Parabellum pistol loads for twenty years now. It meters well enough through the measure on my Dillon progressive press and produces adequate velocity for training and practice loads. It isn't as clean-burning as many newer powders, though. Flake powders are a good choice for many loads where charge density is not an issue, and some may even be useful for reduced-recoil and velocity loads with cast bullets in limited rifle-cartridge applications. If you find that very fine spherical powders leak or cause problems in your powder measure, flake-type powders may be your solution.

Spherical Powders

In general, spherical powders "meter" (the term used for dispensing powder through a volume-based powder measure) to near perfection in even the less expensive powder measures. Some spherical powders are so fine that they will leak out in very small amounts through the various seams or crevices of certain measures, but overall they perform superbly. So why wouldn't we use spherical powder for everything? Spherical powders are double-based powders and most don't have the temperature insensitivity that hunters and some competition shooters demand. With technological advances and market requirements for more stable powder options that still meter well through loading equipment, this is changing, but the general statement that spherical powders tend to be more temperature sensitive is still true. If you're not planning to shoot through huge temperature swings, sub-zero through 100°F, for example, spherical powders make loading much faster and easier by allowing very consistent charges to be dispensed by automated equipment. I use spherical powder in all of my carbine practice and/or blasting ammo and some of my handgun loads. Once I get through the remnants of a big stash of flake-style pistol powder, I'll use spherical powders for all of my pistol loads as well.

Extruded powder sometimes referred to as "Stick" powder.

Burn Rates

Powders are designed with different burn rates for different applications. Pistol powders will be relatively fast, and magnum rifle powders will be relatively slow. All gunpowder will burn with amazing speed when confined inside a chambered cartridge, thus the use of the term "relatively" fast and slow.

A big part of selecting the right powder is choosing a powder of the appropriate burn rate for your application. 90 grains of a slow-burning magnum rifle powder like Retumbo will produce excellent accuracy and velocity in a rifle chambered for the .338 Lapua Magnum cartridge. Put 90 grains of a fast-burning pistol powder in that same cartridge case, and you have a veritable bomb waiting to explode in your face. That huge charge of fast-burning pistol powder will deflagrate with such speed that the force caused by the expansion of hot gasses during combustion will exceed the ability of the barrel's steel to contain that force. The bullet won't be able to move down the bore quickly enough to exit before the pressure overcomes the strength of the barrel steel and a violent rupture will occur. The shooter and others nearby are subject to receive exploding shards of hot metal and gas, resulting in possible death and almost certain injury. Follow the specific guidelines of your loading manual and those provided by the powder manufacturer, and this will never be a problem for you.

All powder manufacturers will be able to provide a burn rate chart for your reference. It is important to note that the steps in these charts are not uniform and all rates are relative. Just because you used a powder in the middle of the chart doesn't mean you can assume the powder directly adjacent will be suitable for your application. We can use these charts for research purposes though. Glancing at the burn-rate chart provided by Hodgdon, I see that my chosen powders for the .308 Winchester cartridge, Varget and IMR 4064, are separated by five powders that I've never used in the .308. One of these powders is Accurate Arms 2520. If Varget and IMR 4064 are unavailable but

AA2520 is on the shelf of my local sporting goods retailer, I can consult the loading data provided by Accurate Arms to discover that AA2520 seems to provide very positive results in the .308. Confident that AA2520 is an appropriate choice, I buy a couple pounds and begin load development in an attempt to replicate the performance I've achieved from my previously selected but currently unavailable powders.

Flake powders are common in shotgun and handgun loads.

Spherical powder meters from a powder measure more uniformly than stick powders.

Relative Burn Rates

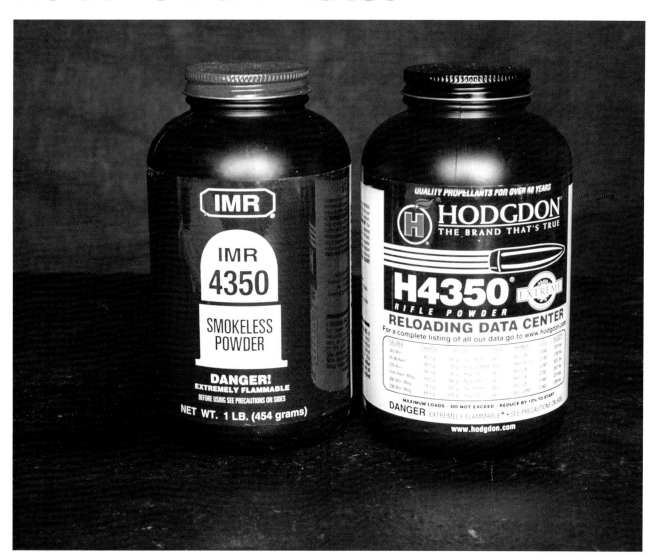

Following is a relative burn rate comparison that ranks powders from the fastest to the slowest burn rates. Handloaders should not use this chart as a reference from which to build loads, since many other variables influence cartridge pressures, such as primers, case volume, bore diameter, etc. Always follow specific handloading data for a particular cartridge and follow the manual's instructions.

1		NORMA R1	Discontinued Norma powder.
2		Winchester WAALite	Sperical 12 gauge powder suited for light loads. Perfect for recoil-sensitive shotgun shooters.
3		VihtaVuori N310	Very fast burning and competitive with Alliant Bullseye, Hodgdon HP38 and Vectan Ba 10. It has applications in a very wide range from .32 S&W Long Wadcutter up to .45 ACP.

4		Accurate Arms Nitro 100	Fast-burning, flattened spherical, double-base shot shell powder that is a clean burning, cost-effective choice for all 12 gauge target applications.
5		Alliant e3	Features reduced charge weights, clean-burning with environmental stability. Premium quality for more consistent and reproducible performance. Named after its core qualities of energy, efficiency, and excellence.
6		Hodgdon TITEWAD	A flattened spherical shotgun powder. TITEWAD features low charge weights, mild muzzle report, minimum recoil, and reduced residue for optimum ballistic performance. Designed for 12 gauge only.
7		Ramshot Competition	COMPETITION is a clean-burning powder for 12 gauge target shooters, and is well suited for many low pressure, low velocity Cowboy Action loads.
8		Alliant Red Dot	The number one premium clay target powder, now 50 percent cleaner.
9		Alliant Promo	Budget-priced 12 gauge target powder that also works in handgun loads. More dense than Red Dot, but same burn rate.
10		Hodgdon CLAYS	Often chosen by competitive target shooters. Works well in shotgun applications, as well as handgun cartridges, such as .45 ACP and .38 Special.
11		Alliant Clay Dot	Economical shotgun target powder desinged to duplicate Hodgdon Clays.
12		IMR Hi-Skor 700-X	This extruded flake type powder is ideally suited for 12 and 16 gauge shot shells. It doubles as an excellent pistol powder for . 38 Special and .45 ACP target loads.
13		Alliant Bullseye	Fast-burning and consistent; economical and accurate. Best known as powder of choice for .45 ACP.
14		Hodgdon TITEGROUP	Spherical propellant designed for accuracy. Because of the unique design, this powder provides flawless ignition with all types of primers. Relatively less powder position sensitivity.
15		Alliant American Select	This ultra-clean burning premium powder makes a versatile target load and superior 1 ounce load for improving clay target scores. Suitable for Cowboy Action handgun loading.

16		Accurate Arms Solo 1000	Fast burning, single-base, flake shotgun powder for trap, sporting clays, and skeet shooting.
17		Alliant Green Dot	Delivers precise burn rates for uniformly tight patterns with lower recoil. Versatile for target and field.
18		Winchester WST	The choice for 12 gauge AA duplicate handloads and standard velocity handgun loads. Ideal for use in .45 Auto match applications. Consistent, clean, and low flash and smoke are benefits to the shooter.
19		IMR Trail Boss	Trail Boss is designed specifically for low velocity lead bullet loads suitable for Cowboy Action shooting. It is primarily a pistol powder but has some application in rifle. It is based on a new technology which allows very high loading density, good flow through powder measures, and most importantly, additional safety to the handloader.
20		Winchester Super Handicap	Super Handicap is the same propellant used in Winchester's Super Handicap ammunition. This slow-burning, high-energy propellant gives the shooter great handicap or long range sporting clays loads at up to 1250 fps with a 1 ⅛ ounce shot charge.
21		Hodgdon INTERNATIONAL	It works in 20 gauge, 12 gauge, 2 ¾ inch light, medium, and heavy 1 ⅛ ounce loads, and high velocity 1 ounce shells. Clean burning and flawless functioning. Available in 14 ounce, 4 pound, and 8 pound containers.
22		Accurate Arms Solo 1250	
23		IMR PB	Named for the porous base structure of its grains by which the burn rate is controlled, PB is an extremely clean burning single-base powder. It gives very low pressure in 12 and 20 gauge shotshell target loads and performs well in a wide variety of handgun loads.
24		VihtaVuori N320	A comparatively fast burning multipurpose handgun powder with burning rate about the same as Winchester 231 or Alliant Red Dot. Currently available reloading data for 9 mm Luger, .38 Special, .357 Magnum, .44 Remington Magnum, and .45 ACP.
25		Accurate Arms No. 2	Fast burning, double-base, spherical handgun powder for a wide range of handgun calibers. Low recoil and low flash for short barrel, concealed carry applications. Non-position sensitive powder and low charge weights for high volume handgun shooters.
26		Ramshot Zip	A clean burning, double-base propellant designed for a wide range of handgun calibers. Low charge weights make it an economical choice for high volume shooters.
27		IMR SR 7625	

28		Hodgdon HP-38	HP-38 is a spherical powder that is great for low velocity and mid-range target loads in .38 Special, .44 Special, and .45 ACP. This high energy powder provides economy in loading.
29		Winchester 231	One of the most popular reload propellants, 231 is a pistol powder ideally suited to the .38 special, .45 Auto, and 9mm standard loads. Consistency, clean burning, and low flash make this powder a choice for any pistol cartridge reloader.
30		Alliant 20/28	Designed to deliver competition-grade performance to 20 and 28 gauge clay target shooters. Extremely clean burning with proven lot-to-lot consistency.
31		Alliant Unique	Versatile shotgun and handgun powder made. Use with most hulls, primers, and wads.
32		Hodgdon UNIVERSAL	Handles a broad spectrum of cartridges for both pistol and shotgun, from the .25 ACP to the .44 magnum and 28 gauge to 12 gauge.
33		Alliant Power Pistol	High-performance semi-auto pistol loads in 9mm, .40 S&W, and .357 SIG.
34		VihtaVuori N330	Burning rate is a bit slower than with N320 and corresponding to Alliant Unique and Vectan Ba 9. Especially designed for 9mm Luger but also suitable for .38 Special, .40 S&W, .44 S&W Special, and .45 (Long) Colt.
35		Alliant Herco	Proven powder for heavy shotshell loads. The ultimate in 12 gauge, 1 ¼ oz. upland game loads.
36		Winchester WSF	Super-Field propellant is the propellant of choice for Winchester 20 gauge AA Target loads. WSF is an ideal choice to maximize velocities in 12 gauge 1 ⅛ ounce and 1 ¼ ounce loads. Super-Field also performs well in 9mm and .40 S&W pistol loads.
37		VihtaVuori N340	An excellent multipurpose handgun powder with burning rate generally about like Accurate No.5 or Alliant Herco. Wide application area covers the following handgun cartridges: 9mm Luger, 9x21mm, .357 SIG, .38 Super Auto, .38 Special, .357 Mag., .40 S&W, 10mm AUTO, .44 S&W Special, and .44 Rem. Mag.
38		IMR Hi-Skor 800-X	This large-grained flake powder is at its best when used in heavy field loads from 10 gauge to 28 gauge. In handgun cartridges, 800-X performs superbly in cartridges like the 10mm Auto, .357 Magnum, and .44 Remington Magnum. Excellent velocity and uniformity translate into top accuracy.
39		IMR SR 4756	This fine grained, easy metering propellant has long been a favorite of upland and waterfowl handloaders. Its top velocities with great patterns are legendary. Like 800-X, SR4756 performs extremely well in the big handgun cartridges. Always a top choice.

40		Ramshot True Blue	Double-base, spherical powder perfect for classic calibers such as the .38 Special, .44 Special, and .45 Long Colt.
41		Accurate Arms No. 5	Fast-burning, double-base, spherical handgun propellant. Wide performance range from target and Cowboy Action applications to full power defense loads.
42		Hodgdon HS-6	HS-6 is a fine spherical propellant that has wide application in pistol and shotshell. In pistol, 9mm, .38 Super, .40 S&W, and 10mm Auto are some of the cartridges where HS-6 provides top performance.
43		Winchester AutoComp	Extremely fine in the .38 Super, 9mm, .45 ACP, and .40 Smith & Wesson race guns. It's just the perfect burning speed to feed the compensators with a higher volume of gas.
44		Hodgdon CFE Pistol	New pistol powder treated with their Copper Fouling Eraser formula. Medium burn rate suitable for a wide range of handgun cartridges.
45		Ramshot Silhouette	A double-base high performance spherical powder, it's an excellent choice for 9mm, .38 Super, .40 S&W, and .45 ACP. It has a low flash signature and high velocity.
46		VihtaVuori 3N37	Listed as a .38 Special target load powder.
47		VihtaVuori N350	This is the slowest burning N300 series handgun powder, which can usually be used in place of Accurate No.7, IMR Hi- Skor 800-X, and Alliant Blue Dot. Appropriate choice for many powerful handgun loads, like 9mm Luger, 10mm AUTO, and .45 ACP.
48		Hodgdon HS-7	Discontinued.
49		VihtaVuori 3N38	Single-based perous powder for handgun loads.
50		Alliant Blue Dot	Designed for magnum lead shotshell loads. Doubles as magnum handgun load.
51		Accurate Arms No. 7	Intermediate burning, double-base, spherical powder suitable for wide range of handgun calibers. For high performance semi-auto handguns and larger magnum handgun calibers.

52		Alliant Pro Reach	Long-range 12 gauge target and hunting load powder.
53		Hodgdon LONGSHOT	Highly versatile spherical powder for heavy shotshell field loads. Great loads in 10, 12, 16, 20, and 28 gauges. Good choice for high velocity pistol, too.
54		Alliant 410	Flake powder specifically designed for the exacting needs of .410 skeet and field loads.
55		Alliant 2400	For .44 Mag and other magnum pistol loads. Originally developed for the .22 Hornet, it's also the shooter's choice for .410 bore.
56		Ramshot Enforcer	Enforcer is a choice for high performance, full-power loads in magnum handgun cartridges.
57		Accurate Arms No. 9	Double base, spherical powder for high power loads in traditional magnums. Can also be used with large magnums such as the .460 S&W and .500 S&W for economical target loads.
58		Accurate Arms 4100	A double-base, slow-burning spherical powder with exceptional metering characteristics. An excellent choice for full power loads in magnum handgun cartridges. It is ideally suited for the .357 Mag, .44 Mag, .454 Casull, .460 S&W, and the .500 S&W.
59		Alliant Steel	New powder for waterfowl shotshell gives steel shot high velocity within safe pressure limits for 10 and 12 gauge loads.
60		NORMA R123	Discontinued NORMA powder.
61		VihtaVuori N110	The fastest burning rifle powder from Vihta Vuori. Similar to Hodgdon H110 and Winchester 296. N110 can be used in small rifle cases like .22 Hornet and .30 Carbine but also in magnum pistol and revolver cartridges like .357 S&W Magnum, .41 Magnum, .44 Magnum, .454 Casull, and .500 S&W.
62		Hodgdon LIL'GUN	Designed to fit, meter, and perform well in the .410 BORE. In addition, LIL'GUN has many magnum pistol applications and is superb in the .22 Hornet.
63		Hodgdon H110	A spherical powder that delivers top velocities and accuracy in .44 Magnum, .454 Casull, .475 Linebaugh, and the .460 and .500 S&W magnums. H110 is a good choice for .410 Bore shotgun.

64	Winchester 296	This propellant was developed for Winchester factory-loaded ammunition for .357 Magnum, .44 Magnum, and .410 bore. It's high loading density provides optimal velocity. The 296 is recommended by Winchester for .410 Bore AA loads.
65	IMR 4227	This is the Magnum Pistol Powder in the IMR lineup. In rifles this powder delivers excellent velocity and accuracy in such cartridges as the .22 Hornet and .221 Fireball.
66	Hodgdon H4227	Also known as AR 2205, distributed by Hodgdon Powder Company.
67	IMR SR 4759	This bulky handgun powder works great in the magnums, but really shines as a reduced load propellant for rifle cartridges. Its large grain size gives good loading density for reduced loads, enhancing velocity uniformity.
68	Accurate Arms 5744	Fast-burning, double-base, extruded powder for a wide range of rifle calibers and magnum handguns. Low bulky density and superior ignition characteristics for rifle calibers and in large capacity black powder cartridges.
69	Accurate Arms 1680	Extremely fast-burning, double-base, spherical rifle powder for large capacity, high performance handgun cartridges. This powder is well suited for low capacity rifle cartridges.
70	NORMA 200	Norma's fastest burning powder suitable for smaller cartridges such as the .222 Remington.
71	Alliant Reloder 7	Designed for small caliber varmint loads. Also good in .45-70 Govt and .450 Marlin.
72	IMR 4198	This fast burning rifle powder gives outstanding performance in cartridges like the .222 Remington, .221 Fireball, .45-70 Govt., and .450 Marlin. Outstanding in cartridges like the .222 Remington, .444 Marlin, and 7.62 X 39.
73	Hodgdon H4198	This propellant has gone through some changes since its inception. The kernels were shortened for improved metering and modified to make it extremely insensitive to hot/cold temperatures. Great in .222 Remington, .444 Marlin, and the 7.62 X 39.
74	VihtaVuori N120	Slower burning powder for small capacity rifle cases and for lighter bullets in many .22 caliber loads. Burning rate is near to Accurate 1680, IMR 4198, and Reloder 7. N120 is suitable also for 7.62x39, .30-30 Winchester, and .444 Marlin.
75	Hodgdon H322	It provides match grade accuracy in small and medium capacity cartridges like the .223 Remington, 6mm PPC, and the 7mm TCU. As a fine extruded powder, it flows through powder measures with superb accuracy.

76		Accurate Arms 2015BR	Fast burning, single-base, extruded rifle powder for small to medium varmint calibers. Recommended for use in large bore straight wall cartridges.
77		Alliant Reloder 10X	For light bullet applications in .222 Rem., .223 Rem., .22-250 Rem., and key benchrest calibers. Also suitable in light bullet .308 Win. loads.
78		VihtaVuori N130	This powder is used in many factory-loaded caliber .22 and 6mm PPC cartridges. Suitable also for lighter bullets in caliber .223 Remington and for straight-wall rifle cases like .45-70 Government and .458 Winchester Magnum. Burning rate is close to Hodgdon H322 and Accurate 2230.
79		IMR 3031	A propellant with many uses, IMR 3031 has long been a favorite of .308 Match shooters using 168 grain match bullets. It is equally effective in small capacity varmint cartridges from .223 Remington to .22-250 Remington, and it's a great .30-30 Winchester powder.
80		VihtaVuori N133	A choice of many benchrest and standard rifle shooters using 6mm PPC. Used also in many loads of .222 Remington, .223 Remington, and in other applications where a relatively fast burning powder is needed, like in .45-70 Govt.
81		Hodgdon BENCHMARK	Ideally suited for bench rest and small varmint cartridges like the 6mm PPC, .22 PPC, 6mm BR, .223 Rem., and .222 Rem. With small, easy-metering granules, competitors will love how it flows through progressive presses.
82		Hodgdon H335	Originated as a military powder, used for the 5.56 NATO, or .223 Remington as handloaders know it. Obviously, it sees endless use in the .222 Remington, .223 Remington, and other small cartridges.
83		Ramshot X-Terminator	A double-base spherical powder with small grain size allowing trouble-free loading in small diameter case necks. It performs extremely well with light to medium weight bullets in .223 Remington.
84		Accurate Arms 2230	Fast burning, double-base, spherical rifle propellant. Designed for .223 Remington but can be used in many small and medium caliber cartridges. Ideal for progressive loading.
85		Accurate Arms 2460s	Fast burning, double-base, spherical rifle powder. Suitable for small and medium-sized caliber applications. Slightly higher loading densities than 2230. Provides an option for shooters to fine tune and optimize loads and combinations with calibers.
86		IMR 8208 XBR	The latest in the IMR line of propellants, this accurate metering, super short-grained extruded rifle powder was designed for match, and AR sniper cartridges. Ideally suited for .223 Remington/5.56mm, and .308 Winchester/7.62mm NATO.
87		Ramshot TAC	TAC is a double-base spherical propellant that sets the standard for extreme accuracy and reliability with heavy bullets in the .223 Remington and match applications in .308 Winchester.

88		Hodgdon H4895	This member of the extreme extruded line powder is suitable for .17 Remington, .250-3000 Savage, .308 Winchester, .458 Winchester, and more. It had its origin in the .30-06 as a military powder and was the first powder Bruce Hodgdon sold.
89		VihtaVuori N530	This is the fastest burning powder in the N500 series and its burning rate is close to Vihtavuori N135 and Hodgdon BL-C(2). Developed especially for the 5.56mm NATO cartridges and does well in .45-70 Govt. loads and light .308 Winchester loads.
90		IMR 4895	Originally a military powder featured in the .30-06, IMR 4895 is extremely versatile. From the .17 Remington to the .243 Winchester to the .375 H&H Magnum, accuracy and performance are excellent. In addition, it is a longtime favorite of match shooters.
91		VihtaVuori N135	An excellent powder for .308 Winchester loads with bullet weight less than 10 grams (155 grains). It will fit applications similar to IMR4064, Hodgdon H4895, or Accurate 2520. Capability for various loads ranging from .222 Remington to .458 Winchester Magnum.
92		Alliant Reloder 12	Discontinued.
93		Accurate Arms 2495BR	Accurate 2495 is a single-base, extruded rifle powder that was developed for the .308 Win and can be used over a wide range of rifle calibers. 2495 is a versatile powder with excellent ignition characteristics that provides excellent accuracy.
94		IMR 4064	One of the most versatile propellants in the IMR line, used for .223 Remington, .22-250 Remington, .220 Swift, 6mm Remington, .243 Winchester Super Short Magnum, .308 Winchester, .338 Winchester Magnum, and the list goes on and on.
95		NORMA 202	Specially developed to provide maximum performance in the .308 Winchester, 202 is a very useful powder for cases with medium capacity compared to caliber, such as the 8x57, 9.3x62mm, and 9.3x74R.
96		Accurate Arms 4064	Accurate 4064 is an intermediate burning, single-base, short cut extruded rifle powder designed for the .30-06 Springfield. Works well in .22-250, .243 WSM, 7×57 Mauser, and the .325 WSM. Popular choice for High Power shooters using the M1 Garand.
97		Accurate Arms 2520	Accurate 2520 is a medium burning, double-base, spherical rifle propellant designed for the .308 Winchester. 2520 is extremely popular with many service shooters. 2520 also performs extremely well in .223 Rem. with heavy match bullets (62 to 80 grain).
98		Alliant Reloder 15	Medium speed rifle powder providing excellent .223 and .308 caliber performance. Broad caliber range; consistent in all temperatures. Suitable for high velocity varmint loads.
99		VihtaVuori N140	A true multipurpose powder, which can usually be used in place of IMR4320, Reloder 15, or Hodgdon H380. Good choice also for .223 Remington, .22-250 Remington, .308 Winchester, .30-06 Springfield, 8x57 IS (8 mm Mauser), and .375 H&H Magnum.

100		Hodgdon VARGET	Small extruded grains for uniform metering, insensitivity to hot/cold temperatures, and higher energy for improved velocities over other powders in its burning speed class. Effective for .223 Rem., .22-250 Rem., .308 Win., .30-06, .375 H&H, and others.
101		IMR 4320	Short granulation, easy metering, and perfect for the 223 Remington, .22-250 Remington, .250 Savage, and other medium burn rate cartridges. Long a top choice for the vintage .300 Savage cartridge as well.
102		Winchester 748	The low flame temperature extends barrel life compared to other similar speed powder. It is ideal for a wide variety of centerfire rifle loads, including .222 Remington, .30-30 Winchester, .30 Winchester, and up to .485 Winchester Magnum.
103		Hodgdon BL-C(2)	A spherical powder that began as a military powder used in the 7.62 NATO, commonly known as the .308 Winchester. BL-C(2) works extremely well in the .204 Ruger, .223 Remington, .17 Remington, and .22 PPC.
104		Hodgdon CFE 223	Introduced in January 2012, this versatile spherical rifle propellant incorporates in its formula CFE, Copper Fouling Eraser. CFE 223 yields top velocities in many cartridges such as the .204 Ruger, .223 Rem., .22-250 Rem., and the .308 Win.
105		Hodgdon LEVEREVO-LUTION	This propellant meters flawlessly, makes lever action cartridges like the .30-30 Win., and yields velocities in excess of 100 fps over any published handloads with even greater gains over factory ammunition.
106		Hodgdon H380	This was an unnamed spherical rifle propellant when the late Bruce Hodgdon first used it. When a 38.0 grain charge behind a 52 grain bullet gave one-hole groups from his .22 caliber wildcat (now called the .22-250), he appropriately named the powder H380.
107		IMR 4007 SSC	Super Short granulated powder with a burn speed that falls between 4320 and 4350. Ideally suited for varmint cartridges like .220 Swift, .22-250 Remington, and .243 Win. Additionally, well suited for the venerable .30-06, the .300 Win. Short Magnum, and the Super Short Magnums .243 and .25 Winchester.
108		Ramshot Big Game	It's an extremely clean burning, double-base spherical powder geared toward .30-06 and .22-250 Rem.
109		VihtaVuori H540	Faster burning powder with a burning rate similar to N140 and close to Hodgdon H414 and Winchester 760. For situations where more power is needed, especially for .223 Remington, .308 Winchester, and .30-06 Springfield loads with heavier bullets.
110		Winchester 760	Combine Winchester components with 760 to duplicate .30-06 Springfield factory load ballistics. The powder has ideal flow characteristics that give it an advantage over other propellants. Excellent for .22-250 Rem., .300 Win. Mag., as well as .300 WSM.
111		Hodgdon H414	This spherical powder has an extremely wide range of use. From the .22-250 Remington to the .375 H & H, it will give excellent results. It is simply ideal in the .30-06. It delivers consistent charge weights through nearly any type of powder measure.

112		VihtaVuori N150	This powder burns a bit slower than N140 and works as well as Hodgdon H414 and Winchester 760. Typically used with heavier bullets in accuracy and hunting loads of cartridges with middle case volumes, like .308 Winchester and .30-06 Springfield.
113		Accurate Arms 2700	Accurate Arms 2700 is a medium burning, double-base, spherical rifle powder that is ideally suited for the .30-06 Springfield and other medium capacity calibers such as the .22-250 Remington, .220 Swift, and the .243 Winchester.
114		IMR 4350	The number one choice for the new short magnums in both Remington and Winchester versions. For magnums with light to medium bullet weights, IMR 4350 is the best choice.
115		Hodgdon H4350	H4350 is ideal in the WSM family of calibers (.270, 7mm, .30, .325). H4350 is the standard in cartridges like the .243 Winchester, 6mm Remington, .270 Winchester, .338 Winchester Magnum, and many more.
116		Alliant Reloder 17	Short Magnum Rifle powder that meters easily and consistently while providing maximum velocity even in extreme weather. Designed for short magnum case capacity. Similar burn speed to IMR 4350. Consistent maximum velocity in extreme weather conditions.
117		Accurate Arms 4350	Accurate 4350 is a short cut, single-base, extruded rifle powder in the extremely popular 4350 burn range. Good in .243 Win. to the .338 Win. Mag. Accurate 4350 is an exceptional choice for the 6mm Rem, .270 Win, .280 Rem., and .300 WSM.
118		NORMA 204	A slow-burning propellant, 204 provides good performance and quite good accuracy in calibers such as 6.5x55 and .30-06.
119		Hodgdon HYBRID 100V	Hybrid 100V has a burn speed between H4350 and H4831. The powder yields superb performance in such popular calibers as .270 Winchester, .243 Winchester Super Short Magnum, 7mm Remington Magnum, .300 Winchester Magnum, and dozens more.
120		VihtaVuori N550	Burning rate is similar to N150 and close to IMR 4350 and Reloder 19. Good choice for more powerful loads for 6.5x55 SE, .308 Winchester, .30-06 Springfield, and many others.
121		Alliant Reloder 19	Provides superb accuracy in most medium and heavy rifle loads and is the powder of choice for .30-06 and .338 calibers.
122		IMR 4831	Slightly slower in burn speed than IMR 4350, IMR 4831 gives top velocities and performance with heavier bullets in medium-sized magnums.
123		Ramshot Hunter	A double-based propellant made for the .270 Winchester, .300 WSM, and .338 Win Mag-type cartridges. Compares to IMR 4350 burn rates.

124		Accurate Arms 3100	A slow burning, single-base, extruded rifle powder specifically designed for large capacity overbore magnum calibers such as the 7mm Rem. Mag., .264 Win. Mag., etc. It provides for high loading densities in these calibers. Will also work well in the new series of WSM and Remington SAUM calibers.
125		VihtaVuori N160	Slow burning powder for magnum cartridges and calibers with large case volume and comparatively small bullet diameter. Burning speed of N160 is close to Reloder 19, Winchester WMR, and the various 4831s. Ideal applications are .243 Win., .300 Win. Mag, .338 Win. Mag.
126		Hodgdon H4831 & H4831SC	It is probably safe to say more big game has been taken with H4831 than any other powder. Bruce Hodgdon introduced this popular burning rate in 1950. A favorite for cartridges like the .270 Winchester, .25-06 Remington, .280 Rem., and .300 Win. Mag.
127		Hodgdon SUPERFOR-MANCE	Delivers striking velocities in cartridges like the .22-250 Remington, .243 Winchester, and .300 Winchester Short Magnum. Velocities well in excess of 100 fps over the best published handloads and even larger gains over factory ammunition.
128		Winchester Supreme 780	Same as powder loaded in Winchester's factory supreme ammunition, such as .243, .270 Win., and .300 Win. Mag. Works well in short mags.
129		NORMA MRP	A very flexible magnum powder, MRP is suitable for calibers with relatively large case volume in relation to the caliber. It has been well-known for many years as a high performance powder in all magnum calibers.
130		Alliant Reloder 22	For big game loads. Provides optimum metering. Powder of choice for .270, 7mm Mag., and .300 Win Mag.
131		VihtaVuori N560	Burning rate is between N160 and N165 and close to Norma MRP and Reloder 22. Powder especially for magnum cartridges from .270 Winchester, 7 mm Rem. Mag., 7 mm Weatherby Mag., .300 Winchester Magnum, and .338 Lapua Mag.
132		VihtaVuori N165	A very slow burning powder for Magnum cartridges with heavy bullets. N165 offers performance equal to Norma MRP and Reloder 22. To be used with heavy bullets in calibers ranging from 6.5x55 SE all the way to .416 Rigby.
133		IMR 7828	This slow burner gives real magnum performance to the large over-bored magnums, such as the .300 Rem. Ultra Mag., the .30-378 Weatherby Mag., and the 7mm Rem. Ultra Mag. This powder is highly regarded by most magnum shooters.
134		Alliant Reloder 25	Advanced powder for big game hunting. Features improved, slower burning and delivers high energy and maximum velocity. Ideal for over bore magnums.
135		VihtaVuori N170	The slowest burning N100 series rifle powder from VihtaVuori and one of the slowest canister reloading powders generally available from any manufacturer. Good performances in most of the belted Magnum cartridges like .338 Lapua Magnum.

136		Accurate Arms Magpro	Slow burning, double-base, spherical rifle powder for short magnums of both Winchester and Remington.
137		Hodgdon H1000	Very slow-burning extruded powder perfect for highly over-bored magnums like the 7mm Rem. Mag., 7mm STW, and the .30-378 Weatherby. With heavy bullets, H1000 gives top velocity in such cartridges as the .270 Win. and .300 Win. Mag.
138		Ramshot Magnum	High performance magnum rifle powder with outstanding performance from the popular 7mm Rem. Mag. and .300 Win. Mag. through the Remington Ultra Mags and .338 Lapua.
139		Hodgdon RETUMBO	Designed for large overbored cartridges such as the 7mm Rem. Ultra Mag., .300 Rem. Ultra Mag., .30-378 Weatherby Magnum, and others. This powder adds 40-100 fps more velocity to these cartridges when compared to other normal magnum powders.
140		VihtaVuori N570	This is the newest member of the N500 series powders and also the slowest burning. The burning rate is near to N170 and faster burning than 24N41. Good for large volume cases like .300 Win. Mag., .300 Rem. Ultra Mag., .338 Lapua Magnum.
141		Accurate Arms 8700	An extremely slow burning, double-base, ball-type powder that is Accurate's slowest powder. It was originally developed for the .50 BMG. Works optimally in heavily necked-down large capacity cartridges such as the .264 Winchester Magnum.
142		Hodgdon H870	Discontinued powder that was originally distributed beginning in 1959 and was surplus 20mm Vulcan Cannon powder.
143		VihtaVuori 24N41	Single-based powder well suited for slow-burning .50 BMG loads, as well as .338 Lapua Magnum.
144		Hodgdon H50BMG	As the name implies, this new generation Extreme Extruded rifle propellant is a clean burning powder designed expressly for the .50 caliber BMG cartridge. Because it shares the same technology as Varget, H50BMG displays a high degree of thermal stability.
145		Hodgdon US869	Magnum spherical rifle powder for heavy bullets in big, overbore rifle cartridges. Great for 7mm Remington Ultra Magnum, .300 Rem. Ultra Mag., .30-378 Weatherby Mag., and others. US869 is superior in the .50 Caliber BMG.
146		VihtaVuori 20N29	Single-based powder well suited for .50 BMG, .338 Lapua Magnum, and .30-378 Weatherby Magnum.

There are kits available that include most of the euipment you will need to get started.

Reloading is a tool-intensive activity. In fact, a flip through a reloading supply catalogue or a browse through a reloading supply website reveals a dizzying amount of choices, some of which are really essential and some of which are a complete waste of your money. One of the goals of this book is to get the reader through the reloading process with a simple, but not necessarily indulgent selection of equipment.

Be wary of anyone suggesting long, laborious, and complicated reloading processes that require you to mortgage your house to buy a truckload of equipment. More than one of the shooting, hunting, or reloading web forums I visit has such information permanently tacked to the top of the reloading forum list. Pages of information exist on how to turn the necks of your brass (to reduce/uniform wall thickness), how to deburr flash holes, how to adjust your sizing dies to achieve infinitely small amounts of shoulder "bump" (resizing), how to "uniform" primer pockets, how you should sort your cases by weight, and how to achieve better accuracy from your rifles by testing handloads where you have varied the bullet seating depth by 0.002"- 0.005" in length. Run. Run away. 99 percent of shooters are completely wasting their time by performing most of those steps. I've tested and proven the futility of much of this work. In most cases, it's just a waste of time.

We will cover what really does matter. Do yourself a favor and ignore everything else for at least the first couple years of your reloading experience.

If you decide to pursue a national-level trophy in any of the Benchrest disciplines and have the skills and equipment to be competitive, some of the minutiae I'm suggesting you avoid might make a small difference for you. For now, stick with the basics we cover in the following chapters, and you will have 95 percent or more of the information you need to be a successful reloader.

Reloading Bench

If you're going to reload, you're going to need a reloading bench. This will be the home for your press, dies, components, trimmers, and all the other tools and toys you will accumulate for various reloading purposes. I've always made my own benches, but there are several options available for purchase as kits or complete benches. Keep in mind that what I'm calling a "reloading bench" is essentially just a workbench like any other, with the only exception being that the height of the bench top may need to be higher for our purposes.

A quick Internet search reveals several pre-made reloading bench options that are available from some of the larger online/catalog sporting goods retailers. These feature heavy wooden bench tops, metal legs, pull-out drawers, and upright pegboard rear panels, some with florescent lights built in. These are great features for a reloading bench! Unfortunately, most of these pre-made "reloading" benches are on the small side of what I prefer. They will be perfectly fine for a new reloader, but (like gun safes) the bench you need today will be outgrown by tomorrow.

In addition to a benchtop, you will need sturdy shelves. Dies, bullets, brass, and powder are all heavy, and you'll be surprised how quickly your inventory grows as you try new combinations of components. Consider the types of containers or storage bins you have available when sizing your shelves. I would suggest electrical power at some point. This can be as simple as running a power strip/extension cord from the nearest outlet. Some reloading processes require examination of small details so you will need plenty of light. I have a bright lamp on each end of the bench, in addition to several overhead lights.

If you plan to build your own bench, there are dozens of plans online that can be adapted to

A sturdy reloading bench made from home-store materials can hold all of your equipment and provide a stable mounting point for your press.

your needs, and most home/hardware stores have workbench kits. An image search for "reloading bench" turns up hundreds of examples, ranging from a repurposed nightstand to some really impressive and expensive custom-built benches with enclosed kitchen cabinet-style storage.

I prefer a 6' minimum bench length. 18" to 24" depth works well. A benchtop height of 42" is ideal for me. I'm 5'10" so you may need to adjust for your height or preferences. That's for a standing reloader, by the way. I sit in front of a computer enough during the week, so I like to reload standing up. There's enough fetching, stacking, and retrieving from shelves in the reloading process that I wouldn't be able to sit for very long anyway. Free plans for a simple but very effective reloading bench you can build from homecenter/lumber yard components for around $125 can be found on my website: *www.gamescoutusa.com*. The bench design is optimized for ease of construction and may require no cutting of lumber on your part. Most of the chain homecenter-type stores will cut your boards and lumber for you. If your local lumber store offers this cutting service, the assembly of the bench is a matter of driving a few nails or turning a few screws.

Other Options

If you live in an apartment or are otherwise restricted on space for your reloading area, a foldup work table can be found for $20 to $30. An online tool retailer lists what they call an "Adjustable Height Heavy Duty Work Station" for $27 that would make a fine bench for the small-space reloader or for use as a portable reloading table at the range. This can be a huge time saver if you live quite a distance from the shooting range where you will test your new handloaded ammunition. Think of the trips back and forth you can save yourself by loading and testing right at the range. Load a few rounds, then shoot a few rounds. You will have immediate feedback on your chosen load combination.

Storage

I can assure you that keeping your bench and related equipment and components organized and accessible can quickly become a challenge. Your local supercenter or discount store will have a range of plastic storage containers that can ease

the pain of organization somewhat. The six-quart size is ideal for holding rifle or pistol brass, bullets, loading dies, and ammunition. They are also pretty good for holding spare rifle or pistol magazines. These containers are stackable and the clear plastic allows you to see what's inside without needing to remove the lid. I generally make a large label for the end of each container with masking tape and a permanent marker to make identification even easier.

The larger tub type bins are great for holding multiples of the smaller bins mentioned above. I'll generally have one of these bigger bins per cartridge. I can keep all of my 9mm brass, for example, in one, two, or three of these bins. Not only do these bins keep my components organized, but they prevent most spills and improve portability. Much of my components "stash" is in the garage, while my reloading bench is upstairs in a spare bedroom. Staying organized with the

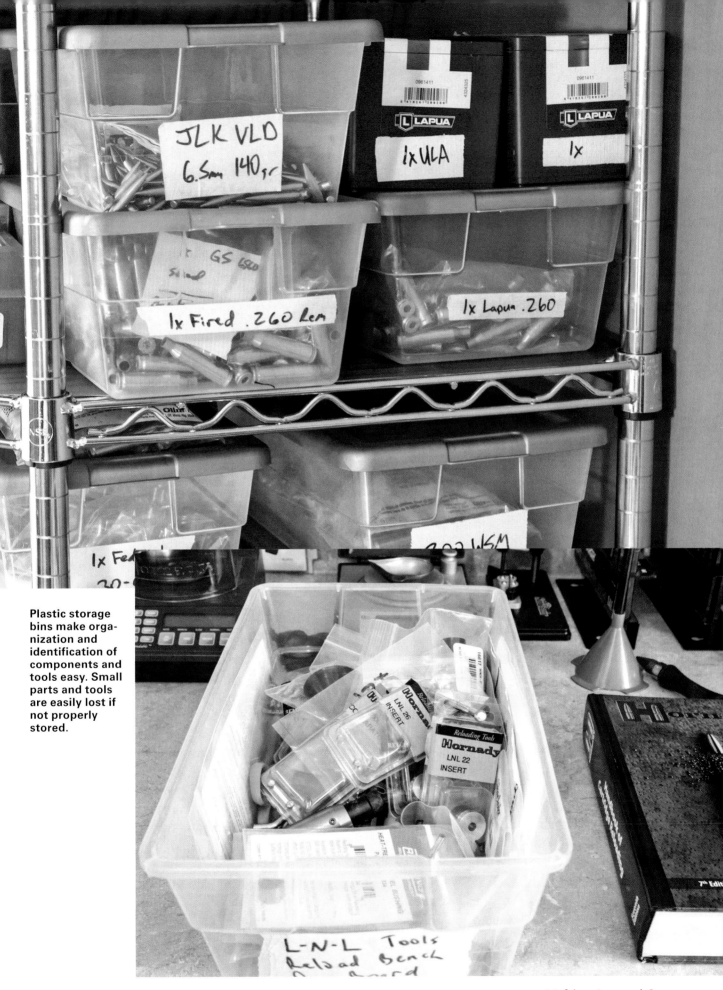

Plastic storage bins make organization and identification of components and tools easy. Small parts and tools are easily lost if not properly stored.

Case cleaning is a messy process and your cleaning area will likely look like this. Don't do this inside unless you like being single or divorced.

use of the various bins makes it easy to grab one container downstairs and have what I need to load for the evening. Although we have already discussed a number of safety issues, it's worth noting that these storage bins help keep children and pets out of our ammo components, too.

Case Cleaning Area

A big part of your reloading efforts will revolve around case preparation. Once fired, your cartridge cases will be oversized, dirty, probably a bit dinged and dented, and in need of some degree of cleaning. For blasting ammo to be used in pistols or carbines, a minimal cleaning may suffice. One reloader I know only cleans his pistol brass every other time he loads it. Though he apparently gets by with that, keep in mind that whatever residue there is left on your cases will soon go into your

expensive dies that are mounted in your expensive press that makes ammunition for your expensive firearms. Even a quick half-hour tumble in some Flitz-treated walnut polishing media will make a big difference in the cleanliness of your cases. Your reloading process will go more smoothly, your bench will stay cleaner, your equipment will last longer, and your loaded ammunition will likely function better in your firearms.

Though I fully understand and appreciate the benefits of the newer types of case cleaning equipment and processes, my old vibratory tumbler still works fine and is relatively easy to use. As a result, my case-cleaning area is fairly simple. I do all my cleaning in the garage where I can easily sweep and clean the concrete floor. A bottle of liquid case polish and a tin of mineral spirits sit on a nearby shelf. A bucket and my media separator sit next to the tumbler. That's it. The main point when choosing your case cleaning area is to be prepared for frequent messes. Trust me when I say you do NOT want to set up your case cleaning area inside the house!

Case Preparation Equipment

Before you load new brass or reload fired brass, you will need to perform a few steps to bring your cases up to a ready-to-load state. Always perform a visual inspection of the outside of the case for cracks or other damage. Small dings and dents are fine. Check for the presence of the extractor rim. Cases will sometimes slip through the quality control checks without an extractor groove being cut properly (or at all). I found an example of this just last night as I sized some 7mm-08 cases. Always check for the presence and condition of the flash hole. I have seen numerous cartridge cases with no flash hole at all! If you load and fire a case without a flash hole, obviously the primer

can't ignite the powder. The primer will dislodge itself from the primer pocket and fall into your action when you eject the faulty cartridge. If you're shooting an AR15, that loose primer will invariably find its way into your lower receiver (between the trigger and receiver) and cause the gun to be totally inoperable until you pull the upper and lower apart and fish out the rogue primer. I have witnessed this many times, and the lack of a flash hole isn't the only reason this can occur.

With new brass I will always size the necks outside, then inside. During shipping, the cases bang against each other and against whatever else is in the shipping box as it gets thrown, dropped, bounced, crushed, and generally abused by the parcel carrier. With the many thousands of pieces of brass I have purchased over the years, I have only had one new case arrive with the mouth so damaged that the case had to be thrown away. Otherwise, a quick trip through the sizing die irons out any issues.

Chamfer

We chamfer the square corner on the inside of each case mouth to ease the entry of the bullet's base into the case neck, reduce the scraping of copper from the outside of our bullet, and help ensure the bullet moves into the neck evenly during seating. This chamfering process sometimes rolls a small burr of brass to the outside of the mouth. We will deburr the outside of the mouth to remedy this. With a little technique during the chamfering process, it is often unnecessary to deburr the outside of the neck. Still, always make a cursory check for burrs. It is possible that large burrs on the case mouth can have a minor impact on feeding and accuracy, especially in custom or "match" rifles built with chambers on the minimum side of the tolerance range. Most factory-chambered rifles have enough room in the neck area of their chamber that we'll never see any negative effects of even moderate burrs.

Note that chamfering and deburring case mouths is almost exclusively done when loading rifle cartridges. Most pistol sizing dies will flare the mouth of the case slightly to eliminate the need to chamfer the inside of cas   e mouth. The crimp that we will apply at the end of the pistol-cartridge loading process generally eliminates the need to deburr the outside of the case. I have yet to experience a single problem related to this in tens of thousands of rounds of hand-loaded pistol ammo.

Chamfering and deburring can be done with simple hand tools or with more expensive and complex (though much faster) electric tools. The simplest are handheld metal tools with an inside-chamfer cutting head on one end and an outside-deburr cutting head on the other. I have used these, and they're fine for small-volume prep work. Models that have a nice handle to grip are easier on the fingers. These will only have

Chamfering the case mouth with a power-driven cutter

Deburring the case mouth

one tool each, so you will need one to chamfer and another to deburr. These are also fine for small-volume prep work and any touch-ups you may need after using fancier electric tools. You may notice that some inside-chamfer tools will be labeled as providing a "VLD" chamfer. The term "VLD" is taken from a particular bullet design, but in this case it relates to the angle of the cutting tool's sharpened surface. VLD chamfers provide a more gradual taper to the inside edge of the case mouth. I haven't found there to be that much of a difference in application between VLD and non-VLD tools, so whichever you find in stock at your reloading supply shop will be fine.

Electric Case-prep Tools

Electric case-prep tools can really reduce the time and elbow grease required to get your cases ready for loading. These motor-driven units range from single tools that are inserted into the chuck of your cordless drill to stand-alone multi-tool benchtop work stations with half a dozen interchangeable brushes, cutters, or reamers. I've used the RCBS version of the latter for years with acceptable results. The motorized rotary tools of this machine handle inside-neck chamfering, outside deburring, primer pocket cleaning, and case-neck brushing in sequence. That's four processes completed after I handle the cartridge case just once. The only knock I have against the RCBS unit is that the chamfer and debur tools aren't as good as I think they should be. With a little improvement to the cutting surfaces on these, they'd be easier and faster to use. Hornady has recently come out with a competing unit. I haven't had the chance to use one yet, but a reloading friend tells me good things about his. There are others on the market, so read some reviews, talk to reloading friends, and make your choice based on what you discover. None of these multi-station units are inexpensive.

Neck turning a rifle cartridge case.

Neck Turning

The process of "neck turning" is very similar to the "turning" process performed on any lathe. If you're a woodworker, machinist, or automotive brake technician, you know exactly what I'm talking about. We'll rotate our cartridge case on its long axis between two points (lathe-style) or on a turning mandrel. A turning mandrel is a carefully sized rod that fits perfectly inside the case neck. The mandrel supports the neck from the inside while the cutter removes material from the outside of the neck. There are several reasons to turn the necks of your cases.

Consistent Thickness

We turn necks to create a consistent neck wall thickness. Consistency on the case neck area certainly is important for best accuracy. If our brass is thicker on one side than the other, the chance that our bullet will enter the rifle's barrel crooked is increased. This can be a result

Close-up view of neck turning tool, cutting bit, and turned neck.

of actual misalignment with the bore (caused by a relatively large discrepancy between the wall thickness on each side of the neck or by an uneven release of the bullet during firing due to inconsistent neck tension). If neck tension is not even around the circumference of the bullet, one side of the neck will release its grip on the bullet sooner than the other side. Keep in mind that we're talking about events happening in milliseconds. It doesn't take much to disturb the path of the bullet when things are happening at such speed. By making a very light "partial turn," we can skim off the high (thick) spots or the high (thick) side, while leaving the thinner areas untouched. This should result in more even neck tension, which will help ensure a consistent release of the bullet and eliminate at least one variable that may affect our bullet to bore alignment.

By taking an aggressive cut, we can remove material from the entire circumference of the neck. This may be necessary in order to use some brands of brass in some custom-chambered rifles. I have a .260 that falls into this category. To use one brand of imported brass, I have to turn necks down to around 0.0145" wall thickness.

When we form brass for one cartridge, say an obscure or wildcat round, we may need to neck turn in order to bring our brass back down to an appropriate thickness. The brass in the neck area is generally thinner than the brass in the case shoulder or body. When we shorten and form cases for a completely different cartridge than we started with, we sometimes end up with a case neck that's made from what used to be the case body. We will likely need to turn the new necks to bring them to the appropriate thickness.

Examples of fully turned (l), partially turned (c), and unturned (r) necks.

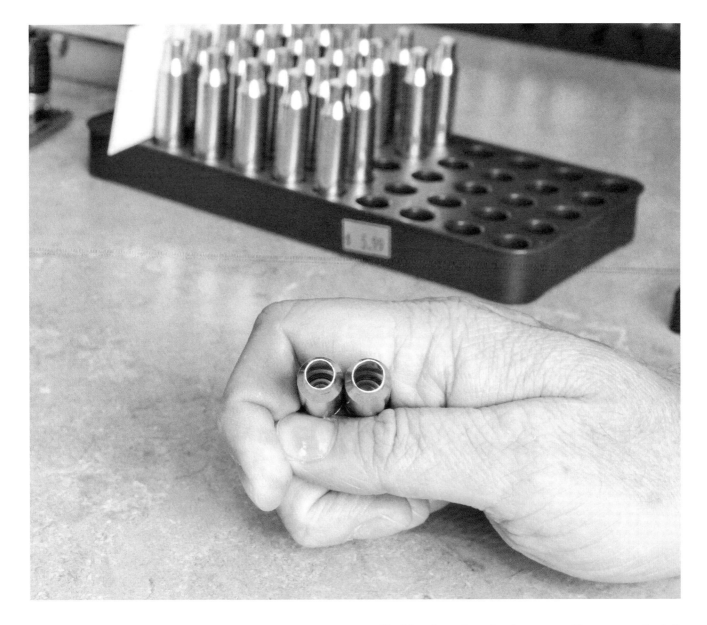

Neck Down

In some instances, brass for less popular cartridges may only be seasonally available or simply hard to find altogether. Often, we can convert cases from another similar cartridge to meet our needs. This is the current situation with 7mm Remington Short Action Ultra Magnum (7mm SAUM) cases. The 7mm SAUM—actually all of the factory-offered SAUM cartridges, for all practical purposes—are dying a free-market death. They just couldn't overcome the head start and performance edge that their rivals, the Winchester Short Magnums (WSM) had in the marketplace. Much of the reason for this shortage is due to a surge in demand for the wildcat 6.5mm SAUM. Shooters create 6.5mm SAUM brass by necking down, trimming and lightly (if at all) turning the necks of 7mm

Necking down is a simple process. The case on the left is now ready to load as a 260 Remington. The case on the right is the parent, a 7mm-08 Winchester. We simply neck down 7mm-08 brass 0.020" to form 260 Remington brass. I prefer to then run the newly formed 260 brass over an expander to smooth out the inside of the neck.

SAUM cases. Since 7mm SAUM brass is being gobbled up faster than it is being manufactured, shooters have begun converting .300 SAUM cases. When we neck down one "step" or roughly 0.020", we can often do no other work on the case necks. I've necked down a couple thousand 7mm-08 cases to make .260 Remington brass with no prep work beyond full-length sizing and chamfering. The same works for the 7mm SAUM to 6.5 SAUM conversion. When we start with .300 SAUM brass, we go from a case neck intended to hold a .308" diameter bullet down to

Neck expanding is perfomed on the press like any other sizing operation. Use the expander to "neck up" brass to form another cartridge or to smooth out the inside of the neck for better accuracy.

one intended to hold a 0.264" diameter bullet, a difference of 0.044" that results in an increase in neck thickness that requires turning. If you think about it, it makes sense. Our sizing die is forcing a large amount of material into a smaller shape from the outside. All that brass has to go somewhere, so the result is a slightly thicker and longer neck.

Neck Up

When we move brass in the opposite direction by "necking up," we may run into a different issue that requires us to neck turn our brass. A "do-nut" occurs when we expand our case necks from the inside, moving part of the shoulder into the new neck area we've formed. If I start with .243 Winchester brass and expand the neck to create .308 Winchester cases, I'll be moving shoulder (thicker) material into, the neck of my formed .308 brass. This will require me to turn my case

necks to remove the ring of thicker material at the bottom of the neck that used to be in the shoulder area of my .243 brass.

Good Thickness

Whatever reason you have for turning necks, a good goal for resulting neck thickness is in the 0.013" to 0.015" range, based on your chamber dimensions and starting thickness. I don't like to remove any more material than is necessary, not only to preserve the integrity of my brass, but also because it is a fairly time-consuming process. Outside of case-forming and chamber-fit issues, I'd venture that 98 percent of reloaders are wasting their time turning necks. Turning 0.0005" off of one side of a case neck won't change the fact that the chamber on a factory-built rifle has twice that much misalignment from the manufacturing process. It also won't change the fact that a factory-spec chamber is so large in diameter that our cartridge has 0.003" in any direction to shift when chambered. Many rifles have spring-loaded ejectors that place constant spring pressure on one side of the case head. This causes the cartridge to be held forward and to turn slightly, taking up any slack or space in the loosely cut rifle chamber. If a chamber is generously cut and an ejector is pushing the cartridge off axis from the bore by 0.002", then it really doesn't matter if we've removed 0.0005" from one side of the case neck. This issue is greatly diminished when we reuse our fired brass. When new brass is fired, it expands to take the shape of our chamber and much of this "slack" is eliminated.

Overall Evenness

Another consideration when attempting to correct brass with uneven neck wall thickness is that the unevenness most likely extends down the entire case length. You can turn the neck, but the body will continue to grow unevenly over multiple firings, pushing back up through the neck eventually. How many firings will this take to become an issue? Who knows? Just keep that in mind. Personally, if I'm turning necks for concentricity and find a case that requires more than a very minimal cleanup, that case either goes in the trash or is segregated for use in practice-only ammo.

On the other hand, if we have a chamber that is relatively snug, our cartridge has little room to move off axis. In this case, the effects of turning brass may justify the effort. If our chamber is perfectly cut in diameter and headspace and is concentric to the bore, unevenness in our neck thickness may be the last area to tackle in pursuit of perfect bullet-to-bore alignment.

In the end, unless you're a dedicated benchrest shooter who competes at a high level, you are FAR better off spending your time learning to observe and correct for wind speed, direction, and changes thereto than you are turning brass

While calipers are not the preferred tool used to measure case-wall thickness, that's what most of us have on the bench. Here, we can see the thin side of a cartridge case with uneven wall thickness.

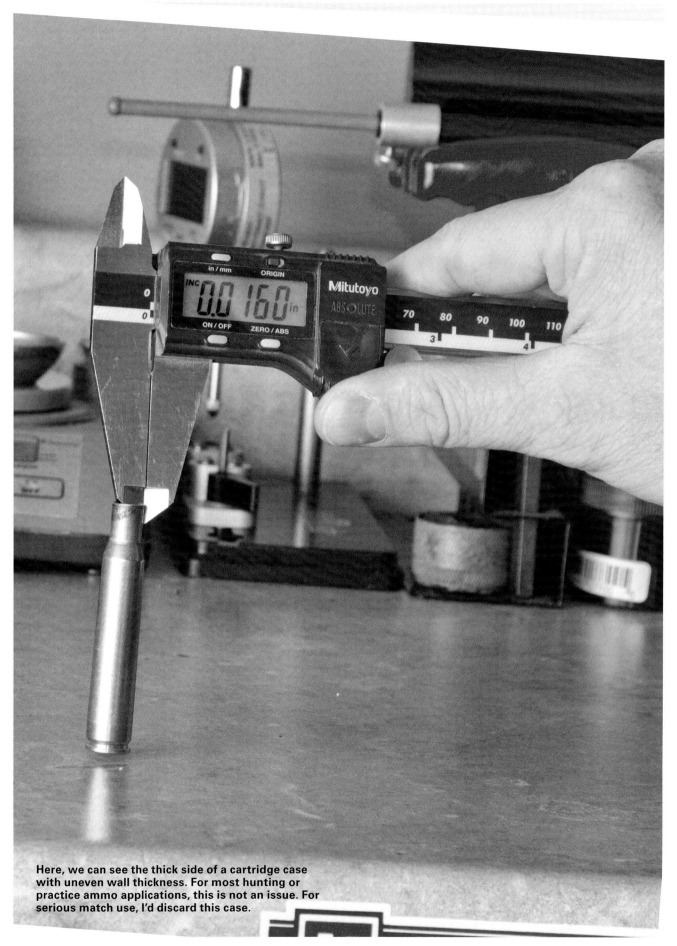

Here, we can see the thick side of a cartridge case with uneven wall thickness. For most hunting or practice ammo applications, this is not an issue. For serious match use, I'd discard this case.

for concentricity.

Case Cleaning

Case Cleaning and polishing is primarily done in one of several ways: vibratory tumbling, stainless-media rotary tumbling, or ultrasonic cleaning. Each has advantages and disadvantages. The most basic and common is the vibratory tumbler.

Vibratory Tumbler

Ground walnut shell or ground corncob "media" is poured into the bowl of the vibratory tumbler. If desired, a liquid polishing additive is poured into the media and allowed to mix with the media while the tumbler runs for several minutes before adding brass to the bowl. I have been tremendously pleased with the Flitz brand brass-polishing additive that's marketed specifically to reloaders. Next, you add your brass to the running tumbler a handful at a time to spread it out a bit in the bowl, close the lid, and allow the tumbler to work. Leave the brass in the tumbler for anywhere from a half hour to three hours, depending on the condition of the brass and how clean and shiny we want it to be. If you forget to turn off the tumbler—before you go to bed, for instance—and it runs for several hours, it is no big deal and won't hurt the brass.

To remove the brass from the tumbler, first pour the brass and media through a strainer-like device called a "separator" into a bucket. The separator will hold the brass while the media falls into the bucket below. Then shake and stir the brass vigorously in the separator to get the last bits of media out of the cases. Really shake and stir aggressively here. Each case needs to be upside down at some point in order for the media to fall free. I highly recommend that you tumble brass with the primers still intact when possible, especially when using corncob media. If you tumble with the primer pockets empty, media can get stuck in the flash holes of your brass. Trust me when I

The vibratory tumbler uses dry media and intense vibration to polish and clean your cases. In this image, we see the last few cases from a batch disappearing into the whirling media.

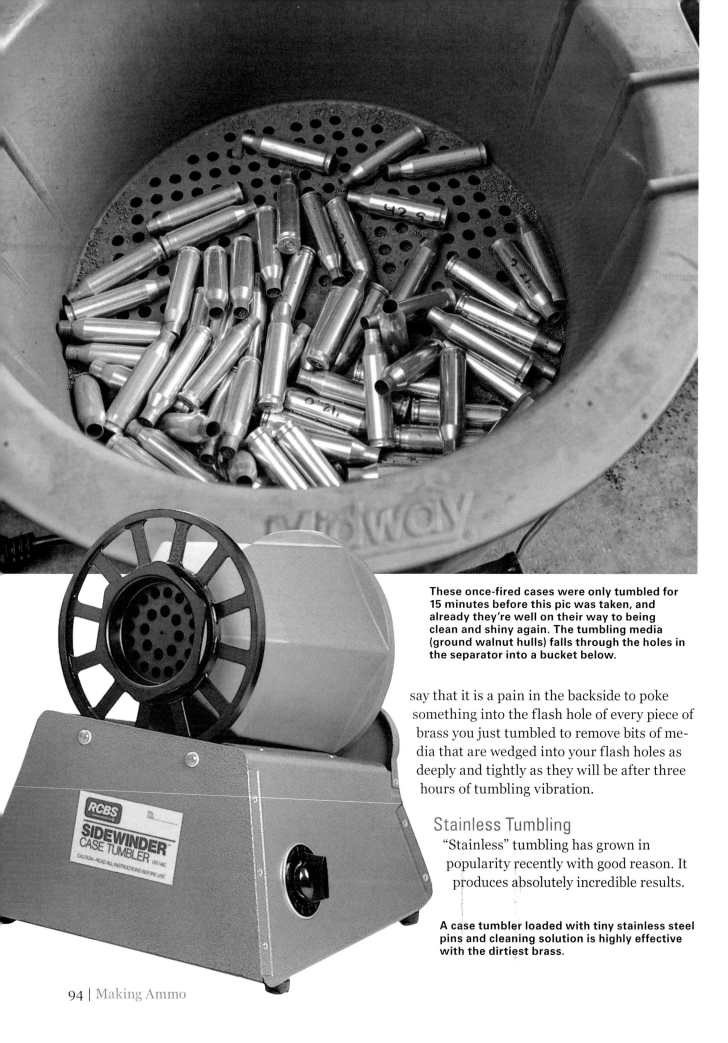

These once-fired cases were only tumbled for 15 minutes before this pic was taken, and already they're well on their way to being clean and shiny again. The tumbling media (ground walnut hulls) falls through the holes in the separator into a bucket below.

say that it is a pain in the backside to poke something into the flash hole of every piece of brass you just tumbled to remove bits of media that are wedged into your flash holes as deeply and tightly as they will be after three hours of tumbling vibration.

Stainless Tumbling

"Stainless" tumbling has grown in popularity recently with good reason. It produces absolutely incredible results.

A case tumbler loaded with tiny stainless steel pins and cleaning solution is highly effective with the dirtiest brass.

Instead of walnut shells or corncob as tumbling media, we use small stainless-steel rods to clean our brass. Obviously, stainless steel rods will provide a much more thorough cleaning than corncob or walnut shell media. Brass cleaned with this process will emerge looking bright, shiny, and brand new, even if it was heavily tarnished or dirty before tumbling.

Note that, unlike vibratory tumbling, stainless tumbling is a wet process. The equipment required is also totally different. We will use a "rotary tumbler" instead of a vibratory model. Our rotary tumbler has a removable drum that we'll fill with water, stainless media, and brass. Some sort of mild soap is often used as well. Once the ingredients are inside, seal the tumbler drum and place it sideways on the rotary base where it will roll until the brass is clean. Think of the rotating drum on a cement truck and you'll have the idea.

The downsides to stainless tumbling are the increased time and difficulty of the wet process. Our brass must be fully dried inside and out before we can begin reloading it. We have to individually arrange our cases so that the water can fall from the inside out of the case mouth or the flash hole. Notice that our flash holes must be open for that to happen. We have to deprime our cases before tumbling in a wet process or we'll trap water in the primer pocket between the primer and case. Depending on your reloading process, this may mean an additional step and additional time to deprime cases. Using a hair dryer to move warm air through the cases can speed up the drying process. If it's a sunny summer day, you can set your upright cases outside and the sun will do much of the work for you. Avoid the temptation to put your cases in the wife's oven, though. Household ovens can easily get hot enough to damage brass, and this approach exposes our food-preparation area and

Ultrasonic cleaners yield good results with dirty brass.

family to potential lead contamination. Neither is good, and both are easily avoidable.

None of this is intended to dissuade you from selecting the stainless media option, of course. There is no doubt that it produces the cleanest and most beautiful brass of any of these methods. Just be aware that the equipment is more expensive, the process is more time consuming, and the clean-up is more involved when making your decision about what brass-cleaning process to select. Also, who says you can only have one? I use vibratory tumbling for the majority of my cleaning, but I will also use stainless tumbling with cases that are really dirty or oxidized, and the additional time is justified.

Ultrasonic Cleaning

Ultrasonic cleaning has been used to clean guns, parts, and tools for many years. Within the last decade or so, it has grown increasingly popular as an option for cleaning brass. Ultrasonic clean-

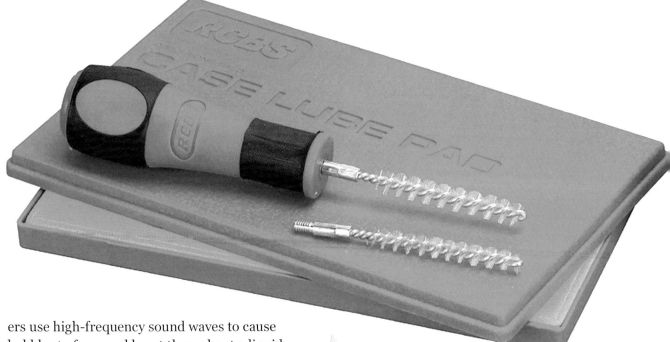

ers use high-frequency sound waves to cause bubbles to form and burst throughout a liquid immersion. Any surface the liquid reaches will be cleaned by millions of bursting bubbles and the resulting heat and agitation they produce. That's only a layman's description, but the effect is that cleaning takes place in all the nooks and crannies both inside and out.

The downside to ultrasonic cleaning is that it is a wet process and most cleaning units targeted at reloaders are small with a very limited capacity. Both of these factors add up to more time and hassle. The more cases you add to the solution, the less cleaning will occur and the longer the process will take.

As with stainless tumbling, ultrasonic is an effective option that can be used when time allows or when brass is extremely dirty. For routine use, I still suggest a vibratory tumbler with ground walnut media. After a few firings or when dry tumbling won't get your cases as clean as you like, break out the stainless or ultrasonic equipment.

Lubrication

Case lubrication is required when loading rifle cartridges. Straight-walled pistol cases like the 9mm Luger or .45 ACP generally don't need to be lubricated in modern pistol cartridge dies with carbide or titanium-based sizing rings. Check the instructions that came with your dies for guidelines. If you haven't purchased your pistol dies yet, look for those that specify no-lube siz-

A small dab of lube outside the case body and inside the case neck is critical to the resizing process.

ing. Standard steel sizing dies will require the use of case lube. Carbide dies are absolutely worth the extra expense if you can swing it.

When you are sizing cases that require lubrication, remember to lube the inside of the case neck when using an expander ball or expander mandrel. If you try to resize a rifle cartridge case without lubrication I can assure you the case will get stuck in your sizing die. Applying too much lube can also cause problems. Too much lubricant build-up in your sizing die can collapse the case shoulder upon resizing. All of that built-up lube has to go somewhere. If the case lube can't escape the path of the cartridge case as it fits firmly inside the sizing die, the case will be deformed and dented. Regardless of the type of lube you select, use just enough to get the job done and no more.

Traditional

The traditional way to lubricate cases is to apply case lube to a flat foam pad and roll the cases across the pad. While this is effective, it is slow and

A gallon-sized zip bag easily contains the mess of applying case lube to a handful of brass.

relatively tedious. Only a few cases can be done at a time, and you will end up with a much lube on your fingers as you have on your cases. A separate step is required to lubricate the inside of the case necks. One thing to note is that the insides of all necks don't need to be lubricated. If you generously lubricate the inside of every third or fourth neck there will be enough lube left on the expander to allow easy sizing of the remaining necks.

Quicker

A quicker way to apply lube is with the use of spray or aerosol case lube. I use an aerosol case lube from Hornady exclusively. I have heard complaints about aerosol lube creating a mess but have a simple technique that is quick, easy, and totally mess free. I use a gallon-size Ziploc bag to contain any spray during lubrication. Grab a handful of brass, drop it in the bag, and generously spray case lube inside. After the first spray, I shake and roll the brass inside the bag to ensure an even coating. I then repeat the process with a second round of spray and roll for a bit of insurance. Since the aerosol lube lands inside some of the case necks, no separate step is required to apply lube there.

The vibratory tumbler and media separator are the basic tools used to clean cases.

Removing Residue

I always tumble cases after resizing to remove any lubricant residue. If I'm working with a small number of cases I may simply wipe the residue away with a cloth, q-tips, and solvent. Some people skip this step, but lubricant residue attracts dust and dirt that will end up inside your loading dies and your firearm's chamber.

This can cause unnecessary wear and has the potential to decrease reliability. By dropping my lubricated cases into the tumbler for 10 or 15 minutes, I can avoid this issue. Throw a spoonful of mineral spirits into your tumbling media along with the lubricated brass to ease this process. Never tumble loaded ammunition. There is a theoretical danger of the tumbler's vibration damaging the structure of the gunpowder. I'm more concerned with the accidental ignition of a primer attached to a case full of gunpowder or the potential for fire in a tumbler full of flammable solids and loaded ammunition.

Case Forming

Case forming operations require thorough lubrication of every case and inside each neck. A tremendous amount of force is required to resize cartridge brass to any significant degree. In addition to becoming stuck in a forming die, cases may be unevenly stretched, deformed, or scratched if inadequately lubricated. Lubricate generously, but watch for and remove excessive lube build-up in your sizing or forming die.

Case Trimming

Case trimming is a necessary but generally unpleasant evil. As we resize cases after firing, the enlarged brass is forced back into a shape and size that will allow it to fit in our firearm's chamber again. As the brass is squeezed to reduce its diameter, the "extra" material is forced to increase the case length. If you look at the inside of your sizing die you can see that there is nowhere else for the material to flow, but up the length of the case toward the open case mouth. This causes cases to grow lengthwise. If not trimmed, cases will get so long that loaded ammo will no longer fit inside your firearm's chamber. At this point, we must either trim or discard cases that have grown too long.

Manual Trimmers

Manual case trimmers have been the basic tool relied upon by reloaders for decades. They're fairly slow, tedious, and almost painful to use, at least in my opinion. I'm not the most patient person when it comes to tedious minutiae so you may not have the same experience. Some people really enjoy the mundane

aspects of reloading as a way to escape from the pressures of work or life in general. If we need to trim small volumes of cases for next hunting season, for example, then manual trimmers are just fine. If we're talking about trimming three thousand .223 cases for next year's 3-gun season, I really don't want to be using a manual trimmer. The time required to manually trim that many cases would be akin to a prison sentence. Not only that, but the jokes

Slow and tedious, manual trimmers get the job done at the expense of time and tedium.

you will have to endure about why you have one forearm that's so much larger than the other may be unbearable.

Manual trimmers fall into a few categories. Some of these include press-mounted trimmers, benchtop lathe-style trimmers, and handheld trimmers. Several of these units can be expanded later to add power-drive capability, either

Some lathe-style trimmers can be motor driven, reducing the effort involved in trimming.

from a cordless drill or a purpose-built power unit from the trimmer's manufacturer. If your reloading budget is tight or you only trim smaller lots of brass, these manual trimmers are just fine. Be sure to read the reviews from customers who have tried each unit and make your purchasing decision after consulting a number of sources. Video reviews of each trimmer, and just about any other reloading tool, are freely available online as well.

Power-Driven Trimmers

Power-driven case trimmers offer a step up in speed and ease of use compared to manual case trimmers. The initial setup process with these can be tedious though. Trim-length adjustments are often accomplished by trial and error only and can be a bit frustrating to setup at first. Once you get things dialed in, you can really zing along pretty quickly. For years I have used a drill press-mounted cutter for my .223 and .308 brass trimming. Recently I added a new version of the .300 Blackout. After the initial setup, I've been able to quickly and easily trim cases with

little trouble. These trimmers, made by Possom Hollow or Little Crow Gunworks, index off of the case shoulder, so be sure to properly resize your cases before trimming. Other similar trimmers are available. As with most tools, a bit of technique and routine help ensure the most consistent results.

Converting

I mentioned earlier that some manual case trimmers can be converted to power-driven case trimmers. This is a good method because it allows the option to do larger volumes of trimming, like .223 brass for your AR15, with the power-drive unit, but still maintains the ability to trim a few cases from your low-volume cartridges without having to charge drill batteries or deal with all the equipment needed for power trimming. If you have enough space for a dedicated trimming setup, this isn't a concern. Still, I have almost always found that having a simpler option to do things manually or with less equipment is a good thing. When I want to load 10 rounds of ammo to test a new powder, I don't

Power-driven trimmers that index the case shoulder size cases quickly and effectively. These trimmers can be mounted in a drill press or in a handheld drill.

break out the multi-station progressive press with the automatic primer and powder systems that have to be filled before use. It's just easier to keep it low-tech in some cases. With a combo manual/powered case trimmer, you can do that.

Trimmer Options

The king of case trimmers at this point is the Giraud Power Trimmer. This motor-driven unit is pupose-built for speed and ease of use. It not only trims cases quickly, but chamfers and deburrs the case mouth at the same time. This is a huge time saver as three operations are performed simultaneously with only one handling of the cartridge case. The Giraud is unmatched for trimming speed. Performance comes at a price though. With a case holder for one cartridge, your initial investment in this unit will be nearly $500. Each additional case holder costs around $30. You have the option to adjust your cutting head each time you change cartridges or you can buy additional cutters so that you have several to dedicate to respective cartridges. This will save time adjusting blades each time you change cartridge-trimming setups, but will cost an additional $45 per. For the ultimate trimming speed and ease, the Giraud is a no-brainer. It's up to you to determine whether or not the cost is justified by your volume and frequency of trimming. One option is to go in together with a number of reloading friends and time-share one of these machines. Since case trimming can be done in large lots, it may be a viable option for you to have the trimmer this month, the next stakeholder the next month, etc. If you are a member of a gun club and can set the trimmer up in the clubhouse to be used by those who have "bought in," that might work as well. If you're the entrepreneur in the bunch, perhaps you can buy the trimmer outright and charge your reloading buddies for time on the machine or sell trimming services via the gun club message board, or the internet. You might have that Giraud paid for and be turning a profit before you know it.

Trim Length

Once you've purchased a trimmer, refer to the reloading manual of your choice, an online reloading site like *www.ammoguide.com* or *www.loaddata.com*, or the load data listed on each bullet manufacturer's website to find the recommended trim length for your cartridge. It isn't necessary to trim to the minimum length, but at least be consistent. When cases exceed their maximum length, the case mouth will make contact with the end of the chamber, potentially causing a number of accuracy-robbing issues. If cases grow much past the maximum length allowed by your chamber dimensions, a dangerous increase in pressure may result during firing or the round simply may not chamber at all, as increased force is required to smash the case into a chamber it no longer fits.

Annealing

Annealing is the controlled process of softening metal through a heating and cooling cycle. Too much heat can soften the metal beyond its usable range. Too little heat and you're wasting time without actually changing the properties

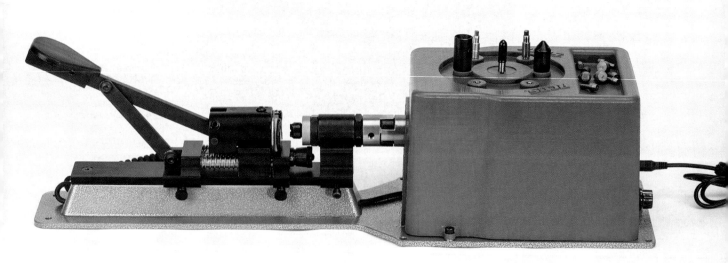

of the metal. Different metals require different temperatures and cooling times. We'll look specifically at cartridge brass. One note on this subject before we go any further: annealing is not a "basic" reloading practice. We'll cover it here for your information but won't go into the detail required to thoroughly teach the annealing process. Don't feel obligated to add an an-

Annealing softens the case neck area to allow it to expand and contract more easily. It keeps the neck from work hardening and cracking, and extends case life.

nealing machine or equipment to your reloading bench any time soon. Rifle cases generally don't need to be annealed until the third, fourth, or fifth firing/loading cycle. Even then, there are sources for annealing services available online that charge a dime or so per case to do this task for you. I've used these services in the past and have been pleased with the results. It's much quicker to box up your brass and ship it off to be annealed than it is to anneal cases with the slower methods available to reloaders. It's also much cheaper than it is to buy the faster annealing tools on the market.

So, why do we need to know about annealing cartridge brass? The most basic reason is to improve our understanding about the components we use to load ammunition. Every piece of brass from every round of rifle ammunition you've ever fired that uses a bottleneck-style case has been annealed at least once, possibly multiple times, during its manufacture. You may have seen the results of the annealing process if you've seen or used mil spec or surplus ammo before. The annealing process changes the color of the cartridge case in the areas where heat has been applied. This results in various colors, from brownish to blue. The off-color area typically extends just down past the case shoulder. What many don't realize is that commercial ammunition with highly polished shiny brass has received the exact same case annealing process. Commercial ammunition manufacturers perform an additional polishing step on each case to remove the discoloration left after the annealing process is complete. Just because you can't see the effects of annealing doesn't mean the cases haven't been annealed.

Reasons To Anneal

Now that we understand that, let's look at the reasons for annealing. Cartridge brass undergoes a process called "work hardening" every time it's "worked" or moved. For the handloader, this is when brass is expanded during firing or sized down during our resizing operations. Every time this cycle repeats, the brass becomes harder and less responsive. Eventually, the brass won't have the "spring" it needs to stretch and contract around a bullet seated into the case neck and our sizing die will stop having the same effect it once did during the resizing process. Eventually, case necks will begin to split and our brass will be fit only for the garbage can.

There is no hard and fast rule as to the number of firing cycles required to make our brass

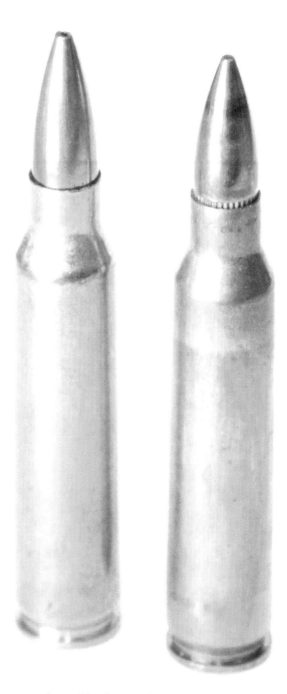

The upper portions of both cases have been annealed at least once during the manufacturing process, but the case on the left has been through a final polishing step to remove the resulting discoloration before being loaded. This is common for commercial ammunition that must look pretty for the customer. The military-grade cartridge on the right has no such cosmetic requirements.

unusable, but it's generally agreed that annealing every third or fourth firing is a good place to start. Some brass may need to be annealed sooner than this, and some may last longer. With regular annealing, minimal sizing, and reloading to sensible pressures, cases can last 10, 15, or even 20 firing cycles. Again, the purpose of

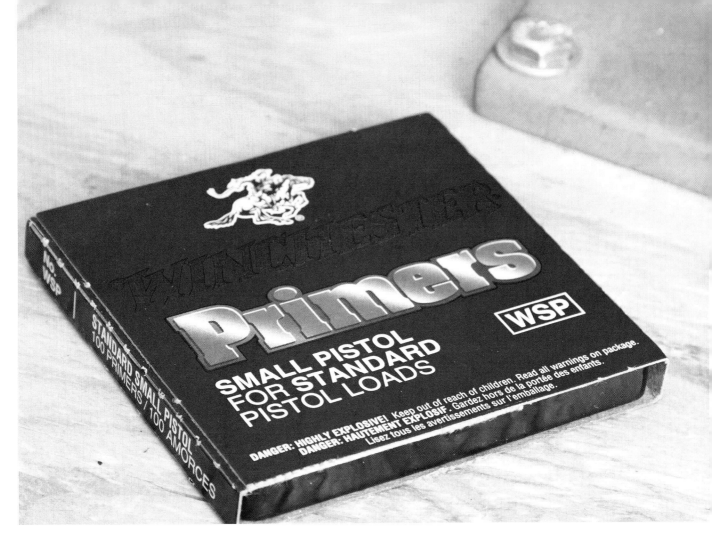

this section is only to increase your knowledge of what annealing is and why it's important to us. When you reach the point of having fired your entire collection of rifle cases three or four times and start to notice an increase in neck hardness, you'll understand why that's happening and where to look for the solution.

The Loading Process

The first components we'll combine to begin the loading process are the case and the primer. The act of inserting the primer into the primer pocket of the cartridge case is generally referred to as "priming." If you're reusing fired brass, obviously you'll have to remove the old primer before you can add the new primer. In some cases, you'll need to remove the primer crimp before you can insert the new primer. This is generally the case if you're using brass from military surplus or high-end tactical ammunition. Military ammo often has crimped primers to ensure

Buy them cheap and stack them deep when you purchase primers and powder. Without them, you can't make ammo.

the primers remain in place during transport and through the violent feed cycles generated by some military automatic weapons.

Deprime Options

We have a few equipment options we can use to deprime our cases. The most common method is to deprime as part of the case resizing process. Most sizing dies come with a "decapping rod" installed. The purpose of this rod is to push the old primer out of the primer pocket. Cases

Here the small decapping rod that removes the spent primer from the fired case is visible protruding through the open bottom of the sizing-die body.

Photo of a RCBS case lube die. Like a universal decapping die, a lube die without lube will allow depriming without resizing the case.

with crimped primers are sometimes stubborn and require significant force to deprime. Have some spare decapping rods on hand. In order to fit through the flash-hole, the decapping rod must be small. If you reload any military brass with crimped primers I can promise you that your decapping rod will break sooner or later!

Universal Die

Several manufacturers make a "universal" decapping die. As the name implies, this universal die is intended for use on any cartridge case. There is no sizing or other action on the case when using a universal decapping die. I find this to be a handy tool when I have really dirty brass that I want to deprime before cleaning. While I don't do this every time I prep a batch of cases, I will sometimes deprime before tumbling the cases so that the tumbling media can clean the inside of the primer pockets a bit. I'll also use this universal die when decapping large lots of fired military brass that has crimped primers. It's much easier and quicker to decap when no sizing is being done on the case. For instance, if I'm decapping rifle brass, I'll have to lube the cases before sizing, so using the universal die lets me avoid this step.

Remove Primer Crimps

Once the cases are deprimed, I can remove the primer crimp, then clean and process as normal. Removing primer crimps can be done by cutting away the crimped area or, preferably, swaging out the primer crimp. Swaging is preferred because it doesn't remove any metal from your cases. The crimped area is simply smashed back to its original shape by the swaging ram. This leaves the primer pocket ready to accept a new primer.

Priming Tool Options

When we're ready to prime our cases, we can use a press-mounted priming system, a hand-held priming tool, or a bench-mounted priming tool. There are advantages and disadvantages to each type. Progressive presses will have a priming system built in or available as an accessory. Some single-stage presses will offer priming capability as well via an add-on system. A bench-mounted priming tool may make sense if you load on a single-stage press, but want to prime large amounts of cases as part of your case prep process. It will be a bit quicker and easier to do this on a purpose-built priming tool where the press frame isn't required. This eases access and speeds the process somewhat.

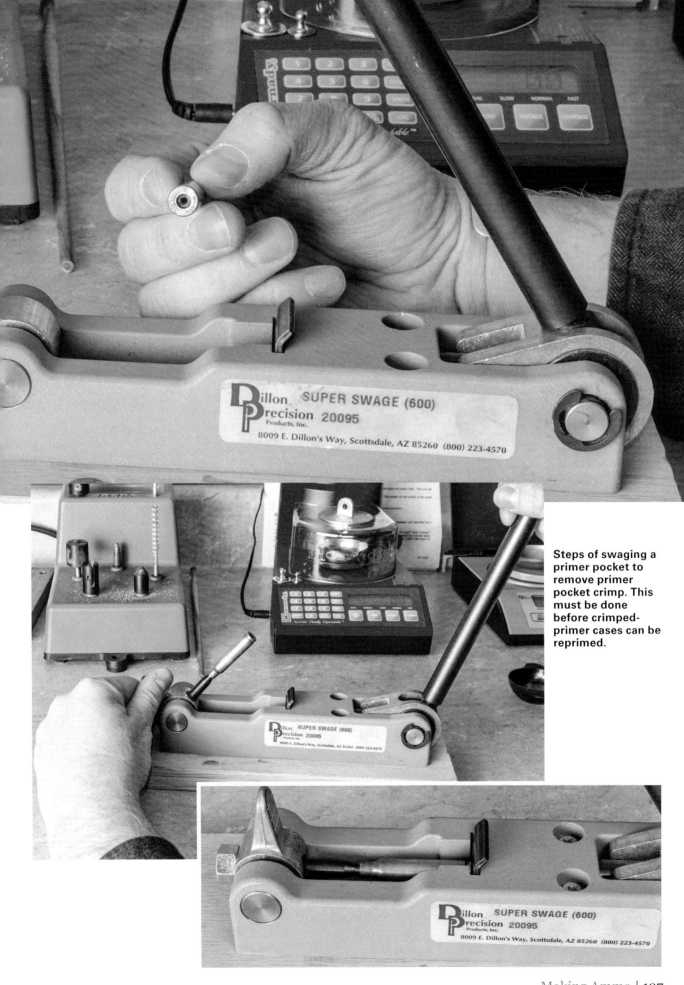

Steps of swaging a primer pocket to remove primer pocket crimp. This must be done before crimped-primer cases can be reprimed.

Priming

My personal preference is to prime on the press when I'm using a progressive press and to prime with a hand-priming tool when I'm loading on a single-stage press. I actually find some amount of enjoyment in hand-priming rifle cases. I really can't explain this, other than to attribute it to the fact that I'm in my "Man Room" at the loading bench without distraction after a day of work or maybe I'm even watching hunting videos on YouTube while I prime. I advise against having a TV or computer feeding you distractions while you load, but it's fairly hard to screw up hand-priming as long as you pay attention during the planning and setup steps and follow through on your data management process. Make sure you have a plan and only put the components required to execute that plan on the bench. The last thing you want to do is put pistol primers in your magnum rifle loads or vice-versa.

My most-used progressive press is a Dillon 550B, which features a pretty good automatic primer-feed system and makes priming on the press easy and quick. I can load up to 400 rounds of pistol or carbine ammo (or prep that many rifle cases) per hour with the priming process happening almost automatically as I load. The Dillon unit comes with several primer tubes that hold 100 primers each. Think of these as speedloaders or high-capacity magazines for your priming system. Pull the retaining pin on one of these primer tubes, and 100 primers fall into the priming system on your press in a second or two. This is really nice!

Primer pockets before (l) and after (r) cleaning (with stainless media, in this example).

Clean Your Pockets

You may find it necessary to clean your primer pockets after residue from a few firings has accumulated. Maybe you just like to clean all parts of your cases before you reload them and want to clean them every time. I don't think it's necessary, but it's perfectly fine if you want to hit every detail. If I find that my primers won't seat below flush or have become more difficult to seat, I'll clean out the primer pocket at that point. Otherwise, cleaning primer pockets just makes them pretty, for the most part. Champion benchrest shooters would argue that point, and I would agree that they're probably right in the context of their game. That's a whole different level of detail and minutiae than we're dealing with at the beginner level of handloading though. If it makes you feel better about what you're doing to clean primer pockets every time you reload your cases, feel free. I'd rather play with my kids or go shooting!

Look closely and you'll see the primer crimp – a narrow ring of pressed-in brass around the primer pocket.

Ways To Clean

When the time to clean primer pockets does come, there are several ways to do it. I mentioned in the depriming section that we could deprime our cases before putting them in our vibratory tumbler and the tumbling media will work on the primer pockets, too. This won't be a thorough cleaning, but it will knock out some of the crusty residue. If we want to conduct a more complete cleaning, there are handheld tools designed just for this task. These will have stiff bristles or metal scraping points that will clear out the hardened residue. Be careful not to overdo it when cleaning primer pockets. If we increase the depth of our pocket too much, our primers may sit deep enough in the case that the firing pin may no longer have adequate reach and force to detonate the priming mixture.

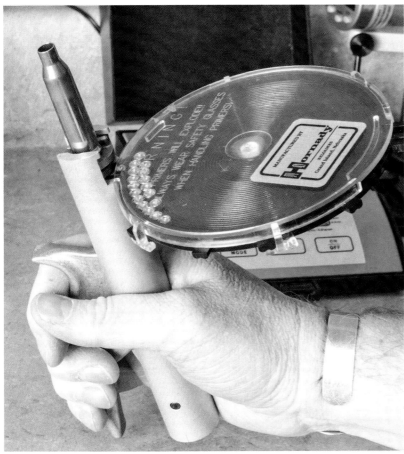

Hand-priming cases is strangely therapeutic. It can also be done from your recliner while you watch the latest action flick or hunting show. It's not as good as shooting, but it's an off-season activity you can enjoy.

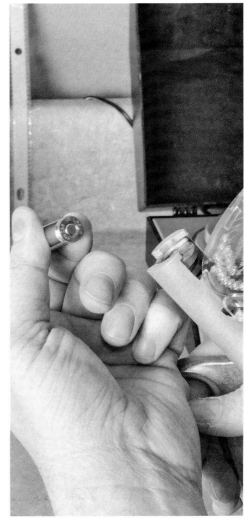

Dies

Sizing dies are used to force the fired cartridge case back to roughly its original size. I say "roughly" original because resized cases will usually be slightly larger through the body and base than unfired brass. That's fine, for the most part. A possible exception is when firing reloaded ammunition in some semi-auto rifles. The violent nature of the firing cycle of a semi-auto may require that we size cases down to the original dimensions of unfired brass in order to ensure reliable operation. If this is the case, we can use what are called "small base" dies to bring the case dimensions down to minimum levels. The downside to this is that we are working the brass more than we would with standard dies. Brass life is reduced through the "work hardening" that occurs every time the brass is stretched or resized. The more stretch or resizing, the more work hardening and less brass life we can expect. The annealing process may allow us to recover some longevity, but that's a whole different operation for discussion separately.

There are several types of sizing dies that can be used together or separately to reform fired cases to your desired dimensions. Some are specific only to rifle reloading, some to pistol reloading, and a few that are common between both. We'll go through these in detail separately.

Full-Length Sizing Dies

"Full-length" sizing dies do exactly what their name implies; they size the entire case from one end to the other. Actually, there is very little sizing that goes on at the base of the cartridge case with standard full-length dies, but the name is fairly appropriate, nonetheless. Full-length dies offer us a one-step resizing tool that brings our brass back into a sized condition, ready to reload.

Rifle Cases

Rifle cases are resized at the neck. The shoulder is bumped back slightly (if it grew to fill headspace), and the body is squeezed back down to near-minimum size. Since the sizing operation works on the outside of the case, most full-length dies also feature an "expander ball" that will open the neck slightly to ensure that the inside diameter of the neck is correct. Think about it;

Hornady full-length sizing die cutaway to show internal parts.

if the case wall of one type of brass is 0.016" thick at the neck and another type of brass is 0.013" thick at the neck, the inside diameter will differ by 0.006". Since it's the inside of the neck that actually holds the bullet in place, we must ensure that the inside diameter is precisely sized. By sizing the neck down more than would be required for the thinnest brass, we guarantee that we can then expand the neck slightly to the appropriate inside diameter. Think of neck expansion as pushing the inconsistencies in our brass from the inside where they will touch the bullet to the outside where they won't.

Pistol Cases

Pistol cases, with a few exceptions, are simply resized throughout the entire length of their straight walls. Since there is no separate neck or shoulder (again, with a few exceptions), this is a very simple and straightforward process. Unlike bottlenecked-rifle brass, straight-walled pistol brass does not need to be lubed before resizing. There are a few bottlenecked pistol cases in use today. The most popular in the U.S. may be the .357 Sig. If using a steel .357 Sig sizing die, we need to lube these cases as though they were rifle cases before sizing or risk having our brass stuck inside our sizing die. A popular option is to size the .357 Sig case in two steps—sizing the body with a carbide .40 Smith & Wesson

Before and after full-length sizing. Notice that we've bumped the case shoulder back 0.002" in this example. That's about perfect for most bolt-action rifles and some generously chambered semi-autos. Some semi-autos may require that cases undergo a more complete resizing process to function properly. In general, try to resize your cases as little as is required to achieve satisfactory function.

(the parent case of the .357 Sig) sizing die that requires no lube and sizing the neck with a standard .357 Sig die.

What full-length dies offer in convenience, they give up in control and flexibility. What we get is what we get. There is very little practical adjustment that can be made to full-length dies. If you're loading small amounts of ammo for plinking or even hunting, standard full-length dies will produce adequate results. Custom full-length dies can be made to your specifications. With this option, we can dictate what neck diameter, what body diameter, and what shoulder position we desire based on the type and thickness of brass we're using. This is a great option that removes some of the negatives of full-length dies while retaining the positive attributes of simplicity and concentricity.

Small Base Dies

"Small base" dies are very similar to full-length dies, but they resize the cases to a greater degree than standard full-length dies do. This is potentially important for shooters loading for semi-auto firearms where the feeding and chambering forces of the action is fast and violent. If the case isn't sized down adequately, feeding malfunctions may result and a case may become stuck in the chamber.

Hybrid Dies

One more option is a hybrid, of sorts. The full-length bushing die features the same fixed body dimensions but allows us to insert bushings into the die neck. These bushings are available in inside diameters varying in 0.001" increments. This allows us to control the amount of resizing (and work hardening) that we perform on our

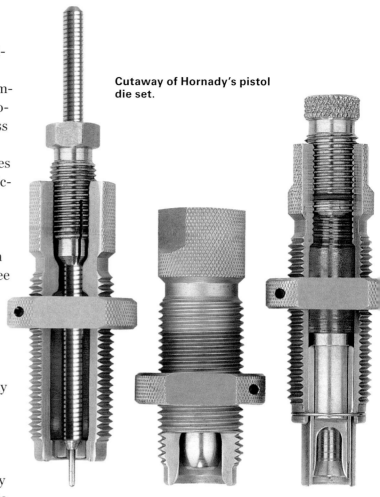

Cutaway of Hornady's pistol die set.

case necks. We can tune the amount of sizing done to the neck to the thickness of our brass and the desired amount of "neck tension" for our loaded rounds.

Neck tension is the interference fit between the sized case neck and our bullet. Shooters hand-loading for hunting or tactical precision rifles generally use 0.002" neck tension. The inside diameter of a resized .30-06 Springfield case, for example, would measure 0.306". The diameter of the bullet is 0.308", resulting in 0.002" neck tension. I've seen some full-length dies produce as much as 0.006" neck tension. Again, for simplicity's sake and general use, that's fine if you just want to load some ammo and shoot. When reloading for pistols, just size (with your die adjusted according to the manufacturer's instructions) and load. Whatever amount of tension your sizing die creates will be fine.

Redding 0.292" neck bushing. Bushing dies allow the reloader to select the exact amount of resizing to be performed on case necks.

Soft-Seating

Competitive benchrest shooters may use almost no neck tension, allowing the rifle's bore to push the long-seated bullet to its final resting place in the case neck. This is called "soft-seating" and should be avoided for any use other than competitive benchrest or similar events. In the event that we need to remove a loaded round from the rifle's chamber without firing it, there is a very high likelihood that a soft-seated bullet will remain in the rifle's barrel when the case is extracted. The entire contents of the case will then spill into the rifle's action. An action full of gunpowder is good for exactly nothing. Even if you can clear enough powder from the action to allow firing to continue, the firing of subsequent rounds may ignite the spilled powder. Spilled powder catching fire inside your rifle's action is not good, to say the least.

Hornady's Match Grade dies incorporate a removable bushing to allow precise neck sizing. This gives control over case neck tension on a seated bullet.

Neck Sizing Dies

"Neck Sizing" dies work on the neck portion of a bottleneck case only. This is useful when reloading the same cases for the same bolt-action or break-action rifle. Remember that each time a cartridge is fired, the case expands to fill the chamber. Unless we resize the case, that near-perfect fit with our chamber will be retained. By sizing the neck only, the case body remains "matched" in size to the chamber walls, providing a very accurate fit. This can lead to an increase in accuracy by reducing the misalignment of our cartridge with the rifle's chamber caused by an undersized case. This technique should only be used on a target rifle

though. Since the chamber-to-case fit is so snug, we may experience more difficult chambering and extraction. If this occurs, we can very slightly size the case body just enough that chambering and extraction are positive. We use a "body die" for this, or we can adjust a full-length die to just begin the sizing process without finishing it.

Body Dies

Body dies size only the body of the case. There is no provision on a body die to resize any portion of the neck. Body dies are useful when we want to have finer control over the sizing process and have elected to size the neck separately in a neck die(s). By sizing these areas of the case separately, we select exactly how much we need to work our brass. If we're shooting a semi-auto .308, we may want to resize the entire case significantly to ensure maximum reliability. If we also load ammo for a bolt-action .308 using the same dies, we can choose to resize the case body very slightly and the neck to an exact thousandth of an inch with a neck bushing die. I prefer to have this ability with almost all my rifle reloading efforts, so I purchase separate neck and body dies.

Expander Dies

"Expander Dies" are another example of breaking down the resizing process into separate steps for greater control of the outcome. When we discussed full-length dies, we detailed the effects of the expander ball on bringing the inside diameter of the neck to a certain size. An expander die performs this step separately through the use of an expander mandrel.

Expander mandrel and expander dies are used to resize case necks from the inside out. This is opposite of the action provided by neck dies and can be used to uniform the insides of case necks where the bullet makes critical contact with the case.

The expander die and expander mandrel are great tools. Full-length dies are nice for their simplicity but don't always give us the exact result we want. The expander inside a full-length die is often the source of concentricity problems. I've witnessed concentricity problems caused by a standard expander ball in a sizing die more than once, though I've seen perfect results from other full-length dies as well. With a dedicated expander, we can select the mandrel that gives us the desired amount of neck expansion. There are usually only a couple of sizes available for each caliber. I've yet to find an instance where one size mandrel or the other fails to give me the results I want. If that ever happens, it is easy to reduce the diameter of a mandrel with a bit of sandpaper and/or crocus cloth. This will give us a decrease in neck expansion but an increase in neck tension. If we intend to turn our case necks, we may need to use these expander mandrels to match our necks to our turning tools. For example, the Sinclair expander die and expander mandrels are intended to be used in conjunction with the Sinclair neck-turning tool.

Steps of forming .300 AAC Blackout cases from once-fired 5.56x45/.223 Remington cases.

Case-Forming Dies

Case-forming dies are used to convert brass manufactured for use with one cartridge into brass that will be used in another cartridge. Case forming is commonly done to create cases for obsolete or hard-to-find cartridges like the 6.5mm Remington Magnum. Common 7mm Remington Magnum cases are formed and shortened to create 6.5mm Rem. Mag. brass when necessary.

Popular Case-Forming Load

The hottest example of case-forming at the moment is with the .300 AAC Blackout (.300 BLK) cartridge. The .300 BLK is the modern version of the .300 Whisper that was pioneered and presented by SSK Industries many years ago. The cartridge case is actually based on the .221 Remington but is most often made today by case forming from readily available .223 Remington donor brass. To accomplish this, .223 cases are chopped off at the shoulder before being deburred, cycled through a .300 BLK sizing die, trimmed to final length, and (ideally) annealed. Some brands of brass may result in necks that are too thick to chamber in some guns. I'd sell that brass (with full disclosure, of course) or pitch it before I turned 1,000 pieces or more of .300 Blackout brass!

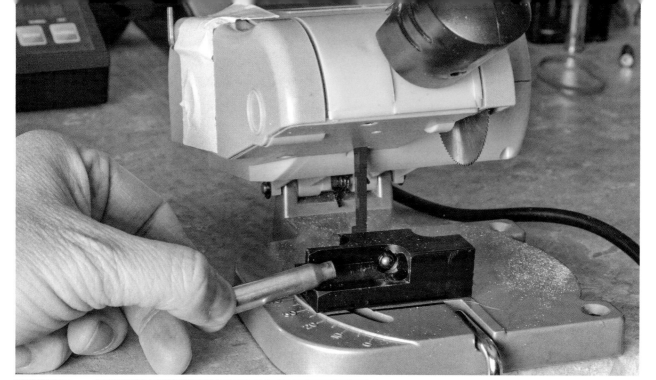

Using an inexpensive mini-chop saw and a case jig to cut down 5.56 brass for forming into .300 AAC Blackout brass.

Purpose Of The Seating Die

The job of the seating die (seater) is to force the bullet to a predetermined position inside the case neck. This must be accomplished in a manner that ensures the bullet is perfectly straight in the neck. The bullet and the case neck should be perfectly aligned. Seaters come in a variety of styles and range from simple and relatively inexpensive to high precision micrometer adjustable units that are much more expensive. Some seaters also perform the crimping function.

Standard Seating Dies

Standard seating dies are available for around $30 and are adequate for all but the most precise ammo-loading requirements. What makes a seater a "standard" model? Standard seating dies are basically a fixed metal cylinder with an internal cavity shaped like the cartridge with which the die is intended to be used. Standard seating dies will have a threaded adjustment stem that you turn to adjust seating depth via a screwdriver slot on top or via a knurled knob on the nicer versions. With no moving parts, standard seaters are simple and inexpensive. The downside is that you get what you get and not much more. Any discrepancy between the resized case and the inside of the seater can lead to minor misalignments between the case and die during seating. The case isn't as firmly held or precisely positioned with a standard seater as it is with a more precise "benchrest" style seater.

Check The Runout

Always check the runout of your completed rounds, if possible. This is true with ALL seating dies, but especially with the inexpensive standard seaters. About a year ago, I found that a standard seating die that I had purchased secondhand produced as much as 0.012" of runout on the seated projectile. The generally accepted norm is 0.003" or less for best accuracy, so my second-hand seater (that I bought used to "save money") was creating four times the acceptable runout! With all the money I wasted on powder, primers, and projectiles shooting crooked ammo, I could have probably bought the best die set made! Truthfully, seaters that won't load to an acceptable level of precision are pretty rare in my experience (if you buy quality seaters). If I'd followed my own rule and checked for runout the FIRST time I loaded with that seater, I could have saved myself some time, misery, and money.

Adjusting

Another compromise you have to make when using a standard seater is ease of adjustment. Since the adjustment stem isn't marked in any way, you have to use trial and error to set your seating depth. This isn't the end of the world, of course, but for a cartridge where you may want to load several different bullets, it can be a hassle. Every time you load a different type of bullet you have to repeat this trial and error process to return to your desired settings.

Benchrest Or Sliding Sleeve-Type Seaters

"Benchrest" or sliding sleeve-type seaters differ in design from standard seaters with the addition of a sliding sleeve in place of the fixed chamber of the standard seater.

Sleeve-type seaters, and certainly those with micrometer adjustments, are worth the higher price they command.

This sliding sleeve is spring-loaded and must be forced upwards by the cartridge case as the press handle is lowered. Where the cartridge-shaped chamber of the standard seater couldn't guarantee a secure fit with the cartridge case during seating, the spring-loaded chamber of the benchrest seater MUST make and maintain firm contact with the case during seating. This helps ensure proper alignment between case and projectile and results in less opportunity for run-out problems. I use the term "benchrest" seater loosely. That's Forster's name for their sleeve-type of seater, but others have this sliding sleeve, too. Hornady's seaters have it, and Redding labels them

Micrometer die (l) and standard seating die (r). The sliding sleeve can be seen at the bottom of the micrometer die.

"competition" seaters. They all help improve alignment of bullet and case by holding both components together in a tightly aligned sleeve.

Micrometer Dies

Micrometer dies take the concept of the sliding-sleeve die and add a REALLY nice feature that makes bullet seating much more convenient and repeatable across many adjustments for different bullets and seating depths. The un-marked stem or knob is replaced with a large knurled knob with laser-engraved markings each 0.001" of adjustment. This allows you to precisely and consistently adjust the seating depth provided by your die. If your current settings result in an overall cartridge length of 2.830", for example, you can turn the dial ten hash marks and the next cartridge will measure 2.820" exactly. You can record the settings you use for different bullets or when loading for different guns and return to those settings even after any number of other adjustments, assuming you haven't changed the position of the die body in the press. This is a huge timesaver when you use the same die for several different bullets or guns.

I load a half-dozen different bullets each for the .260 Remington and .308 Winchester cartridges. I also load these various bullets for more than a half-dozen rifles in these chamberings. Do the math and you can see that there may be 20 or more combinations of bullet and seating depth for each cartridge as a result. With micrometer seating dies, I can record my settings for each combination and return to them quickly and easily when I'm ready for that configuration again.

The downside of the micrometer seater is cost. The least expensive micrometer seaters are more than twice as expensive as standard seaters. The upper end micrometer dies are four to five times as expensive as standard seaters. Still, for those cartridges where you use a large variety of component combinations, micrometer seaters are a must have.

Crimping

Once we've seated our bullets, we have the option of applying a crimp to the case mouth. Crimping's primary purpose is to keep the bullet in place. The violent effects of feeding, chambering, or recoil can sometimes cause a bullet to move within the case mouth, so if we tune our bullet's position to achieve maximum accuracy in a match rifle, we may see inconsistent impacts and lost precision as a result. If we're shooting a hard-recoiling gun, we may experience malfunctions as bullets slip out of case mouths and cartridges become too long to feed properly. Several common types of crimps can be applied. The correct crimp is often determined by the type of firearm we're using.

Bullet Slippage

Bullet slippage in revolvers can actually lock up the cylinder when overlong cartridges prevent the cylinder from rotating. This happens in Ultralight revolvers as the light gun attempts to flee the high-velocity recoil. In hard-recoiling magnum revolvers with heavy bullets, the same can occur. To mitigate this risk, we apply a heavy roll crimp to the case mouth. This prevents bullet slippage and retards the release of the bullet slightly upon firing. The slower release allows pressure to build behind the bullet for a fraction of a second longer, ensuring even and adequate energy to push the heavy bullet down the bore. Revolver cartridges headspace from the case rim, so the case mouth can be heavily crimped without changing that headspace.

Semi-auto pistols headspace on the mouths of their cartridge cases. As a result, we can't use a roll crimp when loading ammo for semi-autos. A taper crimp allows us to hold the bullet in place without eliminating the case mouth termination we need for headspacing. In the following diagram you can see that the walls of the taper-crimped case terminate outside the bullet diameter, but the termination of the case mouth on the roll-crimped case is inside the crimp groove or cannelure.

Most standard seating dies are designed to perform the crimping step if you choose to adjust

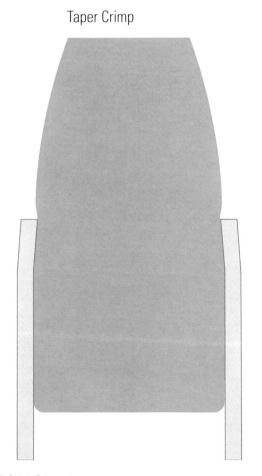

Taper Crimp

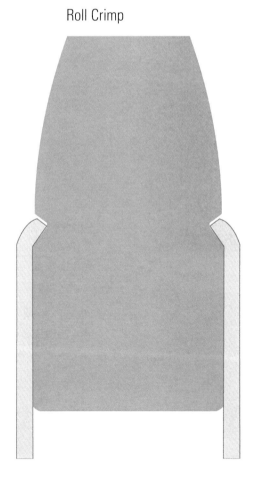

Roll Crimp

them to do so. Crimping is optional in some applications, so follow the instructions included with your die set to ensure a crimp is performed or omitted according to your needs. Seating dies for cartridges that are primarily used for semi-auto pistols will perform a taper crimp, and dies for cartridges primarily used in revolvers are designed to perform a roll crimp. You usually won't have to specify this. There are some crossover applications where a "revolver" cartridge can be fired in a semi-auto and vice-versa. If you're loading for one of those exceptions to the rule, contact your preferred die maker so they can either build you a custom die or provide you with the part number for the die they've already built in that configuration.

Factory Crimp

The so-called "factory" crimp gets its name from the extensive use of a similar crimp applied to factory-loaded rifle rounds. Rather than a roll or taper that's applied from the case mouth end of the cartridge by inserting the cartridge into a crimp die, a factory crimp is applied to the diameter of the case mouth from a nearly perpendicular direction. One advantage of this process is that the crimp can be uniformly applied to car-

tridge cases of somewhat varying lengths with little difference to the resulting crimp tension.

Powder Measures And Charging

Charging cases with powder can be accomplished with the assistance of a number of devices, ranging from incredibly simple (to the point of being crude) to complex, and several steps. While the purpose of all of these devices is to put a predetermined amount of powder into your cartridge cases, the level of accuracy and speed with which this task is accomplished varies significantly. There will be times and tasks that require more accuracy than speed, and sometimes the opposite will be true.

Lee Dipper System

The crude method of measuring powder I mentioned uses the Lee dipper system. When I first saw this process, I actually laughed out loud. I was taken aback at the thought that someone

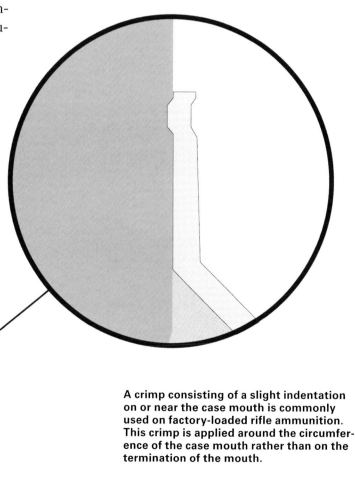

A crimp consisting of a slight indentation on or near the case mouth is commonly used on factory-loaded rifle ammunition. This crimp is applied around the circumference of the case mouth rather than on the termination of the mouth.

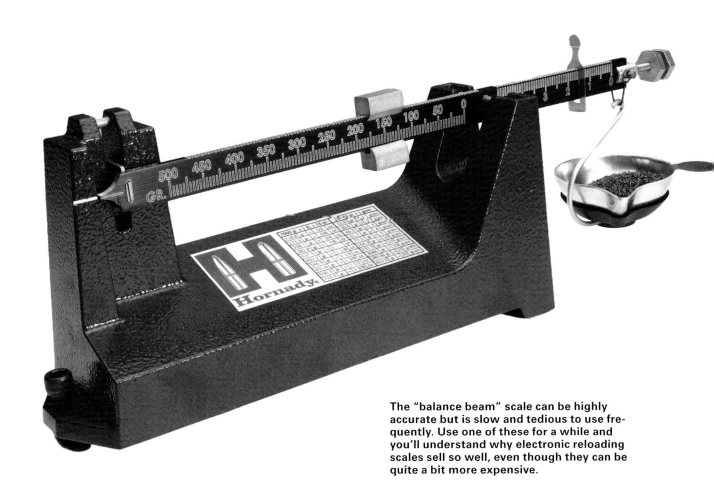

The "balance beam" scale can be highly accurate but is slow and tedious to use frequently. Use one of these for a while and you'll understand why electronic reloading scales sell so well, even though they can be quite a bit more expensive.

The Lee dippers are simple, coarse, but adequate for some applications.

would use a "spoon" to measure charges of a highly flammable and dangerous substance and then fire ammunition loaded with that powder charge in an expensive firearm with the expectation that an explosion and serious bodily harm wouldn't be the obvious result. At the least, I suspected that ammunition loaded with the dipper method must be the source of the old cliché about being unable to "hit the broad side of a barn."

In the twenty years since that first exposure to this system, I've actually come to respect and almost admire it for its simplicity and utility. For those with severely limited budgets, time, or desire to endure an overly complex reloading process, the dippers are a viable option. No, they will never provide the level of precision a competitive 1,000-yard rifle shooter desires, but they aren't meant to. But, for the guy who wants to throw together a dozen rounds for grandpa's old revolver, why not? With spherical or short-cut powders, the dippers can be accurate enough. Always verify the charge provided by your dipper using the powder you intend to load. Different powders have different shapes and sizes, which means that different powders will fill the volume-based dippers differently. Powder "A" may result in a 25.2 grain charge, while powder "B" may result in a 28.6 grain charge using the same dipper. When used with a scale and powder trickler (or by trickling with your fingertips), the dippers can provide a starting point for your trickled load. Technique definitely matters with the dippers. Follow the instructions provided by Lee for best results. If the absolute bare bones reloading setup is all you can afford, don't worry; many thousands of people have made the dippers work until they can do better.

Powder Measure

The term "powder measure" is used to describe manually operated powder-dispensing tools and press-mounted auto-metering tools. For ease, the term may be applied to the electric

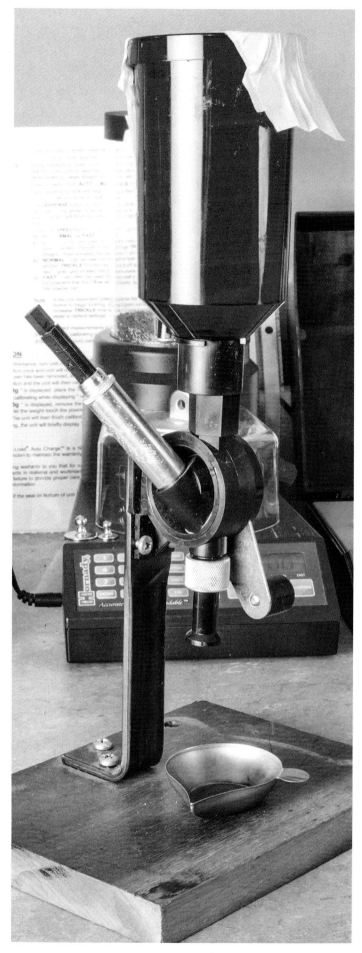

Inexpensive, but surprisingly accurate and durable, the Lee "Perfect" powder measure is a rare example of getting more than you pay for. Though it doesn't throw "perfect" charges every single time, to get appreciably better performance you'll have to spend much more money. You'll have to decide if it's worth it.

Progressive presses like this Dillon 550B have their own automatic powder measures that dispense a charge each time the press is cycled. (Some press components removed for photo clarity.)

motor-driven auto-charging units that are becoming more popular as pricing becomes more competitive, but generally, the term "powder measure" refers to one of the two aforementioned options.

The stand-alone powder measure has a place on almost every reloader's bench. Unless you intend to load solely on a progressive press that has a built-in powder measure, you will want a powder measure eventually. Once again, you can purchase a basic unit at a relatively low price or spend several hundred dollars on a top-of-the-line measure.

Your powder measure will have a large hopper or reservoir into which you'll pour a half-pound or so of powder. Beneath the large reservoir sits an adjustable mechanism that controls how much powder is dispensed with each cycle of the measure's handle. This mechanism is essentially a valve combined with a smaller reservoir that holds one "throw" of powder. Remember the term "throw" as it relates to powder. When we talk about "throwing" powder we mean dispensing one charge from a powder measure. We're not actually throwing powder across the room.

As you cycle the measure's handle, the valve is opened and the smaller reservoir fills with the chosen amount of powder. When the handle is moved back to the start position, the reservoir empties the powder into the pan, funnel, or cartridge case you've positioned below the measure's outlet nozzle. Adjust the amount of powder dispensed with each throw by turning the mechanism according to the manufacturer's instructions.

Though some measures are very accurate and consistent, especially with finer powder, I always throw charges into a pan and confirm the charge weight on an electronic scale when I want to ensure the highest level of precision. The exception to this is when I'm loading pistol or carbine ammo on a progressive press. In that case, loading speed is more important to me than accuracy at 1,000 yards. Obviously, the holdover required to shoot a 9mm that far would be significant! Seriously, if loading blasting or practice ammo, or shooting at short to moderate range, there is no need to weigh charges.

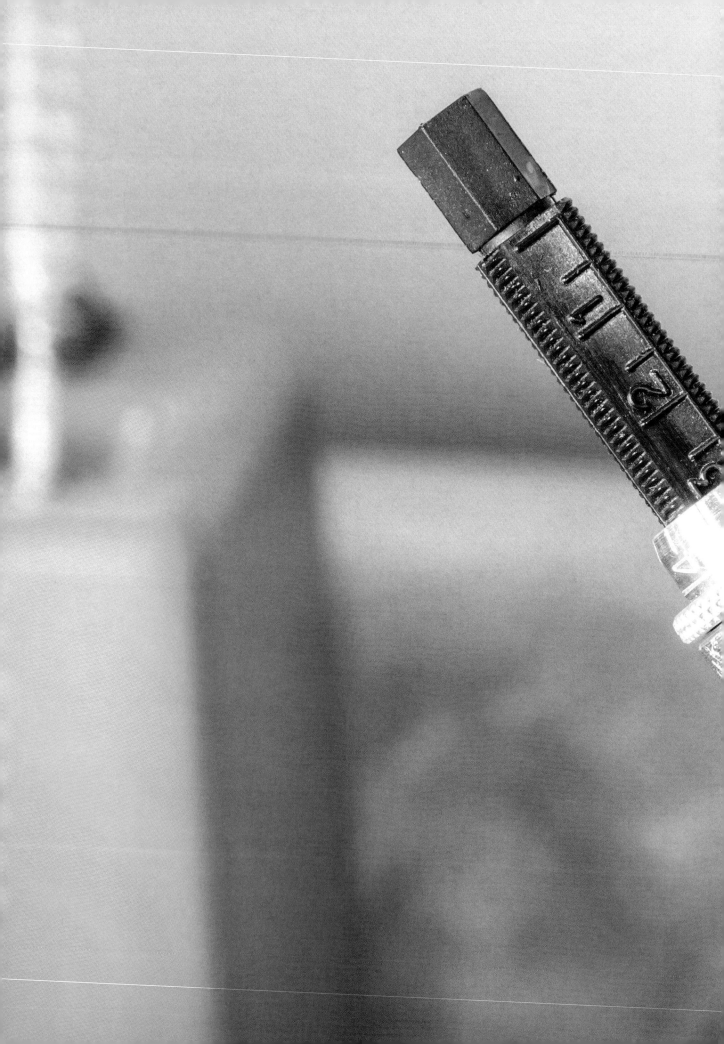

The amount of powder dispensed with each throw is adjusted on this stem.

Throwing Charges

Many successful benchrest shooters throw charges directly from their measure without checking the charge. These guys achieve an incredible level of precision with their system at short to moderate ranges and prove that great results can be achieved right out of the measure. It does take a couple of extra seconds to confirm charge weight on an electronic scale, but it also allows me to easily remove one variable from my loads. Understand that you have options and try both techniques for yourself. Throw a couple dozen charges, weighing each one, and note the difference between the high charge weight and the low charge weight. You may find that your results are perfectly acceptable when you charge directly from the measure.

Charge The Case

Press-mounted automatic powder measures are an incredible timesaver. Loading rates of 400 rounds per hour or more can be achieved with an automatic powder measure on a progressive press. Once the measure's hopper is filled and the charge has been adjusted to the proper weight, all we need to do to charge our cases is cycle the handle on our reloading press. The press does all of the work to charge each case as long as we keep powder in the measure. Remember that the type of powder you're using will significantly impact the level of accuracy you can expect from these measures. Fine flake or spherical powder will provide more consistent charge weights than a long extruded or stick powder. Fortunately, the types of ammo that benefit the most from the higher loading rates of a progressive press are very well suited to those finer powders. Cartridges like 9mm Luger, .45 ACP, and .223 Remington behave wonderfully with flake or spherical powders like Winchester 231, Accurate #5, and (for the .223) Ramshot TAC. Our press-mounted powder measure will produce perfectly acceptable charges for our IDPA, IPSC, 3-Gun, or other training, blasting, and close to moderate-range ammunition.

Throwing a charge directly into a cartridge case
Unless you're loading for an incredible amount of precision, like a 1,000-yard rifle competition, charging from the measure may be perfectly adequate.

Loading Hybrid

What if we want more precision from our progressive press? Do we have to use the automatic powder measure that came with the press? Not necessarily. I've used a simple plastic funnel and manual powder die in place of the powder measure on my Dillon 550 for years. This allows me to individually charge cases after throwing powder from a bench-top powder measure or after verifying each charge on a scale. This is a great hybrid method of loading that allows us to take advantage of many of the benefits of a progressive press while maintaining control of our charges like we would when loading for long-range precision rifles. I personally use this method in conjunction with a bench-top electronic auto-charging system (to be discussed later) when loading .308 Winchester for my AR-platform tactical rifles. I shoot these to 1,000 yards or more, so the consistency gained from the added step of charging individually is worth the little bit of extra time required.

Electronic dispenser/scale combo units are a time-saving luxury worth the price.

Closeup of powder being dispensed by the Hornady Auto Charge.

Electronic Auto-Charging

Electronic auto-charging measures combine the functions of a powder measure, trickler, electronic scale, and computer brain that allows control of dispensing speed, charge weight, and memory. With a little finesse and technique, these auto-charge dispensers are great assets that allow increased speed of loading while retaining the precision of individually weighed charges. Performance comes at a price, of course. The least expensive of these units costs several hundred dollars. If you need to load precisely charged ammunition quickly or in greater volumes, these dispensers can help do just that. As with manual powder measures, powder with smaller granules will dispense more consistently. For example, I can charge IMR 8208XBR to the one tenth of a grain almost every time with my auto-charging dispenser when loading for heavy match loads of 5.56x45 or a 77 grain Sierra Match King. With large stick-type powders like IMR 4064, I often set my dispenser to throw a charge slightly below my target charge weight. I'll then trickle the charge up to the actual target weight, if necessary. Often, the dispenser throws just a bit too much of this long extruded powder and I end up at, near, or even over my desired charge. If I program the dispenser for the exact charge I want, nearly every throw will be over the intended weight when using heavy extruded powders. Technique, practice, and experience will help you develop a system that works for you.

Ready To Load

Matching your ammo performance requirements, loading equipment, and powder selection wisely will increase your success and decrease your frustration with any of these powder measure types. There are a few accessories that may help you develop a more effective charging regimen. These accessories include reloading trays, drop tubes, powder funnels, and index cards.

Reloading Trays

Reloading trays are simple boards or plastic trays with around fifty cup-shaped recesses on one or both sides. These recesses are sized according to the intended cartridge family they're meant to hold. Some universal trays are available with

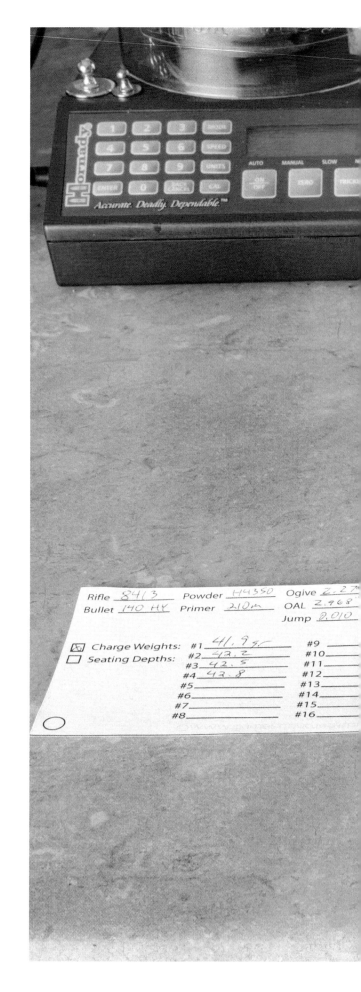

With our loading session planned and components prepared, we're ready to begin. The loading tray positions our cases for easy access and prevents spillage.

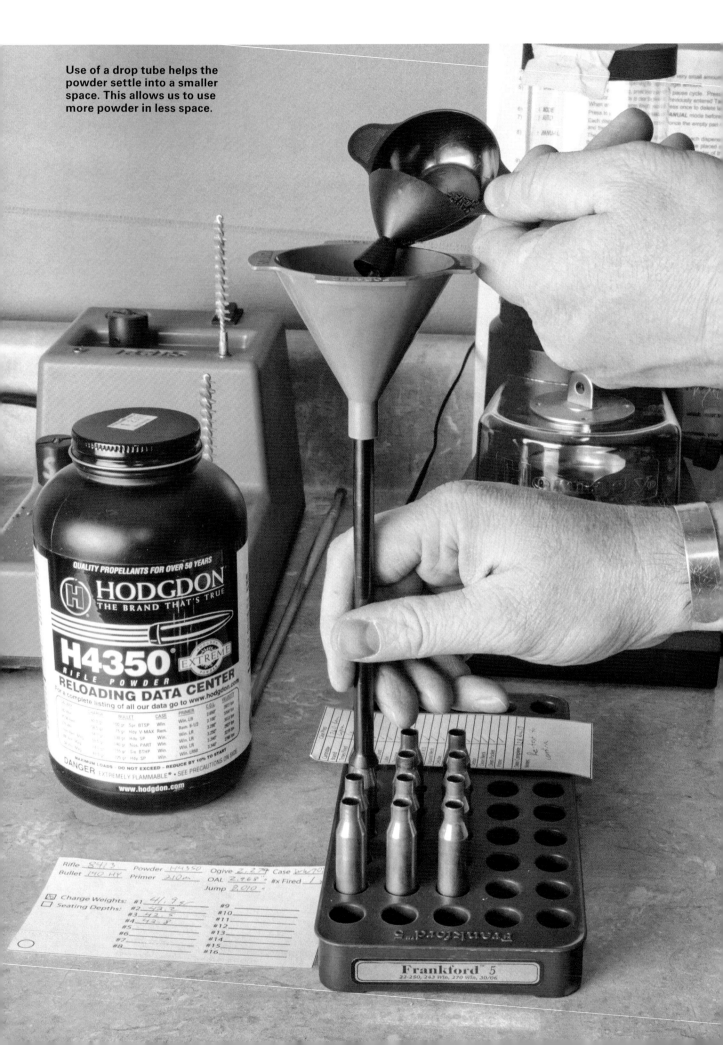

Use of a drop tube helps the powder settle into a smaller space. This allows us to use more powder in less space.

multiple hole sizes on each tray. They all work and personal preference is probably just about the only reason to pick one over the other.

Drop Tube

A drop tube is a valuable tool to have on your bench, especially for the rifle reloader. Some powders work really well in certain cartridges when the powder completely fills the case. Often, the case will be so full of powder that the bullet may be difficult to seat. With the use of a drop tube during powder charging, we can get the same amount of powder in the case, but that powder will fill the case in a more compact manner. Let's say we want to charge a .260 Remington case with 50 grains of extruded powder. Charged

from our powder measure or from the pan of our electronic scale, that 50-grain charge fills the case about two-thirds of the way up the neck. When we pour the same charge through a drop tube, the powder only fills the case to the base of the neck. Clearly, this will allow easier seating of our bullet. I find the drop tube/powder funnel combo so handy that I use it every time I charge a case in a loading tray. Even if powder fill-level doesn't need to be controlled, the drop tube is handy and fast compared to smaller powder funnels.

Index Cards

Index cards are useful for planning your loading session and keeping notes about your loaded rounds, and they fit inside just about any ammo

Basic 3" x 5" index cards are great for the many data management tasks a reloader deals with.

box. They also make a great bench scraper. I use an index card to sweep any spilled powder into my water-filled cup.

Reloading Presses

The basic beginner's option for a reloading press is the "single-stage" press. With a single-stage press, there is only one operation performed with each stroke of the press handle. The press will hold one die at a time and will accept one piece of brass at a time. The basic operation of a single-stage press follows:

1. The proper shell holder is installed into the press ram.

The correct size shell holder for your cartridge is a must.

2. The appropriate die is chosen based on cartridge and task to be performed.

3. The chosen die is installed into the press frame.

4. A case (lubricated, if necessary) is inserted into the shell holder.

5. The press handle is cycled to perform the intended loading operation.

6. The cartridge case (or completed cartridge) is removed to make room for the next case to be inserted.

Inserting a fired case into the Hornady LNL Single-Stage Press. The press has been fitted with a full-length sizing die for the .30-06 cartridge.

The sizing die is squeezing the fired case back down to size as the press ram is raised.

A sized, primed, and charged case ready for bullet seating with a micrometer die.

Beginning to insert the primer with the Hornady hand-priming tool.

Uncrimped case mouth (left) and crimped case mouth (right). It doesn't take much crimp to do the job.

When each loading task is completed on all of the cases you're working with, a new die will replace the former in order to perform the next step in the loading process. Let's look at an example of what these might be if loading for pistols or loading rifle ammo with a full-length sizing die:

1. Resize cases
2. Prime cases (if priming on press)
3. Seat bullet in charged case
4. Crimp case mouth

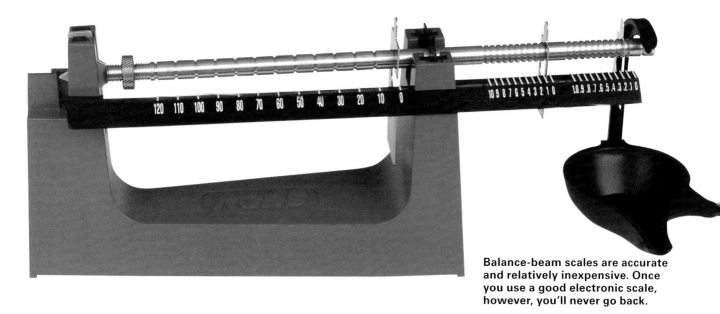

Balance-beam scales are accurate and relatively inexpensive. Once you use a good electronic scale, however, you'll never go back.

Single-Stage Presses

Single-stage presses are fine for low to moderate volume reloading and can be really useful for performing certain tasks a few times when the relative complexity of a progressive press is more of a hassle than a benefit. In addition to simplicity, single-stage presses are vastly less expensive than progressive presses. I've used one of Lee's lower end presses for years to deprime cases and load the least used cartridges in my inventory. When 50 rounds of ammo for a backup hunting rifle will last three or four years, why spend a couple hundred bucks setting up a high-end progressive press? The Lee press cost me around $40 on sale.

Some single-stage presses have removable bushings that allow nearly instantaneous die changes. I use the Hornady version and find it to be a fantastic feature that really improves the utility of the press. Die changes can be made in a couple of seconds without any need to readjust the settings on the dies before use. The downside to this setup is that additional bushings must be purchased for each die you intend to use with this feature. You can unscrew the dies from the bushings each time, but that negates the added speed offered by the bushing systems.

Single-stage presses are generally thought to load more precise ammunition than can be loaded on a progressive press. While I can see the rationale of this line of thought from a theoretical perspective, I wholeheartedly disagree with the assertion that the better progressive presses on the market sacrifice accuracy in a practical sense. I can attest to the fact that match-grade handloaded ammunition capable of high levels of precision can be loaded on a progressive press. If your budget is limited or you don't plan to load high volumes of training or blasting ammo for carbines or pistols, a single-stage press will serve you well. Even if you do jump in with both feet and spring for a $2000 fully progressive automated loader, there will be things you will want to do quickly and without the time and hassle required of a progressive press. Picking up a single-stage will give you the ability to handle these tasks more easily.

Turret Press

A variation on the single-stage press is the turret press. Like a single-stage, turret presses perform one step with each operation of the press. Unlike traditional single-stage presses, turrets hold multiple dies in their rotating toolheads. You can setup all of your dies for a particular cartridge before you begin loading and, on some turret presses, buy multiple toolheads to avoid resetting dies altogether. Some turret presses rotate the turret automatically with each cycle of the handle. This kind of quasi-progressive loading method is useful for some applications. If you're using a full-length die and keeping things pretty simple as far as loading technique goes, you can actually load a round from start to finish with one trip onto the press. In other words, you can prime, size, charge, and seat while only handling each piece of brass once. This is a heck of a lot faster than loading on a single-stage press, which requires you to handle each cartridge case multiple times, not to mention changing dies several times throughout the process.

Other turrets are made with a bit more rigidity in mind and require manual rotation of the toolhead when you want to use another die. This allows you to choose between performing one task or several with each handling of the cartridge case. Sometimes this makes more sense than loading on an auto-indexing turret or progressive press. For example, I like to do rifle brass prep on a progressive press where each cycle of the handle performs several steps. However, I like the finer control and attention I can give each cartridge when seating bullets on a single-stage press when loading ammo for maximum accuracy. With a turret press, I can have the benefit of that same single-stage level of control over bullet seating, but I can save time and handling by rotating the turret to my crimp die without taking the cartridge with the freshly seated bullet off the shell holder and placing it back to the loading block. I'd have to perform those two steps separately with a single-stage press since I can only have one die on the press at a time. That means each round must be handled an additional time when loading with a single stage press.

Basic reloading kits are available from most equipment manufacturers. This is an easy way to get started.

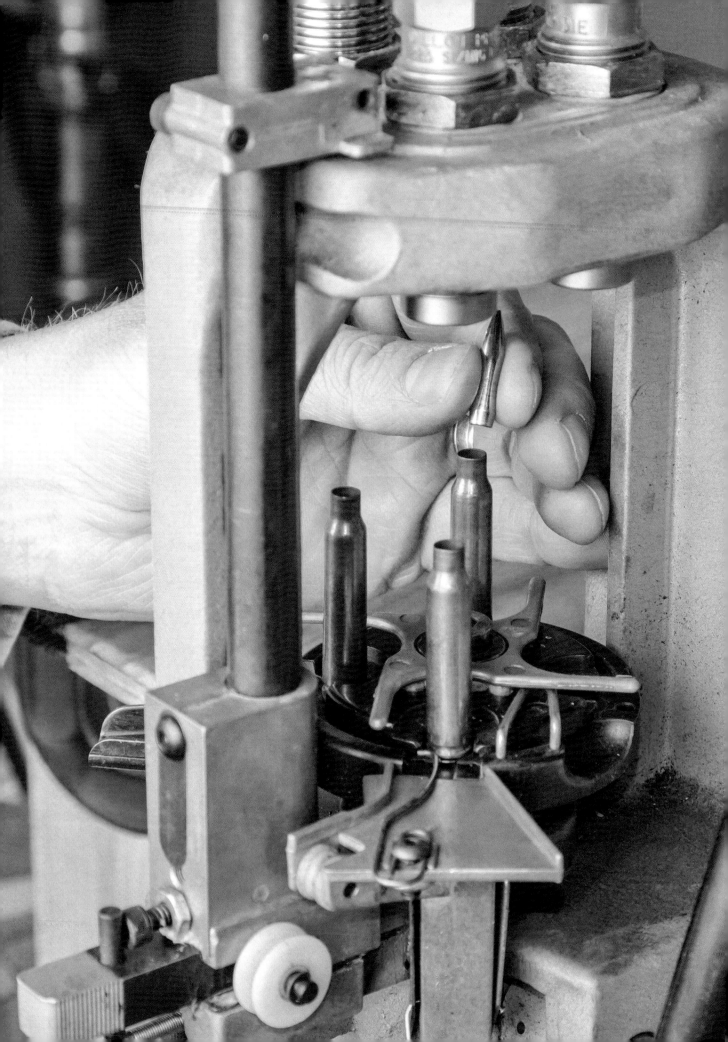

Progressive Presses

Progressive presses perform multiple steps on multiple cartridge cases with each cycle of the press handle. A four-station press, for example, can size, charge, seat, and crimp with one stroke of the handle. The loading speed obtainable with a progressive press is significantly greater than with a turret or single-stage press as a result. If you intend to load pistol or carbine ammo for any of the popular high-volume shooting sports or for training purposes, a progressive press is the only way to load!

Other Progressive Presses

Not all progressive presses are the same, of course. There are relatively simple options that start around the $300 price point, and there are full-blown automated examples that cost more than $2,000. Loading rates vary with price from roughly 400 rounds per hour up to 1,200 rounds per hour. My most-used progressive is a Dillon 550B. I think it is one of the most versatile machines on the market. I do all of my rifle brass prep on the 550 and load all of my higher-volume match rifle ammo on the 550 as well. Though the 550B produces up to 500 rounds per hour in its standard configuration, it still offers a relatively simple and painless changeover to different car-

Multiple operations are performed with each cycle of the progressive press. This Dillon 550B sizes, charges, seats, and crimps in its four stations. (Some press components removed for photo clarity.)

The "toolhead" of this Dillon press holds four dies that perform their function each time the press is cycled. (Some press components removed for photo clarity.)

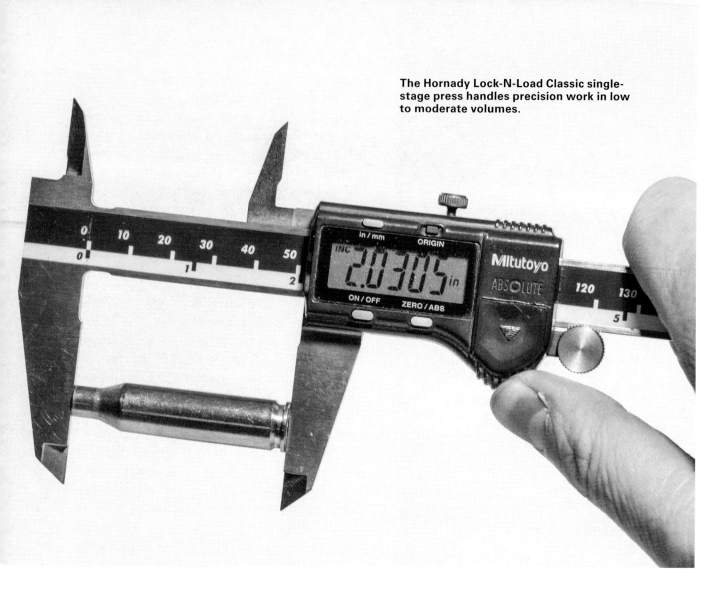

The Hornady Lock-N-Load Classic single-stage press handles precision work in low to moderate volumes.

tridges, assuming each cartridge uses similarly sized primers. If changing primer sizes, the conversion only takes another ten minutes or less.

Perfomance Step Up

A step up in performance with a bit of a step up in price is the Hornady LNL-AP progressive. I call this a step up in performance primarily because of the auto-indexing feature. Unlike the Dillon 550B, the LNL-AP mechanically rotates the shell plate containing the (up to) five cartridge cases that are being worked by the press. On the Dillon, the shell plate must be manually rotated after each press cycle. Manually indexing the shell plate doesn't take a lot of time, but when you have to do it 500 times per hour, the advantage of the auto-indexing machine becomes apparent. The Hornady LNL-AP is slightly slower and more cumbersome to convert to different cartridges, though. Unlike the Dillon's single

toolhead that contains all of the die stations and the powder measure, the LNL-AP has individual bushing-style die locations. Fortunately, the bushings make the task reasonably easy and quick to complete. Both of these machines can be upgraded with a number of additional features, such as automatic case feeders, at an additional cost.

Speed Step Up

A significant step up in speed and price are the Dillon 650 and 1050 machines. These offer loading rates of 800-1200 rounds per hour, depending on the accessories you purchase. With this speed comes increasing complexity and setup, though. Caliber conversions are more involved, more expensive, and more time-consuming than conversions on the simpler machines. If you want to load large volumes of a single cartridge, or a few cartridges, these are the machines for you. If you don't mind the time and complexity of the caliber conversion process, then the loading speed of these machines can justify their cost pretty easily.

Handheld Press

Swinging the proverbial pendulum back in the opposite direction from these blazing fast progressive presses, we find the simple hand-held press. This is a single-stage press that isn't meant to be mounted on a workbench, but hand-held instead. I can assure you that the amount of force that can be applied to these hand presses is significantly less than can be applied to a bench-mounted press. As a result, I avoid attempting to resize large rifle cases on a hand press. Pistol or smaller rifle cases can be resized and completely reloaded (assuming you've primed the cases already) on a hand press. The primary use I've found for a hand press is to adjust the seating depth of my experimental handloads at the range. If I'm conducting seating-depth testing on a new batch of ammo or for a new rifle, I'll load all of my test ammo at maximum length. If the ammo loaded to maximum length

doesn't provide the accuracy I'm looking for, I'll seat the next few rounds a little deeper and try again. This process is repeated until maximum accuracy is achieved. However, if you charge with a battery-powered scale or from the measure, a hand-held press would allow you to load entirely at the range. This can be a huge time saver if you have a long drive to and from the range like many of us do. What may take three separate trips for the loader working entirely at home can be done in one trip for the handloader who is able to experiment with different loads at the range. With all of that said, my one and only press (if I was limited to one) would not be a hand press. The power to handle larger cases just isn't there, and loading to any degree of volume would be painful.

Measuring Calipers

A set of calipers is a must for the reloader. I strongly prefer digital. You will be measuring and checking components, tools, and ammo constantly to ensure you've assembled quality cartridges that will fit and function properly. You can find a perfectly acceptable set at a dis-

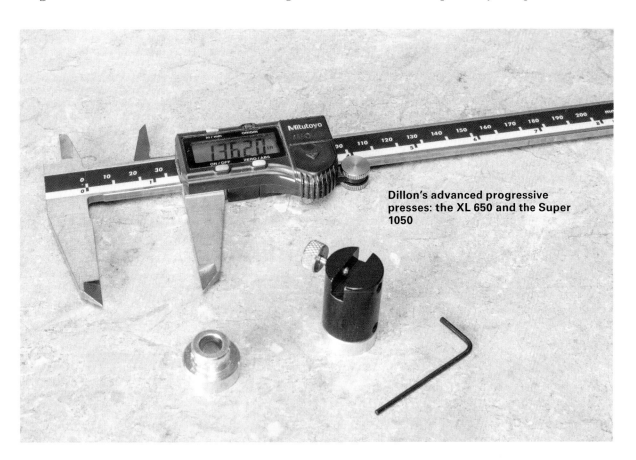

Dillon's advanced progressive presses: the XL 650 and the Super 1050

count tool place like Harbor Freight or you can order them online from Amazon. I have a really nice expensive set and a cheap Chinese set. I can't tell any difference in their performance. Be sure to zero and check them frequently according to the user manual. It is easy to press the wrong button inadvertently and screw up your settings. Go ahead and buy a half-dozen spare batteries, too. I promise you a dead battery in your calipers on Friday night will screw up your weekend of reloading plans if you can't get a replacement.

Comparator

Along with digital calipers, one of the most often used pieces of reloading equipment you will use when loading rifle ammunition with accuracy in mind is an overall-length (OAL) comparator. The comparator assembly consists of a body that attaches to your calipers and various inserts that interchange for use with different bullet diameters. The use of an OAL comparator with the appropriate insert and a set of calipers will allow you to measure the distance from the base of the cartridge to the ogive of the bullet. Knowing and controlling this dimension is critical to achieving maximum accuracy in many cases. This measurement is also likely to be more consistent than an overall length measurement when using open-tipped, hollow-point, or soft-point bullets due to the variations in the meplats of those styles of projectiles.

With the ogive position measurement, we can determine how far our bullets need to move or "jump" before making contact with the lands. We can tune this distance to our particular rifle in order to find better accuracy in many instances. For example, in my Ultralight Arms .260 Remington hunting rifle, I find more consistent accuracy with 0.080" of bullet jump than I do with 0.050" of bullet jump when shooting the Barnes 127gr LRX over H4350. In my mid-weight GA Precision .260 Remington, I find best accuracy at 0.010" jump with the 123gr Scenar over H4350. Each rifle is different and requires an OAL comparator and inserts to work through the process of safely finding the best seating depth. Though some advanced reloaders and competition shooters use zero bullet

jump or actually have the bullet forced slightly into the lands—or "jammed," I strongly suggest you avoid this until you are far down the road of reloading experience for a number of functional and safety reasons. With most semi-auto rifles and some bolt-action rifles, you will run out of room in the magazine before you ever reach the lands. If that's the case with your rifle, you can still benefit from the relative consistency of measurements to the ogive rather than the meplat. Just remember to make sure that your cartridges are short enough to fit within your magazines, like always. Different magazines have different internal lengths. I can load 6.8 SPC ammo to 2.295" in PRI-brand 25-round magazines. If I attempt to use ammo that long in my Barrett 30-round 6.8 SPC magazines, the

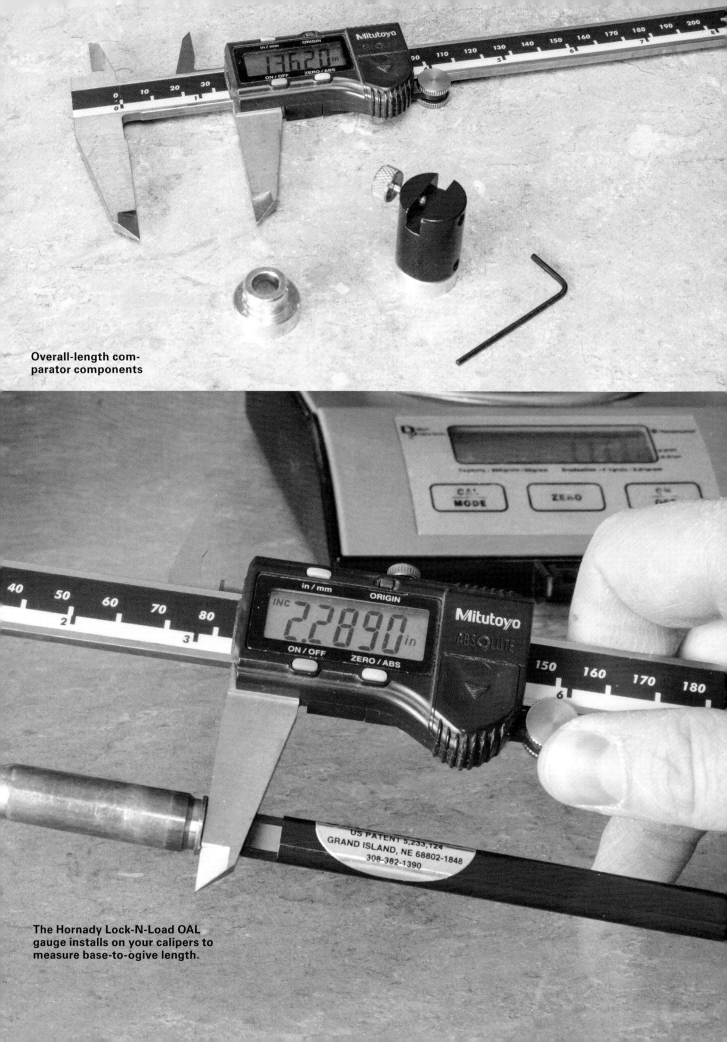

Overall-length comparator components

The Hornady Lock-N-Load OAL gauge installs on your calipers to measure base-to-ogive length.

cartridges won't fit because the magazines are too short inside.

Headspace Comparator

A very helpful, but not necessarily mandatory measuring tool is the headspace comparator. Like the OAL Comparator, the headspace comparator uses inserts of various sizes to measure a specific dimension of your loaded ammo (or your fired and/or resized brass). This dimension is the distance from the base of the cartridge to a datum point on the shoulder of the cartridge case. This is the headspace measurement. With measurements taken before and after resizing, we can determine how much our dies are moving the shoulder during the sizing operation. We only need to bump the shoulder back by 0.001" to 0.003" for fit and function. Any more sizing is only work-hardening the brass and reducing case life. Without the headspace comparator we'd never know what effect our sizing die(s) is having on our cases, and we wouldn't be able to make adjustments to optimize our die settings. As I mentioned, this is an optional tool. A standard full-length sizing die will make perfectly usable ammo almost every time, even if your case life is somewhat diminished in the process by over sizing.

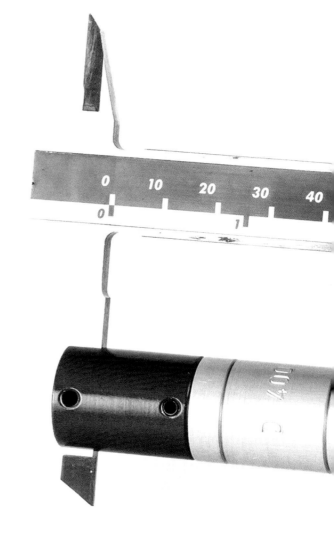

Electronic Scale

One of the most important and beneficial pieces of reloading equipment I've ever purchased is the Dillon Terminator electronic scale. You may have picked up on my affinity for this little jewel as I've spelled out "electronic scale" almost every time the subject of weighing components has been mentioned throughout this book. I made the shift from a balance-beam scale only a couple of years after I started reloading, and I can tell you that the word "relief" isn't adequate to describe how I felt when I ditched the balance-beam scale that got me started. Perhaps balance-beam scales have improved in the fifteen-plus years since then, but I'll ever know. If I had to go back to a balance-beam scale, I might just quit reloading. If you can afford a good electronic scale, they are worth the price! It is important to follow the manufacturer's instructions on setup, use, and maintenance. Until you have the process committed to memory, keep a

printed copy of the use and calibration instructions nearby and follow each step before every reloading session.

Bullet Pullers

Sooner or later (probably sooner) in your reloading career you will disassemble ammo that you've loaded. This could be due to the discovery that you used the wrong powder. You may find that some of the cartridges you've loaded have too much powder for your particular firearm, and some bullets will not shoot well. Eventually, you'll need to disassemble some rounds. To do this, you'll need one of several types of bullet pullers that are available commercially.

Common Pullers

The most common and least expensive type of bullet puller is the impact or kinetic puller. Impact pullers look a bit like an odd hammer. One end of the puller's head has a cap which screws

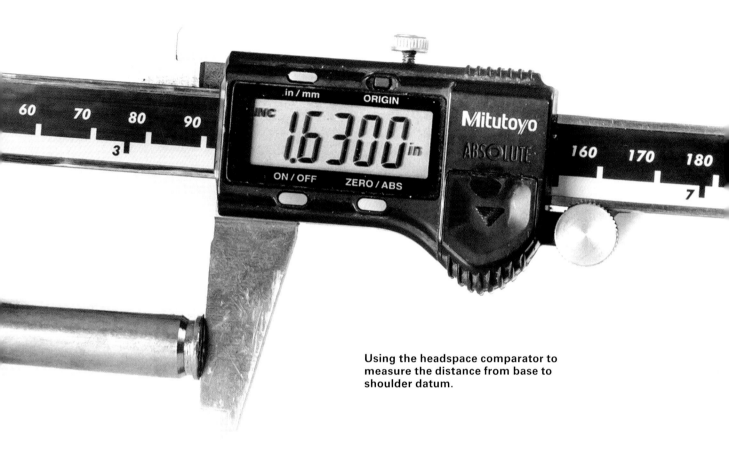

Using the headspace comparator to measure the distance from base to shoulder datum.

This electronic scale by Dillon Precision has served the author well for many thousands of rounds.

into place to retain a collet. The collet holds the cartridge case in place within the inner chamber of the puller head. Insert the loaded cartridge into the collet, screw down the cap until tight, then rap the head of the puller on a hard surface until the bullet comes free from the cartridge case. One or two hard whacks will usually pull just about any bullet, but sometimes it takes a few more impacts to do the job. Now that I have small children at home, I have to be careful when using an impact puller on a stubborn cartridge. When four or five whacks don't do the job, mild profanity is likely to occur. Just an "FYI" so you can be prepared!

Collet Pullers

Another option that is less violent and certainly less noisy than banging a plastic hammer (impact puller) on the end of your reloading bench or garage floor is the press-mounted collet-style puller. These are a very effective way to pull larger numbers of bullets that might lead to a profane cardiac episode if attempted with an impact puller. The Hornady Cam-Lock is an example of this type of bullet puller. The Cam-Lock threads into your press just like any reloading die. You'll need a collet for each caliber of bullet you pull. The collet installs inside the puller die to squeeze the bullet when the locking lever is rotated into the clamp position on the die body.

While these provide much nicer way to pull bullets than the impact puller, the downside is the cost of buying a collet for every bullet diameter you pull. The flip side to that concern is that your bullets are less likely to be damaged during the pulling process. Where an impact puller can leave bullet tips (especially polymer tips) deformed from the violence of the pulling process, collet pullers barely leave a mark if setup properly. Over time, the cost savings of reusing undamaged pulled bullets may provide enough savings to justify the additional expense of multiple collets. A simple trick to reduce the likelihood of damage when using an impact puller is to insert a foam earplug into the bottom of the inner cavity before pulling bullets. Instead

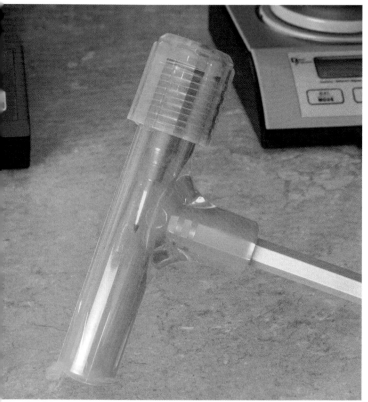

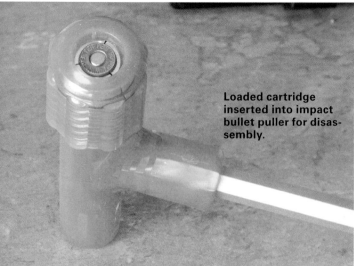

Loaded cartridge inserted into impact bullet puller for disassembly.

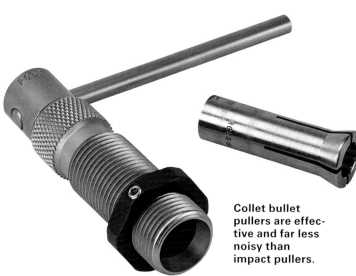

Collet bullet pullers are effective and far less noisy than impact pullers.

A simple squeeze bullet puller works fine for pulling small quantites of bullets.

of the bullet slamming into the face of the impact puller's inner chamber when it is freed from the case neck, the bullet comes to rest on a soft cushy foam earplug. Remove each pulled bullet before disassembling the next cartridge for this trick to work. As you can see, impact pullers and collet pullers have their advantages and disadvantages.

Handheld Pullers

One more type of bullet puller to consider is the handheld squeeze puller. Squeeze pullers can be anything from a pair of pliers (that will destroy your bullets) to purpose-built laser-cut steel models with multiple sizes of apertures for various bullet diameters. I like that these are noiseless, don't require collets, and don't take up much bench space. What I don't like is the tendency to scar bullet jackets—even if only cosmetically, and the significant amount of grip strength that is required to pull any number of rounds. I fully intend to put my kids to work

doing brass prep and pulling bullets as soon as they're capable of doing so without injuring themselves. The squeeze pullers I've used require so much grip strength that there is just no way my kids can use them successfully.

Jokes aside, even when I'm pulling bullets with a squeeze puller, I notice that slippage increases after a few dozen bullet pulls. Slippage means jacket damage, so I limit pulling sessions to a couple of dozen or so cartridges. As long as I don't let a big pile of defective cartridges accumulate before I pull them apart, this isn't much of an issue. In fact, there have only been one or two times in roughly twenty years of reloading where I've needed to disassemble more than a couple dozen cartridges. The compact size, lack of moving parts, and quickness of setup offered by the squeeze puller result in me usually grabbing it first.

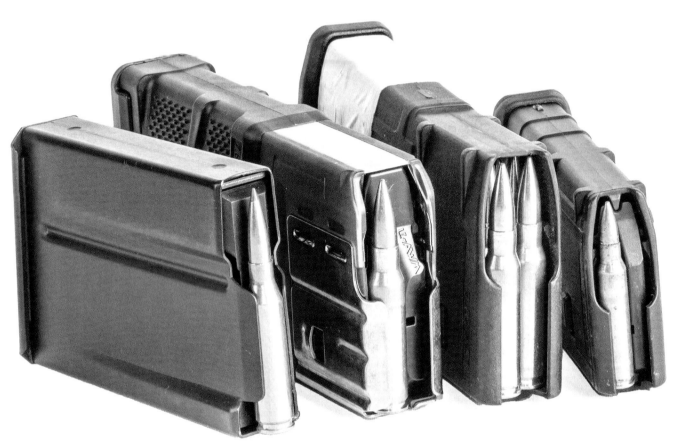

Magazine Length

Unless you're loading for a single-shot pistol or rifle, you will need to make sure that the longest cartridge you load is short enough to fit inside the firearm's magazine. If you don't mind loading rounds by hand, or "single loading," this won't matter.

Take the Remington 700 bolt-action rifle, for example. The Model 700 is offered in two action lengths; "long action" and "short action" are the commonly used designations for these. The short-action rifles are offered in cartridges like the .308 Winchester and its derivatives. Magazine length for a short-action Model 700 is usually limited to approximately 2.82". That means that our loaded rounds must be shorter

than 2.82" in order to fit inside the rifle's internal magazine. This is primarily important because of function-related issues. There are length-related measurements that affect accuracy, but the overall length of our cartridges matters most for functional issues. There is no standard magazine length used by all manufacturers, so check your particular firearm's magazines to determine what your length limitations are.

Magazine length often determines your maximum usable cartridge length.

Jump Measurement

Another length measurement we need to inspect, primarily when loading rifle ammunition, is the distance from the base or head of the cartridge to the ogive of the bullet. This length measurement can be very critical to the accuracy results we'll achieve with our handloaded ammunition and should be planned for during our preparations for loading. The location of the bullet ogive on our loaded round tells us how far the bullet must move or "jump" before it engages the rifling inside the barrel. There is no hard rule that tells us what amount of jump will yield the best results and every firearm is totally unique. I have rifles that like as little as 0.010" of jump and rifles that like more than 0.100" bullet jump.

We measure the ogive position of our loaded rounds with a set of calipers and and a comparator body and insert. Each bullet diameter will require its own comparator insert or the selection of the appropriate aperture on a multi-caliber comparator tool. These multi-caliber tools often look like a large hex nut with a hole through each facet. Each of these holes will be used for a different caliber projectile. Either comparator style is just fine for our purposes. Make sure your rounds will fit inside your magazine. You may also need to adjust the overall length of your loaded rounds to ensure reliable feeding in some semi-auto or even bolt-action rifles.

Seating Depth

The distance the bullet travels before engaging the rifling can affect the amount of pressure our ammunition generates. As the bullet touches the rifling, a certain amount of force is required to "engrave" the bullet, or cause the bullet to deform in order to move inside the rifling. As we

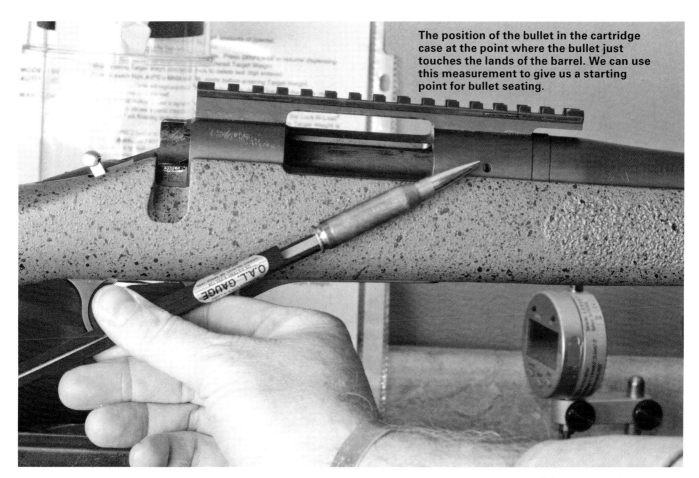

The position of the bullet in the cartridge case at the point where the bullet just touches the lands of the barrel. We can use this measurement to give us a starting point for bullet seating.

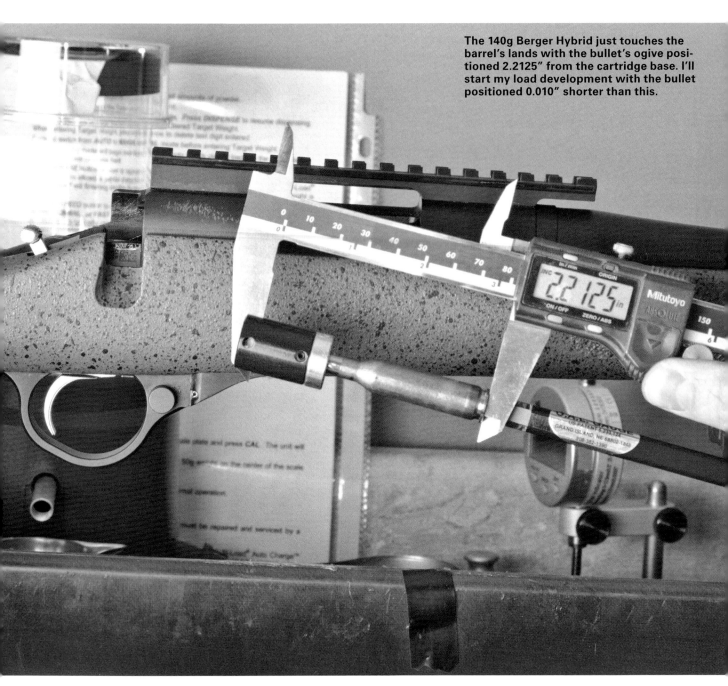

The 140g Berger Hybrid just touches the barrel's lands with the bullet's ogive positioned 2.2125″ from the cartridge base. I'll start my load development with the bullet positioned 0.010″ shorter than this.

illustrated in the Components and Nomenclature section, the diameter of the area between the lands is actually *smaller* than the diameter of the bullet. This is why the bullet must be engraved or forced into the lands. If we seat our bullets away from the lands, the bullet gets a running start, so to speak, that makes the engraving process somewhat easier and quicker. This means the bullet keeps moving through the engraving process quickly. If we seat our bullets so long that they are touching the rifling before the shot is fired, there is no opportunity for the bullet to get a running start before the engraving process begins. As pressure is building behind the bullet

during this engraving process, the relatively slow speed the long-seated bullet is traveling when it enters the rifling allows much higher pressure to build. This can lead to a dangerous situation where pressure builds too quickly and to a higher level than our components can tolerate. We can remedy this situation by using a smaller powder charge when we need or want to seat our bullets to touch the lands. Loading to the lands isn't dangerous or wrong; in fact, it can potentially provide the highest accuracy in some cases. You need to be aware of pressure issues when seating to touch the lands and adjust your charge accordingly though.

Bullet position in the case can be tuned for best accuracy in many instances. Long and heavy bullets may reduce powder capacity in smaller cartridge cases.

Weight

Once we've completed our cartridges and are in the final inspection stage, we should pick a few rounds at random to weigh on our reloading scale. This is a good check for the presence of a powder charge and/or primer. More than once during my first years as a rifle reloader, I discovered a cartridge that had been skipped during the powder-charging process by performing a weight check on my rounds. When this happens, go back and check every round thoroughly. If you missed one step on one cartridge, there is a good chance you missed another step somewhere else. Pull apart and reload any rounds that you suspect may be improperly loaded.

Primer Seating

Check the head of the cartridge to ensure the primer is seated smoothly without dents or damage. Make sure the surface of the primer is slightly below flush. If a primer is seated above the case head, there is a possibility that one of two bad outcomes will result. It's possible that the firing pin will use its energy to finish seating the primer in the pocket, the way it should have been in the first place, instead of smashing the cup into the anvil and detonating the priming compound in between. In other words, the entire primer moves forward under the pressure of the firing pin. This will result in a "click" when you expect a "bang."

The other possibility is that the primer will be detonated by the bolt or breechface or another object that makes violent contact with the surface of the protruding primer before the cartridge is seated fully in the chamber. This can

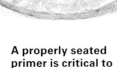

A properly seated primer is critical to reliable ignition.

be highly dangerous with injury or death a possible outcome. Think of the primer as similar to the gun's trigger. Nothing should come into contact with the cartridge's primer or the gun's trigger until your sights are on target and you intend to fire a shot.

Case Concentricity

Merriam Webster defines the root word concentric as "having a common center" or "having a common axis." Both definitions are decent ways to think about the term concentric or concentricity from a reloader's perspective. Basically, we don't want the bullet to be crooked in the case neck. We want the bullet to be perfectly straight or in line with the case neck.

Checking for concentricity is a bit outside of what the new reloader needs to be concerned with, at least initially. This is one of those things that you may never need to do. It is also one of those things that can explain accuracy issues that otherwise might leave you scratching your head and wondering why your handloads won't shoot well. I have one set of dies that will not load ammunition to an acceptable level of concentricity. As a result, the shot groups of ammo loaded with these dies—the seating die in particular—are far larger than they should be. Typically, two shots will be fairly close together and a third shot will be up to two inches away. When I first observed this grouping phenomenon in ammo loaded with these dies, I suspected an incompatible load, a bedding issue, or a scope and/or mounting issue with the rifle. Fortunately, before I spent too much time on that investigation, I checked the ammo on a concentricity gauge. What I found was revealing. About half of the ammo loaded with this die set achieved an adequate level of concentricity of less than

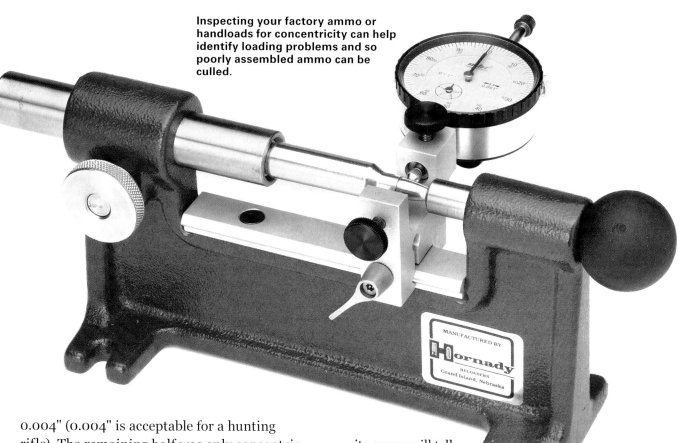

Inspecting your factory ammo or handloads for concentricity can help identify loading problems and so poorly assembled ammo can be culled.

0.004" (0.004" is acceptable for a hunting rifle). The remaining half was only concentric to within 0.009"-0.017". That's a huge failure. Match ammo is generally held to a concentricity standard of 0.002".

With that said, when you're just starting out, do not feel pressure to spend $100 on a concentricity gauge. If you're only going to load pistol and carbine ammo for blasting or even 3-Gun competition, don't *ever* worry about it. If you intend to load for 1,000-yard competition, F-Class, or benchrest, then you'll want to put a concentricity gauge on your Christmas list at some point down the road.

To check for concentricity, you'll need to follow the setup instructions that come with your gauge. Once you understand the workings of your tool, you may want to check multiple measurements. To verify the concentricity of your case necks, place your gauge probe about three-fourths of the way up the neck towards the case mouth on a resized case. Measuring the exact location on the neck where you place your probe to the thousandth of an inch isn't critical, but try to be as consistent as you can. The importance of this measurement is to identify any issues with your case-sizing operation. Do your sizing dies bring cases back into a nice perfectly round shape? If not, the concentric-

ity gauge will tell the tale. I have identified neck runout problems caused by the expander ball in a standard full-length sizing die with this method. Once the expander ball was removed, cases resized without the expander displayed almost no runout. My full-length sizing die for the 6.8 SPC creates 0.002" neck runout with the expander ball installed. I don't like it, but that isn't enough of an issue to warrant concern. I'm loading small volumes of hunting ammo for that rifle and consider it a 300-yard gun at most. At ranges that short, it isn't worth the time and trouble to chase runout with a separate expander die and mandrel or buy one or more new dies that may work better.

Semi-Auto Extras

Other issues to be aware of are the special needs of semi-auto rifles like the AR15. Semi-autos do not have the same amount of mechanical force and/or leverage for chambering and extracting that bolt-action rifles have. Slightly oversized ammunition that can be forced into and out of the chamber in a bolt-action rifle will cause a stoppage in a semi-

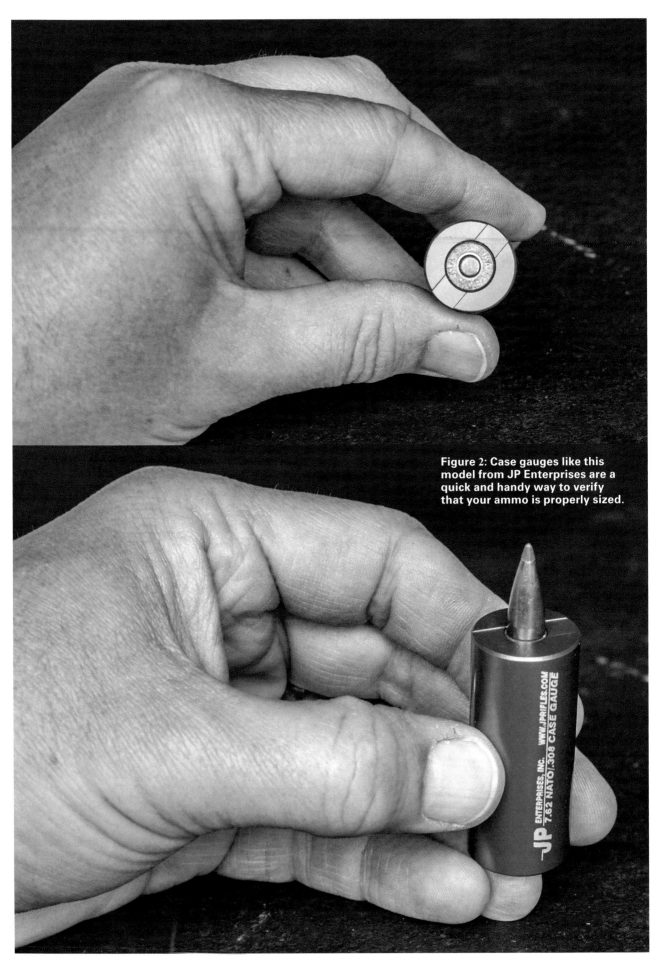

Figure 2: Case gauges like this model from JP Enterprises are a quick and handy way to verify that your ammo is properly sized.

auto. If the cartridge case hasn't been adequately resized, the bolt may be prevented from going into battery. Even if the bolt is able to slam into place chambering the round, extraction of that round may be impossible once it has been lodged in place by the forward momentum of the bolt carrier group. Many years ago when I first started loading .223 ammunition for my AR15s, I encountered this problem. Once I adjusted my full-length seating die to size the case down a bit more, the problem went away. JP Enterprises makes a really handy cartridge inspection gauge that will tell you if your ammo is properly sized for function in your .308 or .223 semi-auto. If the cartridge is too large, the loaded round won't seat fully into the chamber within the gauge. Using a tool like this to ensure that your ammo will chamber and extract is FAR safer than chambering a loaded round then attempting to extract it when you're standing beside your loading bench at home. Not only is the risk of damage to your rifle eliminated, but the risk of accidentally putting a hole in the ceiling (or worse) is also eliminated. Once a live cartridge gets stuck in your rifle's action, the task of dislodging it with a dowel or cleaning rod is not a pleasant one. The case gauges are worth the roughly $30 price!

Seating Concentricity

Once you've verified that your sizing dies have produced the desired results, you should check the concentricity of your seated bullets. This is the measurement we'll refer to when we discuss ammunition concentricity. Place the gauge probe just in front of the bullet's bearing surface and rotate the cartridge on the concentricity gauge. Any movement displayed on by the gauge indicates the amount of misalignment between your bullet and the cartridge case. As mentioned previously, match ammo should be kept to approximately 0.002" concentricity. The more perfectly our bullet is seated in the case, the more perfectly our bullet will be aligned with the bore of our rifle (if we've done everything else right, at least).

Visual Inspection

Major ammunition manufacturers perform a visual inspection on every round of ammunition. Think about that for a second. With millions of rounds of ammunition flowing through some of the larger plants annually, a set of eyes goes over every round of loaded ammo. Don't skip this step. This may be your last chance to catch something that's just not right before it's too late. A dent in a case neck, a primer incorrectly seated or missing altogether, a bullet with a deformed tip, or an improperly applied crimp on the case mouth are just some of the factors that can cause issues. These issues can be caught with a quick visual inspection before firing. I also suggest you adopt this practice with factory ammo.

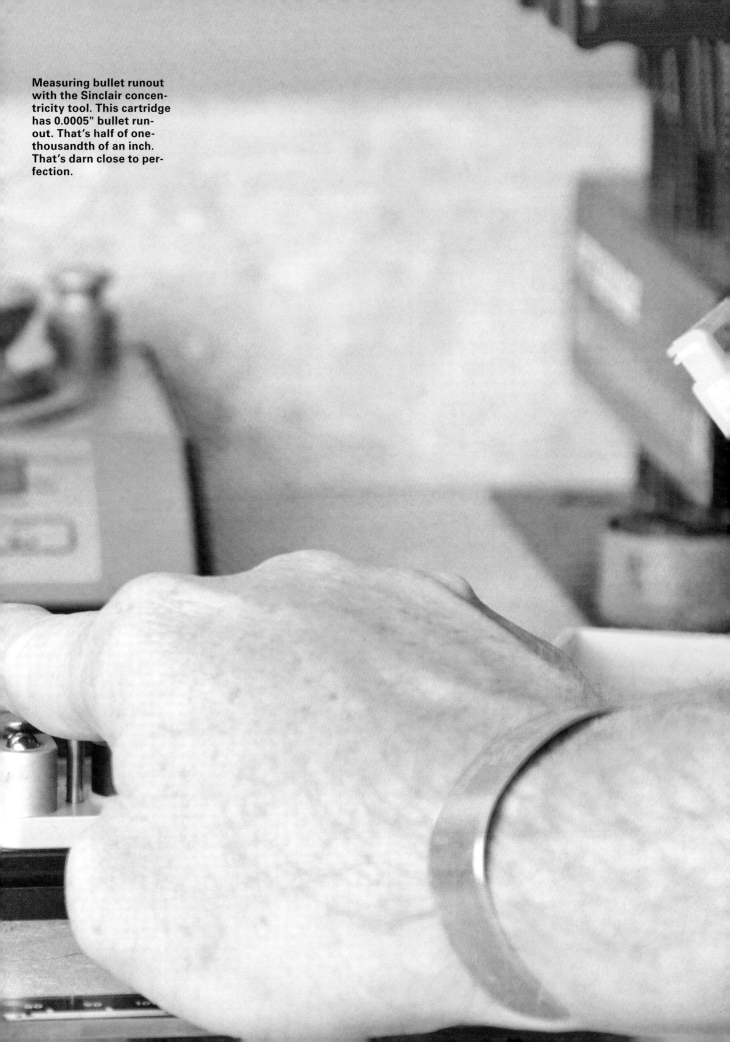

Measuring bullet runout with the Sinclair concentricity tool. This cartridge has 0.0005" bullet runout. That's half of one-thousandth of an inch. That's darn close to perfection.

Data management

Reloading Data Manuals

Every reloader needs at least one reloading manual on the bench. Three or four are even better. One of the most important lessons I hope you'll learn is that cross-referencing data is a must. One source may claim that 29.5 grains of powder is a maximum load while another source may claim that 29.0 grains is max. Why the discrepancy? Every firearm is different. Every rifle barrel is different, every pistol barrel is different, and every gun has its own little peculiarities. Cartridge cases from one manufacturer are not the same as those made by another company. Even different lots of cases from the same manufacturer will have variations that can affect pressure and performance. Primers from Winchester and primers from Federal or CCI will have different burn characteristics and create different pressure curves. Different types of actions can handle different amounts of pressure. Some manuals provide baseline data for the lowest common denominator, so to speak.

Cross-Reference

By cross-referencing several sources and paying attention to the details provided with the data, you can begin to formulate a plan that will allow you to safely load ammunition for your own set of components and firearms. If three out of four manuals list 45.0 to 45.4 grains of powder as a maximum load, you should be asking questions about the one manual that lists 47.0 grains as max. Are they using a different primer, different cases, or different overall length? What velocity does their max charge produce? The smart reloader will discard the outlier data, at least initially. Load to the charge range that multiple sources agree on. Let your results and common sense guide you from there. Another 50 or 100 feet per second of muzzle velocity is not worth damaging your gun or yourself.

Online Sources

In addition to printed reloading manuals from the big players in the ammo and handloading business, we are blessed (and cursed) with an astounding amount of information that is available online with the click of a mouse. Some online sources are fantastic tools for the handloader, but some contain information that is questionable, if not outright dangerous. Cross-referencing online info with data from other sources is a must, once again. As you gain experience and knowledge, you'll be able to pick up on things that make you question info that just doesn't seem right or seems inconsistent with what you've learned through your own reloading efforts. For example, load data for the .350 Remington Magnum can be found online from some very reputable sources that, I'm sure, have validated the data completely and verified it's safety in whatever barrel was used to develop the data. If I use that data in my 1980s-vintage Remington Classic, however, I'll need a hammer to open my rifle's action after firing the first round. While the online data suggests that charges of 64gr of a particular ball powder are safe, I see MAJOR signs of pressure at only 62gr of that powder. Cross-reference, verify, check, and always work up, regardless of the data source.

Record Keeping

If you stick with reloading long enough, sooner or later you'll find yourself drowning in data. It is frustrating as hell to find a box of handloaded ammo on your bench (even worse when this happens at the range) and have no idea what powder, primer, or bullet weight make up the cartridges inside. You may be able to Sherlock Holmes your way to an answer, or you may end up pulling apart the rounds and dumping the powder and possibly the primers just to reload the cases again with a known powder and charge. Over the years, I've developed an excellent system for managing reloading data that will keep you out of trouble and help you organize the information you need for the short and long term. In the years prior to my adoption of these methods, I loaded the same combination of components more than once during load testing/development. In other words, I did the same thing that I'd already done all over again without realizing it. I'm glad no one else knew about this. The results the second time weren't any better than the first! Because I did not keep good notes on the loads I'd already tried, I totally forgot about having already tested a powder/bullet combination that seemed like a sure thing. I eventually found a good load for the rifle with a different powder. The embarrassment from this episode, not to mention wasted time and money, were

View History Bookmarks Window Help

cout.com/forums

NCAA   NFL   MLB   FANTASY   WARRIOR   ⊕ SCOUT   HUNT   FISH   HOME   MORE ⌄   SIGN IN   SIGN UP

SNIPER'S HIDE

FORUMS   CONTESTS   BENEFITS   ARCHIVES   FAQ

7 CONCEALED CARRY MYTHS:

Sad Truth on why Most CCW Americans are Training to make good Victims:

Menu ⌄

come to Sniper's Hide, Online community for the Serious Tactical Marksman.

	TOPICS	POSTS	LAST POST
g	22	101	24 minutes ago by: Lakeray

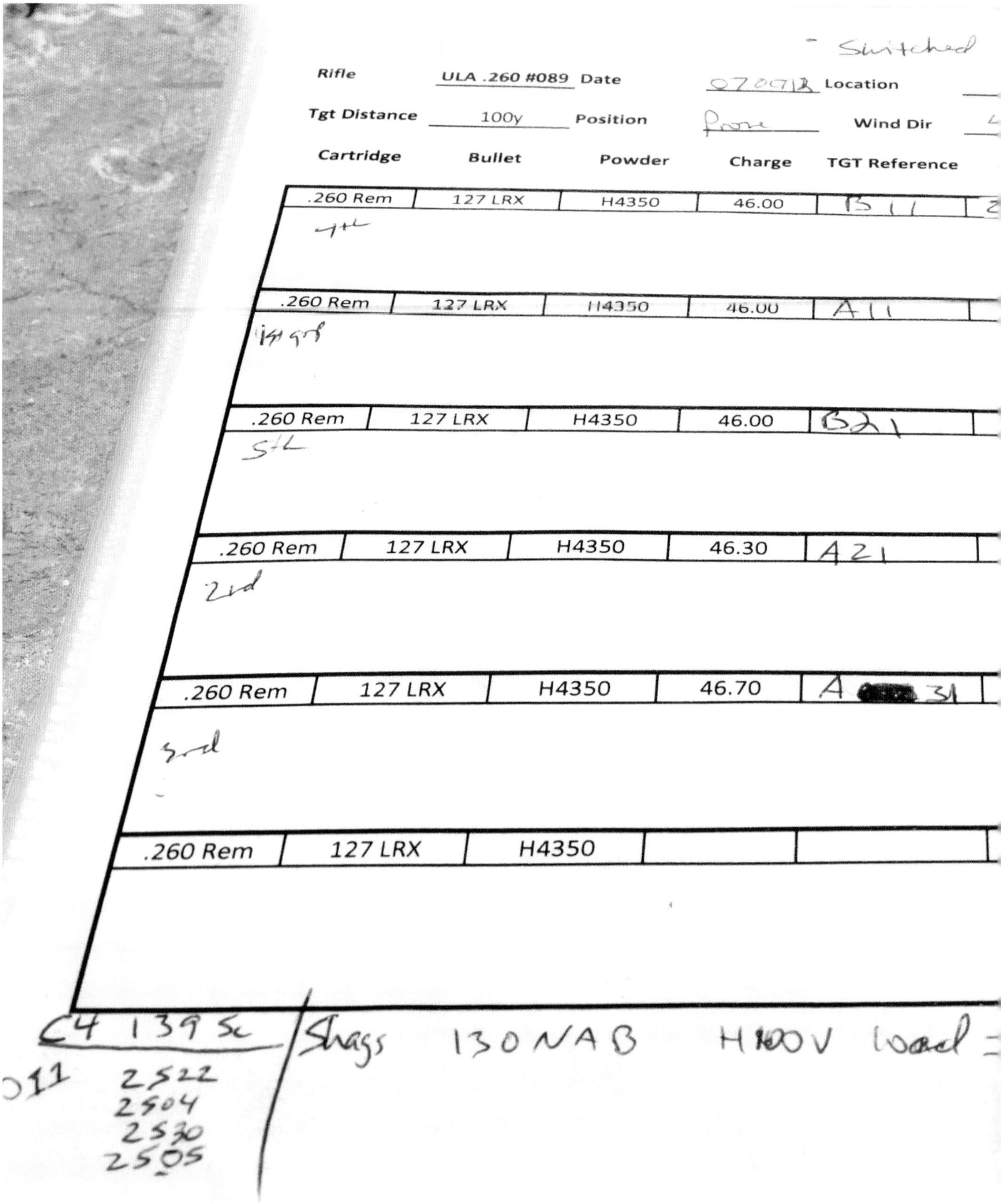

Rifle	ULA .260 #089	Date	070712	Location	
Tgt Distance	100y	Position	Prone	Wind Dir	

Cartridge	Bullet	Powder	Charge	TGT Reference	
.260 Rem	127 LRX	H4350	46.00	B11	
4th					
.260 Rem	127 LRX	H4350	46.00	A11	
14 grs					
.260 Rem	127 LRX	H4350	46.00	B21	
5th					
.260 Rem	127 LRX	H4350	46.30	A21	
2nd					
.260 Rem	127 LRX	H4350	46.70	A ● 31	
3rd					
.260 Rem	127 LRX	H4350			

C4 1395c / Snags 130 NAB H100V load

011 2522
2504
2530
2505

Recording the performance of your ammo is a critical step to avoid repetition of failures and loss of data on your successes. Notes like, "Today's results suck!" are a bit vague, but backed up with real numbers, they can create an adequate picture of how a certain powder or bullet works in your gun.

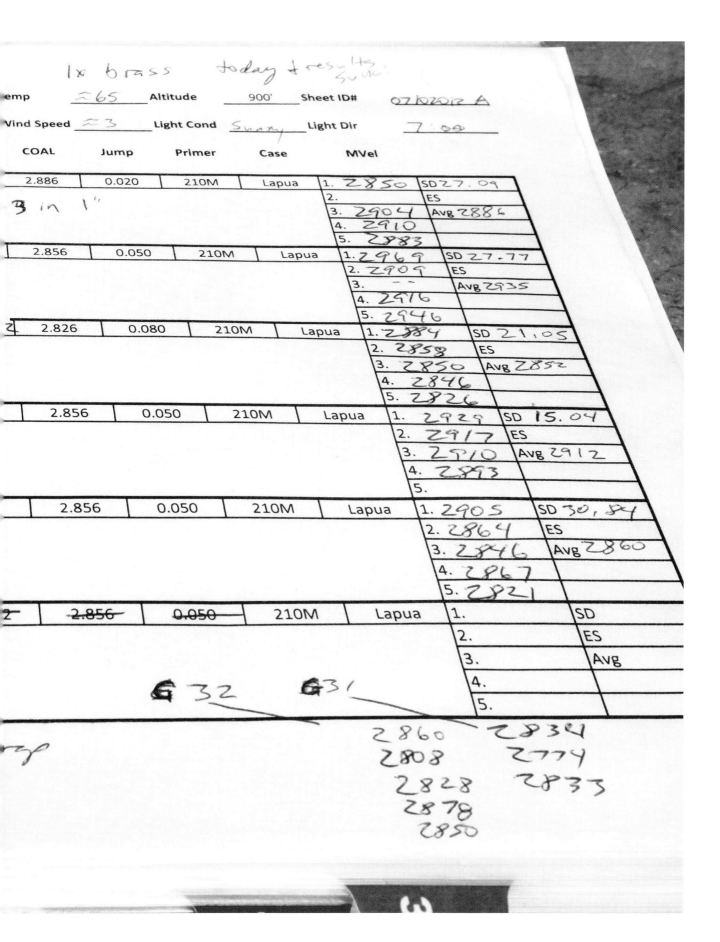

1x brass   today + results: sub.

Temp ≈65   Altitude   900'   Sheet ID#   07/02/2012 A

Wind Speed ≈3   Light Cond Sunny   Light Dir   7:00

COAL	Jump	Primer	Case		MVel	
2.886	0.020	210M	Lapua	1.	2850	SD 27.09
				2.		ES
3 in 1"				3.	2904	Avg 2886
				4.	2910	
				5.	2883	
2.856	0.050	210M	Lapua	1.	2969	SD 27.77
				2.	2909	ES
				3.	--	Avg 2935
				4.	2946	
				5.	2946	
2.826	0.080	210M	Lapua	1.	2884	SD 21.05
				2.	2858	ES
				3.	2850	Avg 2852
				4.	2846	
				5.	2826	
2.856	0.050	210M	Lapua	1.	2929	SD 15.04
				2.	2917	ES
				3.	2910	Avg 2912
				4.	2893	
				5.		
2.856	0.050	210M	Lapua	1.	2905	SD 30.84
				2.	2864	ES
				3.	2846	Avg 2860
				4.	2867	
				5.	2821	
~~2.856~~	~~0.050~~	210M	Lapua	1.		SD
				2.		ES
				3.		Avg
				4.		
				5.		

G 32   G 31

2860   2834
2808   2774
2828   2833
2878
2850

a big part of what led to the techniques we will cover in the following paragraphs. Imagine how frustrating it is to have a target or multiple targets with several REALLY good groups on them and not remember what rifle, bullet, powder, primer, and ogive length turned in such a stellar performance. How can you recreate that load if you don't know what it is? We're all way too busy these days and the complications of work, family, keeping up a home, and all the other demands on our time require us to be smart and efficient with our reloading efforts and the data that we need to be successful. My way certainly isn't the only way, but it is easy, efficient, and effective.

On The Bench

The first group of data you need to manage relates to your rifle or pistol. I recommend you organize a 3-ring binder with clear sheet protectors to hold your data pages. Start with a blank page and list the make, model, chambering, and serial number of the gun. Write down any notes about the gun provided by the manufacturer. Glock pistols aren't intended to be used with lead bullets, for example. Next, add things like the internal length of the magazine. If we're loading to the overall length listed in a reloading manual, we can be fairly sure our rounds will fit in the magazine. If we're looking for better accuracy from our handloads, we may want to adjust the length of our rounds to find the bullet position our rifle prefers. How long can we load our rounds and still fit into the magazine? We need to know. Some handguns won't tolerate certain types of hollow-point bullets. Adjusting the length of loaded rounds can have an impact on feed reliability, especially in some older semi-auto handguns. We need to note the constraints of our system.

I'll generally pick the top three or four bullets I want to use in a particular gun. Check with the manufacturer of your chosen projectile for their recommendations on length and loading. Make note of that information. When loading for rifles, I'll measure the length at which my chosen selection of bullets makes contact with the barrel's lands, as we discussed in the Tools and Process section. I'll write this base-to-ogive measurement and the corresponding overall length next to each bullet on the list. With these two mea-

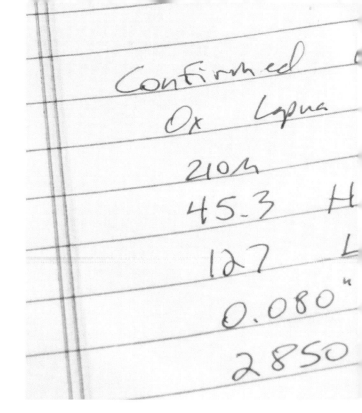

surements, I can determine whether or not I can seat the bullet to touch the lands and still have a cartridge that fits inside the magazine. If not, I can determine how much jump the bullet will have before engaging the rifling. This may help me narrow down my bullet selection to one or two that are best suited to my rifle.

Load Development Process

As I go through the load development process, I'll use the case prep cards, load plan cards, and range data sheets we'll discuss a bit later. Once I settle on a load or two that I intend to use with a particular gun, I'll write every available bit of data on that load on the gun's logbook page. Think of it as the "home page" for that gun. All of the core data we need regarding that gun is kept on this page. Once I validate and confirm my load, I'll cut that confirmation/zero target out and tape it to my home page. This is intended to represent the culmination of all of my load development work, my trials and errors, and my tweaking until I achieved the best possible performance from my handloaded ammunition. For some loads that I don't build with any great

Examples of data that should be recorded in your benchtop data binder: rifle data sheets with base-to-ogive measurements, completed range data sheets, load-development targets, and confirmed final load data for each rifle or pistol.

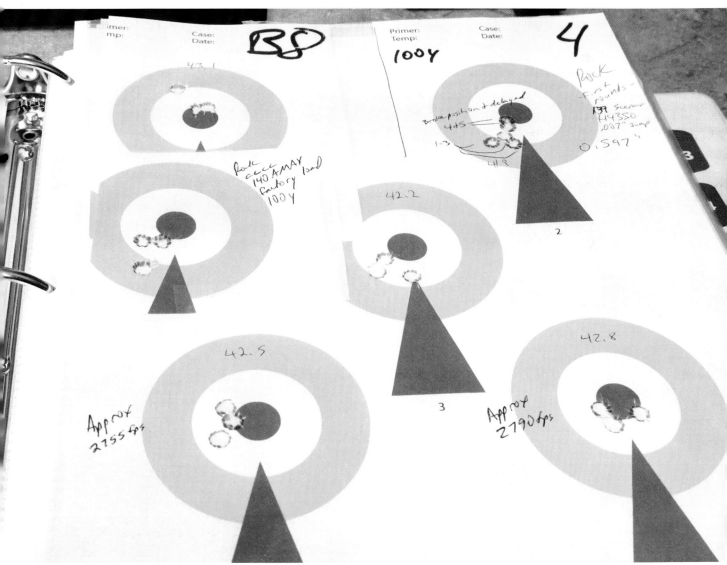

frequency, hunting loads for a backup rifle, for example, I may completely forget my load information by the time I shoot all of the ammo I loaded last. Three years ago, I went through the load development process for a .300 Winchester short Mag elk rifle, adjusting this and trying that until I got the charge, seating depth and primer just right. The rifle consistently shot under ¾" at 100 yards with very consistent velocity numbers. With the load confirmed out to 650 yards, I loaded 150 rounds of my final combination of components. I still have over 100 of these rounds and, after three years, I couldn't tell you what the ogive measurement, primer type, powder charge, or muzzle velocity is from memory with any certainty. Fortunately, I don't have to because every detail has been recorded in my logbook on that rifle's "home page."

Benchtop Binder

I also suggest you keep a benchtop binder with load data for your favorite cartridges copied from several loading manuals. Instead of thumbing through four different 700+ page load manuals to check the recommended max charge weight for a 175gr Sierra Match King over Reloader 15, you can flip through three or four tabbed and labeled pages in your binder. If you don't own a

A three-ring binder makes it easy to keep your important or often used loading data and records handy.

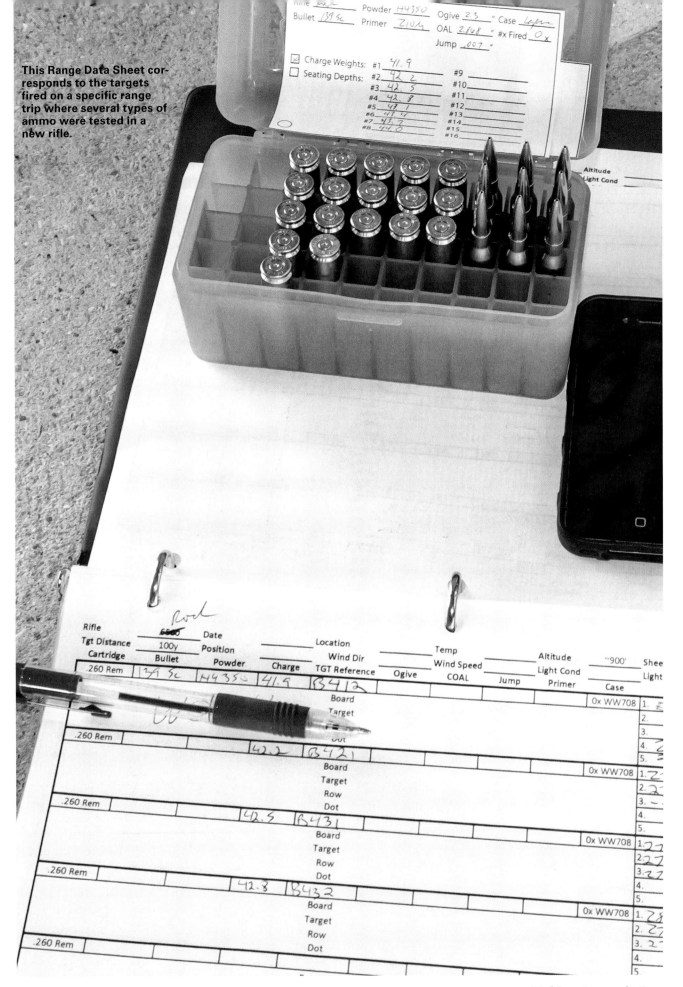

This Range Data Sheet corresponds to the targets fired on a specific range trip where several types of ammo were tested in a new rifle.

.256 Newton, a .22 Zipper, or a .296 Whiz Bang, why deal with a huge manual listing load data for those and other cartridges you don't care about?

Range Binder

The final binder I suggest you build will go to the range with you. Here you'll note the cartridge, the rifle or pistol you'll use, the powder type, the powder charge, the bullet type and weight, the overall length, and any other factors you believe are important. Remember, it's better to have too much data than too little. Write it down & thank me later.

As I fire my test loads, I'll add chronograph data for each shot and make notes about feeding, firing, extraction, ejection, and accuracy. Does my resized brass chamber easily, or do I have to force the bolt closed? Does my empty eject vigorously? Do I experience stovepipes, double-feeds, or other malfunctions? Did I make a bad shot on that last round, or is the ammo just not shooting well? Did I find the Holy Grail and shoot a half a dozen consecutive groups under a half-inch at 300 yards? Write it down. What target am I shooting at with this load? How many times have I retrieved targets at the end of the day and had no idea which groups were shot with which gun/ammo? Don't ask. That's just one more of many lessons that led to the development of my data management system.

I've kept every target I've fired during load development for the last several years as well. Matched with my range data sheets and rifle data sheets, I have an incredibly complete picture of how each firearm performs, what sort of accuracy I've found, velocities achieved, the bullets or powders that seem to consistently perform well in many similar guns, and what components don't seem to perform at all. An assessment of this data can lead you to eliminate some components from your load-development process and focus on others altogether. For example, one Nosler 165gr Accubond load I've tried has been a sub-MOA performer in three different factory-built hunting rifles I've played with over the last several seasons, but that doesn't mean anything when I start testing with a new rifle. Every rifle is different. Still, because of past successes, I will try that load

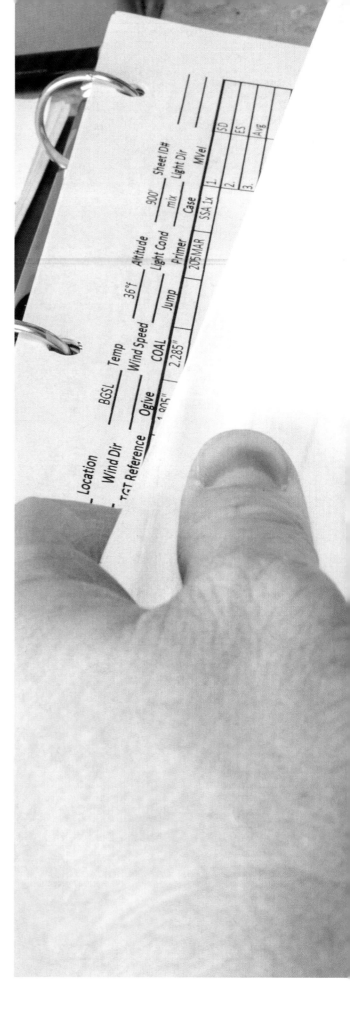

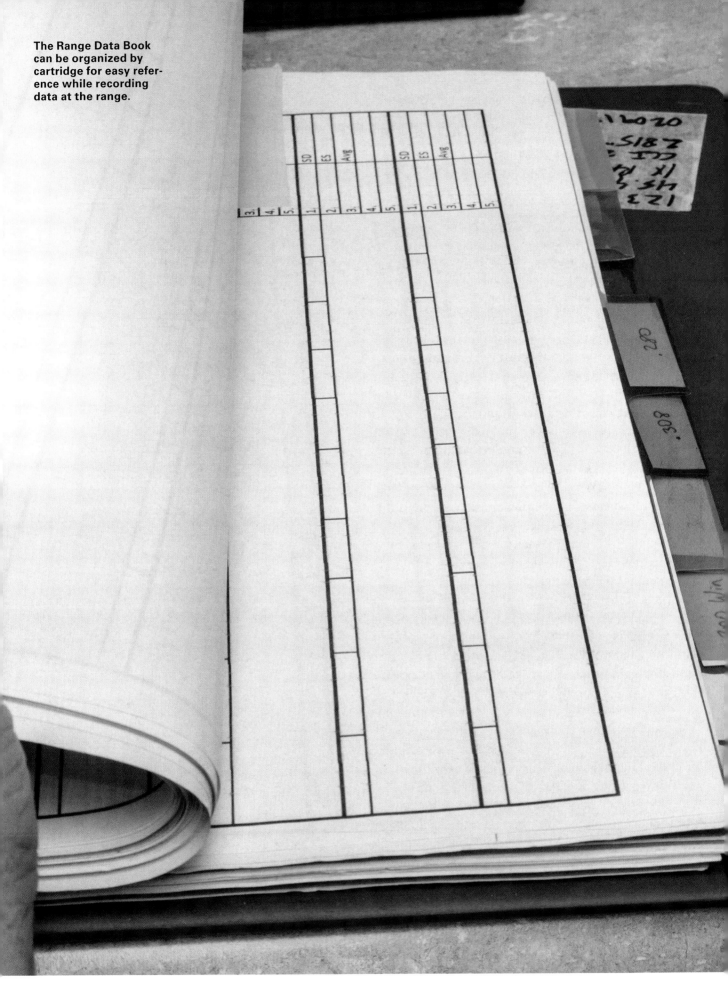

in the next .30-06 that finds its way home with me. Because I've kept detailed notes, I've identified a consistent load that works in at least three rifles. This may keep me from reinventing the wheel again and again with my handloads. Keep these targets in sheet protectors behind your rifle data sheet for easy access.

Case Prep Card

Reloading requires a quantity of components greater than the amount of ammo you intend to keep on hand. You may have 200 rounds loaded and ready to shoot but have another 100 empty cases prepped and ready to load and 150 spent cases ready to be cleaned, resized, primed, and reloaded. Over time, you'll accumulate brass from different manufacturers, different numbers of firing and loading cycles, brass that "belongs" to different pistols or rifles, and brass that may be designated for practice only. You'll eventually find a box or tub of brass and wonder "Has this brass been resized? How many times has this brass been fired? Which rifle was this brass fired in?" My favorite is, "Where the hell did this come from? I don't remember buying any of this brand."

Imagine you're loading for six different cartridges. The management of all of the components you'll need just got significantly more complex. To deal with this, I've stolen some inspiration from my limited experience in manufacturing. I've created a case-prep card that is much like a "work order" or "job traveller" form. The card lists the cartridge, the brand of brass, the number of times the brass has been fired (not loaded), the gun the brass was fired in, and each step that I perform on my brass next to a check box that I mark as each task is completed. I found this tool to be so valuable in improving my organization and efficiency that I had custom-printed notepads made based on the template I've refined over the years. I've made these available on my site (*www.gamescoutusa.com*) for those looking for the utmost in speed and

Case-prep cards are an invaluable tool that allows the reloader to organize and keep track of batches of brass and what preparation steps have already been performed. You can easily make these yourself or purchase them from *www.gamescoutusa.com*.

Case Prep	
Cartridge	2̲6̲0̲ Rem
Brand	Lapua
Times Fired	1x
Rifle Used	8413
Tumble	✓
Wipe Dust	✓
Lube	✓
Deprime	✓
Size Neck	✓
Size Body	✓
Size Neck (#2)	✓
Expand Neck	✓
Delube	✓
Trim	—
Chamfer	✓
Debur	✓
Clean Neck	✓
Clean Pocket	✓
Prime	✓
Date Complete	28 Dec
Notes:	Re test confirm

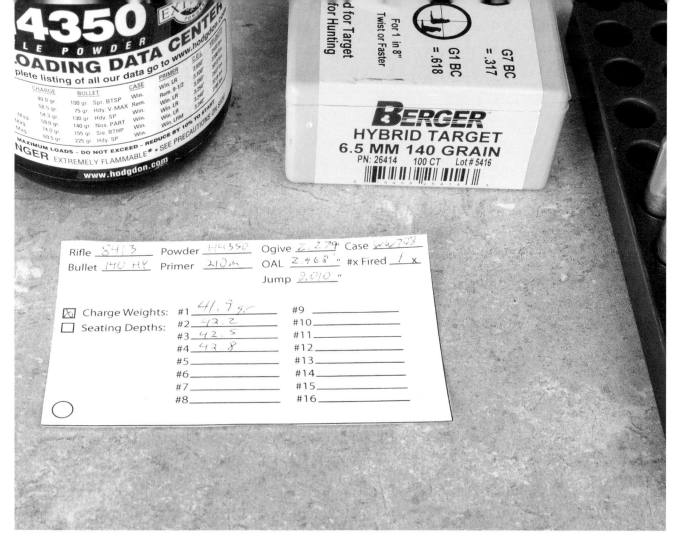

Load-planning cards ensure that you have a completely thought-out plan *before* you begin handling powder, primers, or other components. This increases efficiency and safety by reducing opportunities for mistakes to happen. You can easily make these yourself or purchase load-planning notepads from *www.gamescoutusa.com*.

ease. The pages are sized to fit inside an ammo box. Just peel off a page, mark each step, and never get your components mixed up again.

Load-Planning

Once my cases are prepped, I'll move on to plan my next batch of reloads. For years, I'd walk up to the bench with a general idea about what I wanted to do, but no real plan. As such, loading was much more frustrating and inefficient than it needed to be. Now I make a plan FIRST, before ever touching any components, tools, or dies. I can't begin to tell you how helpful this is. Start with a 3"x5" card. Write the cartridge, rifle, or pistol (I use the last four digits of the serial number), bullet, powder type, powder charge, primer, seating depth (to both ogive and overall length), case brand, and the number of times the case has been fired across the top of the card. Next add the powder charge(s) and number of rounds you intend to load.

With this information, you know what brass to retrieve from the shelf, what primers to pull from the box, what powder jug to grab, and what bullets to pick up. You also know how much of these components you'll need. I recently made a load-plan card for an AR15 in 6.8 SPC and discovered that I didn't have enough bullets left to complete that plan. A quick check of my components inventory confirmed that I had everything else I needed, so I made an order for more bullets. Instead of wasting time getting started, filling my powder measure with powder, filling my priming tool with primers, prepping brass, then discovering mid-stream that I was out of bullets, I was able to move on to another task that I could complete that day. Regular 3"x5" index cards work very well for this. Over the last few years, I've refined these load-plan cards to the point that I've made them available as custom-printed notepads on my website, *www.gamescoutusa.com*, for those looking for the utmost in speed and ease. With these pre-printed cards, you can simply fill in the blanks and be confident you haven't forgotten any critical info.

Boxed To Shoot

When I'm loading test ammo for development purposes, I'll take the time to actually mark each loaded round with the powder charge for that particular round. Since I'll be putting up to ten different charges in one ammo box, there is a danger of these rounds becoming mixed up should the box come open in transit or should the rounds spill out for some reason. If a box of mixed ammo spills and you have no idea what powder charge each round contains, you'll have no choice but to disassemble every round, dump out the powder, resize the necks, and reload the cartridge. I recommend the 50-round plastic boxes from MTM Case-Guard.

Once my rounds are labeled, I'll put the case prep sheet and load plan sheet in the box with my ammo. These sheets will stay with the ammo through the range trip and stay with the empty brass once the ammo is fired. Finally, I'll stick a big piece of masking tape across the top of the box with the basics of the loads inside and the gun the ammo is intended for.

Loaded, labeled, and ready. We'll use our load-planning card to label our ammo box. The case prep card is inside to be kept until rounds are fired and brass is ready to be cleaned and reloaded.

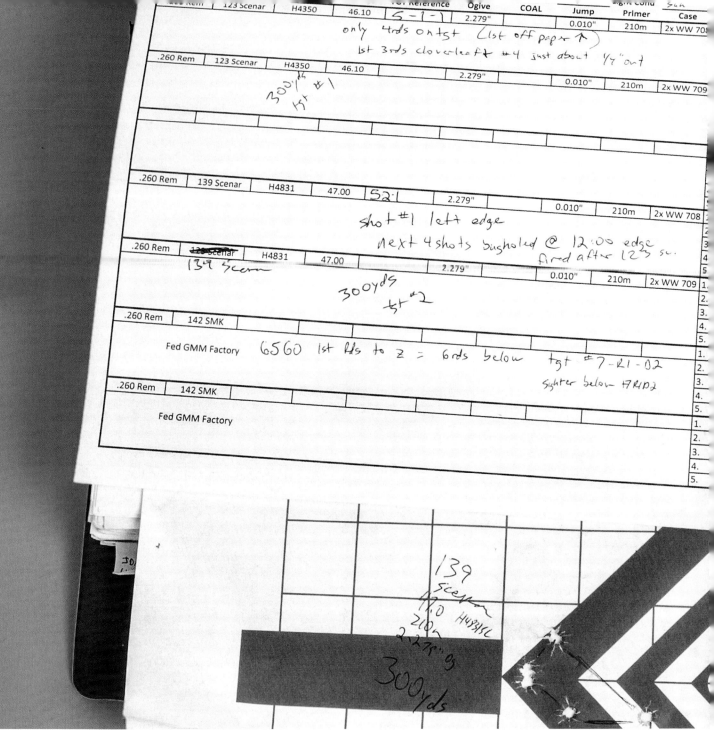

	123 Scenar	H4350	46.10		5-1-1	2.279"			0.010"	210m	2x WW 70

only 4rds on 1st (1st off paper ↑)
1st 3rds cloverleaf #4 just about ¼"out

.260 Rem	123 Scenar	H4350	46.10			2.279"			0.010"	210m	2x WW 709

300yds #1
1st

.260 Rem	139 Scenar	H4831	47.00	52.1		2.279"			0.010"	210m	2x WW 708

shot #1 left edge
next 4 shots bugholed @ 12:00 edge
fired after 123 sc.

.260 Rem	~~123 Scenar~~ 139 Scenar	H4831	47.00			2.279"			0.010"	210m	2x WW 709

300yds
1st #2

.260 Rem	142 SMK										
	Fed GMM Factory			6560	1st rds to z = 6rds below	tgt #7-R1-02					
						sighter below #R1D2					

.260 Rem	142 SMK										
	Fed GMM Factory										

Range Data Sheet

Once you've completed your load plan and your rounds are boxed or bagged and ready for the range, you need to begin planning your range session. The first step is to make a range data sheet. This will mirror the load-plan sheet you've just placed inside your ammo box. Write the cartridge, rifle or pistol, bullet, powder type, powder charge, primer, seating depth to both ogive and overall length, case brand, and number of times the case has been fired across the top of the card just like your load plan card.

Range data book with range data sheet, ammo boxed with load-planning card, and case-prep card, all with corresponding information. This sort of organization makes reloading so much more efficient and effective. Handle this data casually, and you'll soon be frustrated and retracing your steps after your results are forgotten or lost.

Next, add a spot for distance to target, target number, date, weather conditions, shooting position, and any other information about your range plan. Add places for chronograph data, notes on load performance, and notes on feed-

ing, extraction, ejection, and any other issues you may observe while at the range. Finally, punch three holes across the top of the pages and place them in your range binder. Ready-made range data sheets are also available as custom-printed notepads on my website, *www.gamescoutusa.com*.

The purpose of this book isn't to sell you case prep notepads, load plan notepads, or range data sheet notepads. These are offered very inexpensively as a convenience to folks like me who are way too busy and want to maximize loading and shooting time. You can certainly make your own and print them at home with the info in this section. My point is to emphasize the need for data management for the reloader. This is an area that no one considers at first but that can quickly become overwhelming and confusing. Take a few steps to plan what you're going to do and record what you've done on the bench and at the range, and you'll save yourself time, money, and frustration. You'll be a much happier and more effective reloader in the long run.

Data At The Range

When you arrive at the range, your range data sheet and the target you're shooting will be used to record what happens. Make sure to note the date, distance, gun, load, and weather conditions (temperature at a minimum) on the target itself. If using multiple targets, label each target with a number, a letter, or some other identifier. I label target boards with letters and individual targets printed on regular 8.5"x11" paper with numbers. If a target has multiple bullseyes, I'll number those, too. This will take an extra minute or two but is absolutely worth the effort. The results of your loading efforts are of no value if you can't keep them organized. What powder charge shot best? What seating depth resulted in the greatest precision? When you zeroed your scope last year, was your point of impact "dead on" at 100 yards or did you zero an inch and a half high at 100 yards in anticipation of that big hunt out west? If you don't keep that information straight, you may just have to redo the entire process.

This record-keeping system is also beneficial once you've settled on a load(s). For the match shooter that burns through 1,000 rounds in a season, these records can paint a very clear picture of the condition of a rifle's barrel as it wears. What started out as a gun capable of putting five rounds into half an inch with consistency may now struggle to keep five rounds in three quarters of an inch. Maybe it has even opened up to an inch or more. With thorough recordkeeping, we can evaluate the condition of our rifle or handgun system based on real numbers and evidence from our range sheets and corresponding targets.

The classic Winchester Model 70 chambered for the classic .30-06 cartridge.

Now that we've learned about the tools, components, and pitfalls of reloading, it's time to put that knowledge into practice. We'll go through three distinct examples that will cover most general reloading applications. We'll load 9mm Parabellum (Luger) ammo for use in a semi-auto pistol on a Dillon progressive press, .30-06 ammo with a simple full-length die setup for Uncle Larry's hunting rifle, and match-grade .260 Remington ammo for tactical competition rifle.

In each example, we will follow our load-planning and data-handling routine as outlined in earlier chapters. We'll prepare only the components we need for each loading application to reduce the chance of powder or primer mixups. Once we've completed each batch of ammo, we'll clean up our mess and put away the remaining components to make room for the next group of cartridges. The starting point for each example

will be a bag of once-fired brass that has been tumbled clean and inspected for damage or defects. Any load data provided is only listed as an example and should not be used for loading purposes. Consult your load manuals and remember to cross-reference when choosing your own data!

Example #1:
Loading Basic .30-06 Hunting Ammo On A Hornady LNL Single-stage Press

Step 1. Prepare load plan card for our selected components. We'll load a total of 18 rounds for this initial workup.

    a.   Chosen bullet: 168gr Boattail HP

    b.   Chosen primer: Federal #210 Large Rifle

    c.   Chosen case: Federal .30-06

    d.   Chosen powder: Hodgdon 4350 (starting 4% below max and working up to Hodgdon's max load in 0.5gr increments)

    e.   Overall length: 3.340"

Step 2. Prepare your range data sheet according to your load-planning card.

Step 3. Check batteries in calipers.

Step 4. Install full-length sizing die, set up according to manufacturer's instructions in our reloading press.

Step 5. Lubricate 18 cases for sizing.

Step 6. Place lubed cases in loading tray.

Step 7. Size all 18 cases. Ensure primers pop free from each case.

Step 8. Tumble resized cases for 15-20 minutes to remove sizing lube. (Some people skip this, but it drives me nuts to have sticky cases. It's up to you.) Wipe the media dust off the cases to keep your dies cleaner.

Step 9. Check case length and trim, if needed.

Step 10. Chamfer and debur case mouths.

Step 10. Place cases back into loading tray in six rows of three cases each.

Step 11. Fill powder measure with H4350. Follow safety and handling guidelines from powder-handling section of this book. Adjust measure until desired *starting* charge is dispensed.

Step 12. Position bullet box on bench for easy retrieval.

Raising the ram to full-length size a fired .30-06 case.

Step 13. Fill priming tool with two rows of Federal #210 primers. (These are packaged in rows of 10 so you'll end up with a couple of extras if you don't count them out on the front end.)

Step 14. Prime cases using Hornady hand-priming tool.

Step 15. Charge first row of three cases (I start at the far row and work closer.) with our *starting* charge.

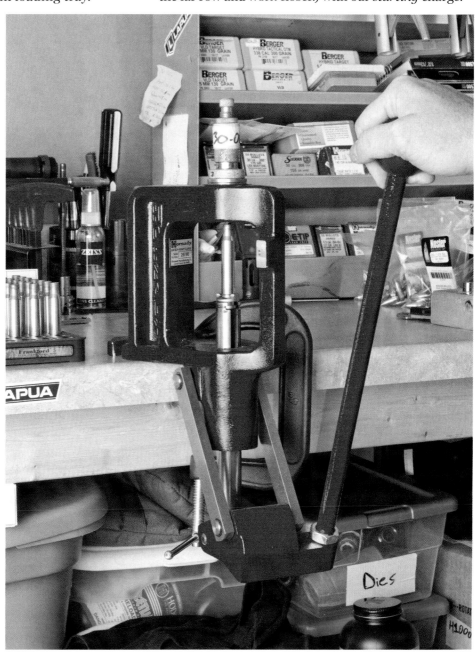

Step 16. Increase powder charge by 0.5gr.

Step 17. Charge next row of three cases with new powder charge.

Step 18. Repeat steps 17-18 until all 18 cases have been charged. You'll have three cases in each row with six different charges. The highest charge will be the max recommended load we've been provided by our data sources.

Step 19. Seat bullets in all 18 cases, being careful to place each completed cartridge back in the same spot on the loading tray from which it came. Check the base-to-ogive length on every round at this stage of development. You should be able to hold +/- 0.003" or close to that.

Step 20. (Optional) Crimp all 18 case necks. This can be done during seating or with a separate crimp die.

Step 21. Use a permanent marker to mark every case with the powder charge it contains. This will save you from heartache if your ammo spills on the way to the range.

Step 22. Inspect every round thoroughly according to the steps in the "Inspection" portion of this book.

Step 23. Place your cartridges into an ammo box or case. I'd also suggest taping the lid shut or wrapping a rubber band around the box during transport.

Step 24. Clean up your area and put away all components. Make sure leftover powder goes back into the appropriate container and is stored properly.

Example #2:
Loading Match-grade 260 Remington Ammo On A Hornady LNL Single-stage Press

Step 1. Prepare load-plan card for our selected components. We'll load a total of 24 rounds for this initial workup.
   a. Chosen bullet: 140gr Boattail HP
   b. Chosen primer: Federal #210m Large Rifle Match

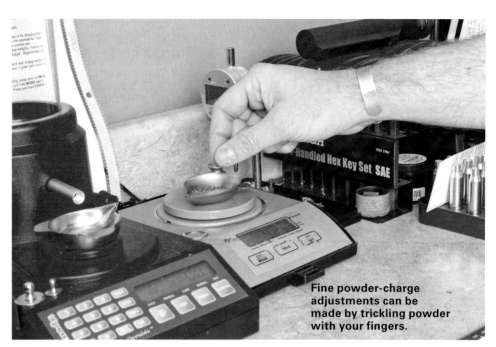

Fine powder-charge adjustments can be made by trickling powder with your fingers.

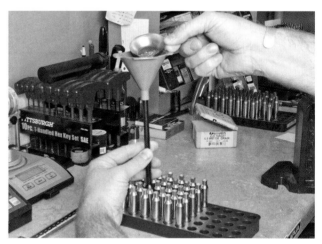

A funnel/drop-tube combo is one of the handiest things to have on your bench.

   c. Chosen case: Lapua .260 Remington, lightly neck turned
   d. Chosen powder: Hodgdon 4350 (starting 4% below max and working up to Hodgdon's max load in 0.3gr increments)
   e. Base to Ogive: 2.300" (0.010" jump in this rifle)

Step 2. Prepare your range data sheet according to your load-planning card.

Step 3. Check batteries in calipers.

Step 4. Lubricate 24 cases for sizing.

Step 5. Place lubed cases in loading tray.

Step 6. Install neck-sizing die #1, set up according to manufacturer's instructions in our reloading press, and size all cases. We're stepping our neck diameter down in two increments.

Partial resizing will occur due to the 0.295" neck bushing. Leave the decapping mechanism installed in this die to deprime during this step.

Step 7. Install body die, set up according to manufacturer's instructions in our reloading press, and size all cases. You want 0.001" to 0.002" resizing at the shoulder. Use your case comparator to set the desired level of resizing performed by your die.

Step 8. Install neck-sizing die #2, set up according to manufacturer's instructions in our reloading press, and size all cases with a 0.291" neck bushing. This will take the inner neck diameter down 0.001" to 0.0005" below where we want it to be.

Step 9. Install neck expander die with 0.263" expander mandrel and size all case necks. With the typical 0.001" springback after sizing, this will give us our final inside neck dimension of 0.262" for perfect neck tension on a 0.264" bullet.

Step 10. Tumble resized cases for 15-20 minutes to remove sizing lube. As I mentioned in example #1, some people skip this, but I can't stand to.

Step 11. Check case length and trim, if needed.

Step 12. Chamfer and debur case mouths.

Step 13. Brush inside of case necks clean.

Step 14. Place cases back into loading tray in eight rows of three cases each.

Step 15. Fill powder measure with H4350. Follow safety and handling guidelines from powder-handling section of this book. Adjust measure until desired *starting* charge is dispensed.

Step 16. Position bullet box on bench for easy retrieval.

Step 17. Fill priming tool with three rows of Federal #210m match-grade primers. (These are packaged in rows of 10 so you'll end up with six extras if you don't count them out now.)

Step 18. Prime cases using Hornady hand-priming tool.

Step 19. Charge first row of three cases (I start at the far row and work closer.) with our *starting* charge, verifying every charge to exactly equal to our intended charge weight. In other words, +/- 0.0gr.

Step 20. Increase powder charge by 0.3gr.

Step 21. Charge next row of three cases with new powder charge.

Hand priming is another option and provides a more precise and tactile feel to the primer seating process when compared to priming on a press.

Once you've primed and charged the case with powder, bullet seating comes next. Using a micrometer seating die provides more precision

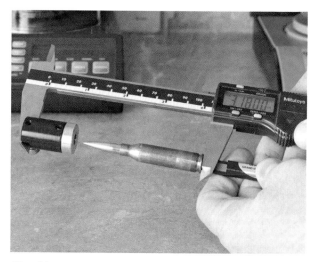

Checking runout ensures that bullet seating depth matches your rifle's chamber.

Step 22. Repeat steps 20-21 until all 24 cases have been charged. Verify every charge to exactly match our intended charge weight. You will have three cases in each row with eight different charge weights 0.3gr apart. The highest charge will be the max recommended load we've been provided by our data sources.

Step 23. Seat bullets in all 24 cases, being careful to place each completed cartridge back in the same spot on the loading tray from which it came. Check the base-to-ogive length on every round at this stage of development. You should be able to hold +/- 0.002" or better with match-grade bullets.

Step 24. (Optional) Crimp all 24 case necks. If you're going to crimp match-grade ammo, I strongly recommend doing so with a separate factory crimp die. With very few exceptions, I only crimp match ammo that I will use in a semi-auto rifle.

Step 25. Mark every case with the powder charge it contains. This will save you from heartache if your ammo spills on the way to the range.

Step 26. Inspect every round thoroughly according to the steps in this book. If any rounds are found to be more than 0.003" out of concentricity, place those rounds aside and reload that charge weight into another case. If that charge weight is determined to be safe, you can shoot the "crooked" rounds for practice. If not, pull the rejected rounds apart to salvage the components.

Step 27. Place your cartridges into an ammo box or case. I also suggest taping the lid shut or wrapping a rubber band around the box during transport.

Step 28. Clean up your area and put away all components. Make sure leftover powder goes back into the appropriate container and is stored properly.

Example #3:
Loading 9mm Parabellum (Luger) On A Dillon 550B Progressive Press

Step 1. Prepare load-plan card for our selected components. We'll load a total of 45 rounds for this initial workup–3 groups of 15 rounds in 0.2gr increments up to the max-recommended charge.

   a. Chosen bullet: 115gr FMJ
   b. Chosen primer: Winchester Small Pistol
   c. Chosen case: Mixed brass
   d. Chosen powder: Winchester Super Field
   e. Overall length: 1.150"

Components to load 9mm practice ammo.

Step 2. Prepare your range data sheet according to your load-planning card.

Step 3. Check batteries in calipers.

Step 4. Sort through brass to ensure that all cases are 9mm Luger. It's easy for .380 or .40S&W cases to blend in. Discard any cases with obvious defects.

Step 5. Install sizing die, set up according to manufacturer's instructions in toolhead station number one.

Step 6. Install powder die, powder funnel, and powder measure according to manufacturer's instructions in station number two.

Step 7. Install seating die in station number three.

Step 8. Install crimp die in station number four.

Step 9. Install shell plate and locator buttons for 9mm cases.

Step 10. Fill powder measure with chosen powder.

Step 11. Adjust powder measure to dispense *starting* charge. You will need the small powder bar for this. Insure that's installed in the powder measure.

Step 12. Verify that small primer system is installed. If not, replace large primer system components with small primer system components or use a handheld priming tool.

Step 13. Fill primer pickup tube with 45 small pistol primers.

Step 14. Empty primers into primer feed tube and reinstall low primer warning rod.

Step 15. Place bin of cartridge cases on bench to the right side of press.

Step 16. Place bin of bullets on bench to left side of press.

Step 17. Insert a clean 9mm case into station one. Cycle the handle all the way down and then all the way back up, pushing firmly forward at the top of the stroke. This forward push seats the primer. Remember that part!

Step 18. Rotate the shell plate clockwise.

Step 19. Insert another case into station one and cycle the handle all the way down and back up. Remember to push firmly forward at the top of the stroke.

Step 20. Rotate the shell plate clockwise.

Step 21. Insert another case into station one. You now have a primed and charged case with flared case mouth in station three, a resized and primed case in station two awaiting powder, and fired case in station one.

Step 22. Place a bullet on the case in station three and cycle the handle all the way down and back up. Remember to push firmly forward at the top of the stroke to prime.

Step 23. Rotate the shell plate clockwise.

Step 24. Repeat steps 21-23 until you've loaded 15 rounds with the starting charge. Your cartridges will be crimped in station four.

Step 25. Adjust the powder measure to dispense an additional 0.2 grains of powder with each cycle.

Step 26. Repeat the entire process until you've loaded the next 15 rounds at our mid-range charge.

The shell plate on this progressive press has four stations and is manually indexed after each press cycle.

Step 27. Adjust the powder measure to dispense an additional 0.2 grains of powder with each cycle. We've now reached the maximum recommended powder charge.

Step 28. Repeat the entire process until you've loaded the last 15 rounds at our max charge.

Step 29. Mark every case with the powder charge it contains: A, B, C, or 1, 2, 3, etc. This will save you from heartache if your ammo spills on the way to the range.

Step 30. Inspect every round thoroughly according to the steps in this book.

Step 31. Place your cartridges into an ammo box or case. I'd also suggest taping the lid shut or wrapping a rubber band around the box during transport.

Step 32. Clean up your area and put away all components. Make sure leftover powder goes back into the appropriate container and is stored properly.

Congratulations! You've just loaded your first batch of ammo!

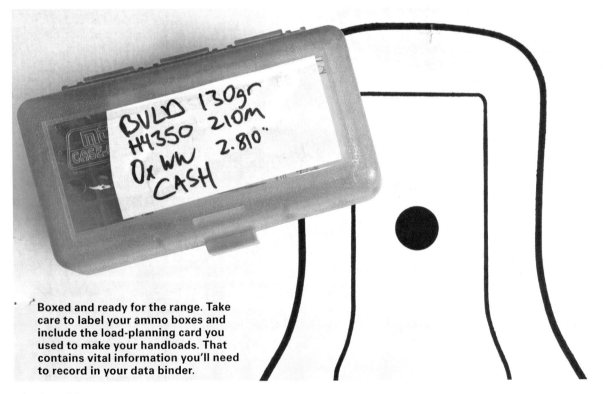

Boxed and ready for the range. Take care to label your ammo boxes and include the load-planning card you used to make your handloads. That contains vital information you'll need to record in your data binder.

Range Equipment

Shooters today can benefit from an array of range equipment that allows us to establish a solid and consistent firing position, gather and record information about our handloads, and analyze the results afterward. I've included a general list of items you'll want to take to the range to help achieve consistent results in Appendix C. This list is primarily aimed at rifle shooters, but there are a number of important items for pistoleros as well.

Rifle shooting and fieldcraft require a variety of equipment.

You may not "need" a rest to shoot a pistol, but it only takes one bullet through the display of your expensive chronograph to make you think otherwise, especially if you let your friends do the shooting. (Want to guess how I know this?) A few of the more critical items follow:

Portable Chronograph

A portable chronograph can be the reloader's best friend. While it is possible to back into an accurate muzzle velocity figure with extended-range trajectory information, it is much easier and faster to shoot through a chronograph. Prices on portable chronographs start at around $100 and go up from there. In general, you get what you pay for, and to really get a high-quality unit you'll need to spend roughly $400-$600. Don't let that get in your way. You can get perfectly usable information from the

If your budget allows, add a chronograph to your reloading equipment inventory.

personal chronographs I've used over the years and the professional-grade computerized indoor ballistics laboratory of a well-known firearms manufacturer. A comparison of the *average* muzzle velocities displayed by the personal chronographs and the ballistic lab were pretty darn close. The individual numbers varied much more so. The small personal chronographs are also vastly more susceptible to errors caused by improper setup, lighting conditions, alignment, and positioning (and sometimes unexplainable gremlins). For example, the 300-grain Lapua Scenar loaded in Lapua factory ammunition leaves the muzzle of a 24-inch barrel at an average muzzle velocity of 2,720 feet per second according to the professional chronograph system. According to measurements taken from two different portable chronographs over a period of several years, the same ammunition averages 2,710 feet per second. Looking at the tiny difference between those average velocity numbers, it's hard to find a justification to pay for an expensive professional chronograph. Where the high-end system delivers for us is in the accuracy of individual shot data. The small portable chronographs display wildly varied numbers with far greater frequency than the big money systems. So much so that velocity extreme spread and standard deviation numbers and any data taken from a small sample size (30 rounds or less) are potentially suspect and may be almost worthless.

How is it possible that we can have such inaccuracy with portable chronographs, yet come to almost exactly the same conclusion (average velocity) in the end as the much more expensive systems? Error works both ways. Sometimes the portable chronograph provides a reading that is way too high. Sometimes it provides a reading that's way too low. With enough data over time, the errors seem to average out. Now, when I say "enough" data, I'm talking multiple dozens or hundreds of rounds through the chronograph. With attentive setup and use, we can achieve an acceptable result with the inexpensive portable chronographs that are available to reloaders for the cost of a couple of good steak dinners.

less expensive units as long as you follow their setup instructions and use them properly. You should understand their limitations though. The least expensive chronographs are good for one thing—to provide a general idea about your average muzzle velocity. I've conducted comparative tests between the inexpensive

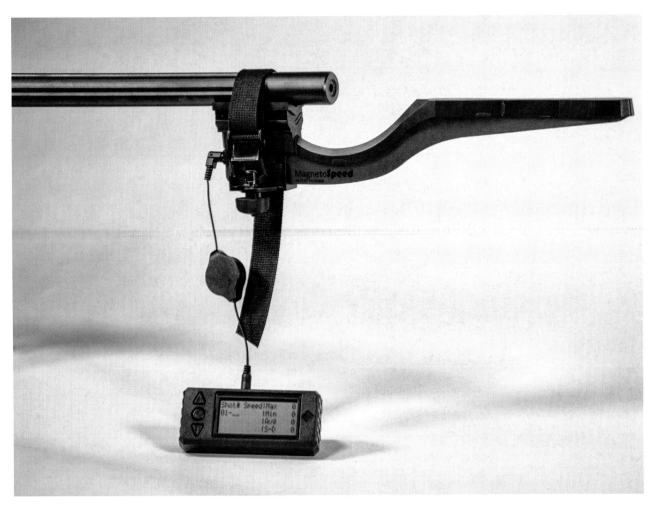

MagnetoSpeed

There is a fairly new type of chronograph on the market that I've not had the chance to test yet. It is called the MagnetoSpeed. The MagnetoSpeed actually attaches to the barrel of your rifle, revolver, or other non-reciprocating barrel/slide-equipped handgun. (It is not compatible with semi-auto pistols.) A "bayonet" protrudes forward of and below the muzzle to measure the bullet's velocity upon firing. Reports on this new style of chronograph are overwhelmingly positive. The one catch is that your shot-group size may or may not change due to the weight of the MagnetoSpeed hanging on the barrel. Your point of impact may also change. If you can live with the idea of doing accuracy testing and chronographing separately as a worst-case scenario, the MagnetoSpeed is worth looking into. I see real value in the ease of use offered by this system and will be evaluating one ASAP.

Recommendation

Here's my bottom-line recommendation based on nearly two decades of chronograph use: when your budget allows, spend $150-$250 on a portable chronograph. You may be able to find a shooting partner that will split this cost with you. Follow the manufacturer's instructions carefully every time the unit is used. Find a reference load that you can use to establish a baseline of performance with your chronograph and firearm. I recommend a first-quality military cartridge such as M118LR. Military rifle ammo has to perform within a specific range of velocity and consistency. This makes it a good candidate for our reference load. There is also a tremendous amount of data available for this cartridge online that will give you an idea of how accurately your chronograph is measuring velocity. Remember to refer to data obtained from a barrel of the same or very similar length as your own. If you can't find data from the same barrel length as your rifle, you can adjust up or down by 20 to 30 feet per second per inch of barrel length. That's a general rule of thumb

The MagnetoSpeed chronograph attaches directly to the rifle's barrel.

for standard centerfire rifles. For centerfire pistols of typical service-pistol size (roughly four to five-inch barrel length), you can double the anticipated velocity gain/loss per inch as a general guideline. The changes per inch usually aren't linear, but we're talking guidelines here. With some additional research, you may be able to find data directly related to your gun. Don't be afraid to call the manufacturer either. Their tech support people can give you specific info about the performance of their products. Fire at least a dozen rounds of your chosen reference load through your new chronograph, and average the results.

Velocity Average

Once you've confirmed that your chronograph is measuring a known load at a reasonably accurate velocity, you can then begin to trust the average of your velocity readings. Remember to fire as many rounds as possible, at least a dozen but preferably up to three dozen, to get a large enough data set to create a reliable average.

Basic chronographs are relatively inexpensive and can provide valuable, though imperfect, data to the reloader.

If you see any signs of high pressure or other problems with your handloads, stop shooting. If your results with a reference load don't match the data you've obtained in your research, call the chronograph manufacturer for their recommendation.

Another Confirmation

Another way to confirm the average velocity readings you get from your chronograph is to shoot at an extended but exact range of at least 300 yards, preferably 500 yards or more. Starting with a well-established 100-yard zero (or the zero of your choice), fire at a large paper target placed at a precisely known distance. Your target will need to be 24" tall for most modern centerfire rounds at 300 yards. At 500 yards, I'd suggest a target at least 6" tall. Notice that I didn't mention anything about adjusting your scope or aiming point to "hold high" before firing. Place your aiming point near the top of the target board and fire as accurately as you possibly can. Pick a day with calm winds for this, if possible. Horizontal dispersion won't change the results of what we're doing, but it makes it harder to keep our rounds on paper and measure the results. If you don't have a target board large enough for this test, you can adjust your scope to half or even three-quarters of the estimated drop provided by your ballistics program. Make it a round number, and write that number on the target. Once you fire your group, you can add the measured drop to the amount you adjusted for a total drop number. We will have added in some level of error by doing this, since most scopes have at least minor adjustment errors throughout their range, but we should be close enough for our purposes. I like a 3-inch target dot if I'm firing at 300 yards, a 5-inch target dot if I'm firing at 500 yards, and so forth. Use what your eyes will allow you to aim at consistently.

Once you've fired 5-10 rounds at your distant target board, it's time to measure drop and do some math. Measure the vertical distance from the center of your aiming point to the center of your shot group. Disregard the horizontal spread for this task. Input the known factors

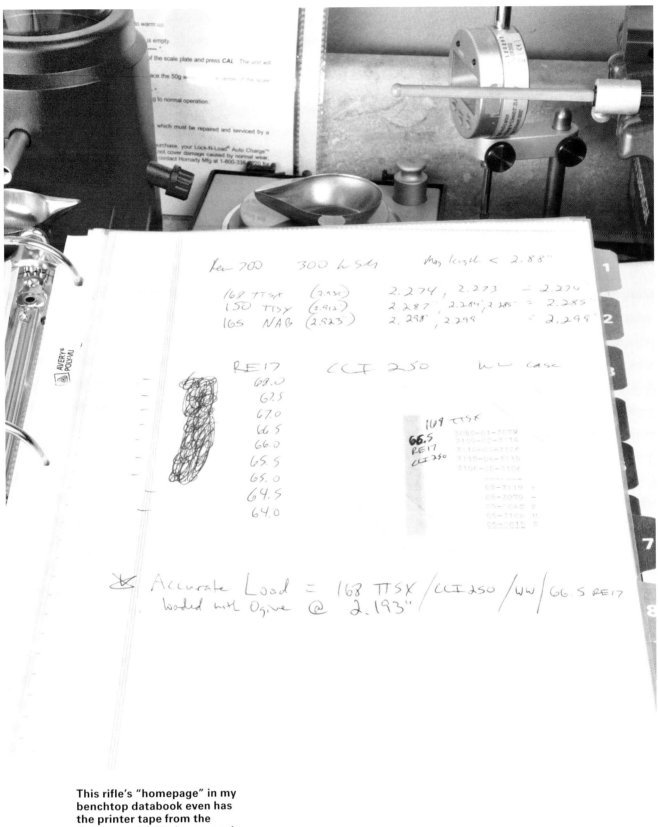

This rifle's "homepage" in my benchtop databook even has the printer tape from the Oehler model 35 chronograph used to measure the velocity of the rifle's preferred handload. Mean velocity, high and low velocities, and standard deviations are listed.

Establishing real numbers based on the actual drop of your bullet at a known distance is more valuable than computer predictions or hobbyist-level chronograph data.

of your load into a ballistic calculator. JBM is a free online tool that is fantastic for this. It can be found at *www.jbmballistics.com*. Enter your bullet type, weight, estimated muzzle velocity, ballistic coefficient (see the bullet manufacturer's website), temperature, altitude of the firing location, sight height above bore (using a ruler will get you close enough in this case), and all of the other inputs required in the "Simplified" Trajectory Card page on JBM. Be as accurate as you can with these inputs. Make sure you select "inches" as one of your output units. (There are two drop-down menus near the bottom of the page labeled "Column 1 Units" and "Column 2 Units.") Once you've entered all of your inputs, click the "Calculate" button.

Upon calculation, JBM will output a chart of trajectory and windage data based on your in-

puts. If you've entered everything correctly and the estimated muzzle velocity is fairly close to reality, you'll see a measurement in inches that's pretty close to what you measured on your drop target. If not, adjust the muzzle velocity input and recalculate until the drop predicted by JBM and the actual drop you measured are the same. Again, the farther you can stretch your shot distance on this test, the better your results will be.

Scenario
Here's an example that will hopefully help this make sense: firing a .224" diameter 77grain Sierra Match King from an 18" barrelled AR-15 that has been zeroed and confirmed at 100 yards, I measured 14" of drop from my point of aim to the center of a five-round shot group fired at a target 300 yards distant from my firing position

(the farthest distance available at the time). The vertical size of this group was less than 3", so I'm pretty confident about the accuracy and consistency of this load. Inputting the environmental conditions for the day, the sight height, bullet type, ballistic coefficient (I use the JBM bullet library for this.), and an estimated muzzle velocity of 2750fps, I see a predicted drop of 12.4" at 300 yards. Since my real world drop of 14" is greater than the 12.4" prediction from JBM, I know that my velocity estimate is too high. After confirming the accuracy of the rest of my data inputs into JBM, I begin adjusting my velocity estimate downward. With an adjusted muzzle velocity input of 2655fps, JBM predicts exactly 14" of drop. Through this process, I've determined that my 77gr Match King load produces a muzzle velocity of approximately 2655fps from my 18-inch barrel. Though not as fast as I'd hoped, 2655fps is right in the velocity range that should be expected from the combination of components I've chosen, and I'm impressed with the accuracy of this load. Based on this test, I'll select this as my standard load for this rifle. When I have the chance to test this load at greater distances, I'll rerun the numbers to refine my estimated velocity if I notice that the predicted elevation corrections from JBM aren't mirroring my real world results.

Chronograph Extras

Always have a spare battery for your chronograph in your range bag. Always take the instruction manual, too. Gasp! I know it's a violation of the Man Code to read the instruction manual, let alone take it with you to the range, but once you've gotten lost in the chronograph's menu functions a few times and come home in defeat because you can't figure out how to fix it, you'll understand why I say this. A cheap camera tripod is a must if you'll be firing off a bench. Your chronograph will have a threaded socket for mounting to the tripod head. You'll use the tripod to elevate and adjust the chronograph to the correct height for use. The alternative is to place the chronograph on the ground and fire over it from the prone position. I frequently do this for ease of setup, but it does become a literal pain in the neck after a long day of load testing.

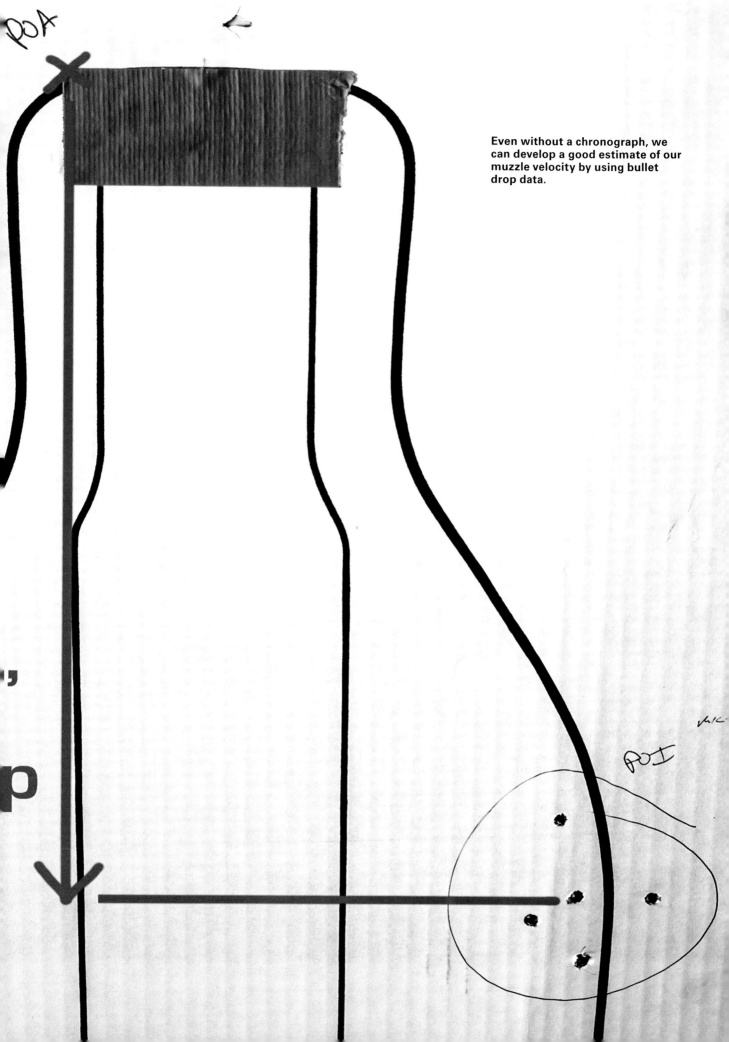

Even without a chronograph, we can develop a good estimate of our muzzle velocity by using bullet drop data.

A solid position with front and rear rests is critical for accurately gauging the performance of your handloads.

Rifle Rest

A steady firing position is critical to your ability to accurately evaluate the results of your load development efforts. You must have a solid rest for the front and rear of your rifle. Options for the fore-end rest include the use of a bipod, sandbags, or an adjustable benchrest-style rest. My personal favorite is a heavy sandbag up front and a well-made squeeze bag like the "Brick" from Armageddon Gear under the buttstock. I strongly suggest bags that contain a hard pelletized fill material or even sand. There are some bags on the market that contain a semi-squishy fill that won't allow you to get a good solid rest position.

I'm primarily a "field" shooter. In this case the term "field" shooter refers to engaging targets at unknown distance in the natural terrain from whatever position the terrain allows. This describes the hunting style of most western hunters and eastern still-hunters as well. It also describes the format of a particular style of rifle competition that has really exploded in popularity over the last decade. Examples of this match style include the "Steel Safari," the Thunder Beast "Team Challenge," and the "Snipershide Cup" (2012 forward). For these types of events, a bipod and rear squeeze bag is a must. If you're going to buy gear for the range, keep in mind the other uses you may have for that gear down the road. You won't be hauling an adjustable benchrest-style rifle rest around with you on an antelope hunt or in the Steel Safari. That's just food for thought that may save you a penny on gear.

Spotting Scopes

Spotting scopes (spotters) are another optional but very nice accessory. Typical riflescopes offer top magnification ranges of 9x-24x. Typical spotting scopes offer magnification ranges of 45x-60x power or more. Spotters also benefit from much larger lenses than riflescopes. All else equal, this creates the potential for greater

The use of a portable "rear bag" like this one from Armageddon Gear provides significantly increased stability in field positions or on the bench. Stabilization of the rear of the rifle is often overlooked by shooters of almost all experience levels.

resolving power and, to some degree, bright-
ness. With a decent spotter, you can see bullet holes in your target from several hundred
yards away without leaving the bench. I use the
descriptor "decent" for a reason. While it's possible to spend $5,000 plus on a Ziess-Hensoldt
military-grade spotter, it isn't necessary. I use
one of Bushnell's "Elite Tactical" models for
short-range work with fine results. This unit
cost less than $375 and included a carrying case
and a small, inexpensive benchtop tripod. It
won't win any awards for optical performance
compared to a Zeiss, Swarovski, or Leica Spotter, but it doesn't need to. I can throw the Bushnell in the range box and let it bounce around
in the back of the truck without losing sleep at
night. The old adage, "Buy nice or buy twice," is
absolutely true, and I'm not suggesting you buy
cheap (i.e., low-quality) gear. Just be aware that
you can purchase an acceptable spotting scope
without spending a small fortune. Read the
online reviews and discussion available on any
of the poplar shooting and optics forums before
buying, and you should be fine.

Safety Gear

Following safety guidelines when using firearms
is always important, but we compound the need
for safety when we introduce our first reloaded
ammunition. If you've followed the steps in
this book and the instructions provided by the
manufacturers of your reloading equipment and
if you've cross-referenced data from several reliable sources, there is very little chance that you'll
have any problems. Pick a day when you can fire
your first rounds without other shooters nearby.
You need to be able to use all of your senses to
notice anything that may seem wrong with the
sound, look, or feel of your initial firings.

Always have a medical kit with you when you
use firearms. Have at least some knowledge of
self-aid procedures in case of an accident. We've
seen a huge surge in the availability of "battlefield medicine" courses available to civilians over
the last decade. TAKE ONE! I guarantee there
will never be a time when you think, "I wish I

**Though not mandatory, a good spotting scope can be
a tremendous time saver and should be on your wish
list of range gear.**

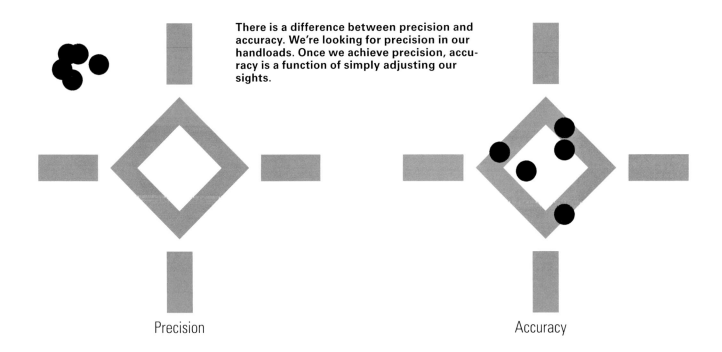

There is a difference between precision and accuracy. We're looking for precision in our handloads. Once we achieve precision, accuracy is a function of simply adjusting our sights.

Precision

Accuracy

had less training, knowledge, and experience." It may save your life or someone else's. Always wear your safety glasses and ear protection when using any firearm. Be sure you have a safe backstop behind your target to prevent rounds from traveling beyond your range. There is never any justification for neglecting these steps. Every time you fire a gun without hearing protection you permanently damage your hearing. Every time you shoot without eye protection, you risk losing your sight forever. Don't be a fool.

If there are other people using the range, wait until they take a break before firing. Let them know that you're firing your first handloaded ammunition, and they will probably back up to give you room on their own. If not, ask if they will give you some space as you fire your first rounds. Some ranges have pretty strict procedures that may make it difficult for you to be the only one firing. Talk to the owner or the range officer. Explain your situation and ask for two minutes to fire your first rounds without other shooters firing nearby.

First Rounds

Fire several rounds of factory ammo before firing your handloads. Notice the sound and feel of these rounds as you fire. Observe the ejection of the brass if you're shooting a semi-auto. Pick up your brass and look at it closely. Look at the neck, case body, cartridge head, and primer in detail. Keep those cases for comparison to your handloads later. Look at the target. What sort of accuracy did you achieve? Mark the holes in the target and label them as having been made by your factory ammo. If you shoot more than one type of factory ammo (this is recommended), be sure to mark the groups made by each type separately. Ideally, you'll put up a new target before you fire your reloads. You want to be able to compare the results of each type of ammo later.

The moment of truth has arrived. Firing that first round of handloaded ammunition can be a stressful event. Regardless of the homework we've done leading up to that moment, it is natural to have some trepidation before pulling the trigger. You've researched load data, selected the right components, followed the manufacturer's instructions for use of all of your equipment, used sound component storage and handling safety procedures on the bench, managed your loading data with precision, inspected your firearm for safety, and assembled a well-made group of cartridges. It's time to fire your first handloaded rounds with confidence!

Getting Ready

Insert a single handloaded cartridge of the starting (the lowest charge of your chosen powder recommended by your reloading manual) powder charge into the magazine, cylinder, or chamber of your firearm. Remove all other ammunition from your firing position. Check your eye protec-

My good friend and publisher, Jay Langston, and 1911 guru Bill Wilson after a successful boar hunt.

tion. Check your ear protection. Inform anyone nearby that you're firing your first handloads. Before, during, and after you fire the first shot of your handloaded ammo, you need to look, listen, and feel for anything unusual, problematic, or potentially problematic. If you notice anything unusual or different than you saw, felt, or heard when you fired the baseline rounds of factory ammunition, STOP shooting. More than likely, your trigger squeeze will result in a satisfying BANG and a big smile on your face. Once you realize that you're still alive and all eyes, fingers, and toes are intact, congratulate yourself!

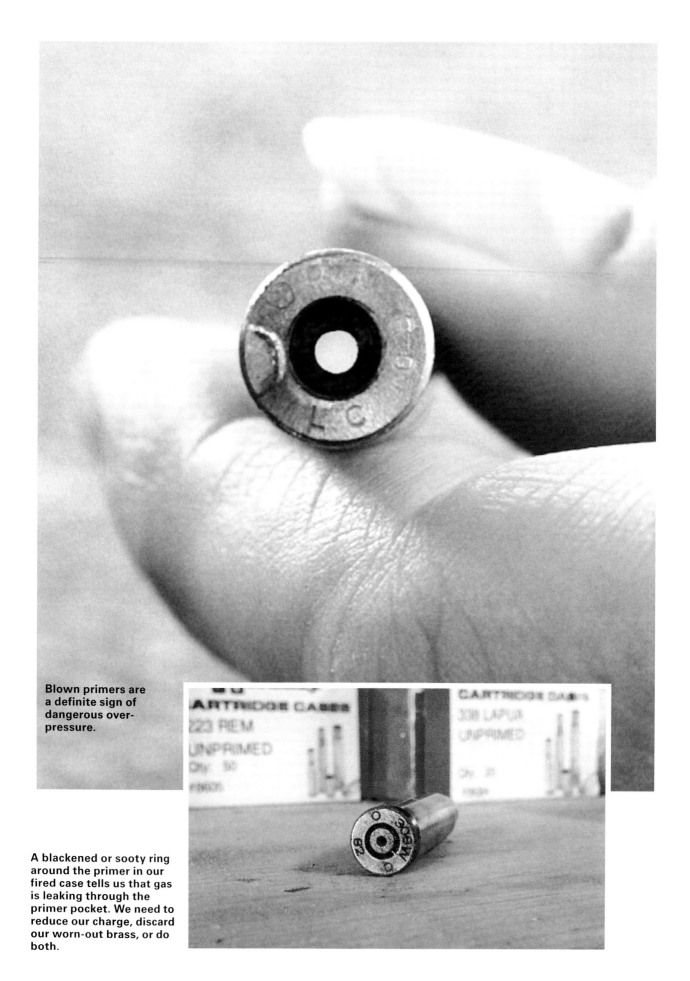

Blown primers are a definite sign of dangerous over-pressure.

A blackened or sooty ring around the primer in our fired case tells us that gas is leaking through the primer pocket. We need to reduce our charge, discard our worn-out brass, or do both.

Inspect Everything

Before firing any more rounds, it is critical that you check yourself for extra holes, check your firearm for signs of damage or unusual wear, check the target for the presence of a new bullet hole (yes, seriously), and check your fired brass for any damage including cracks, splits, holes, loose primers, or signs of gas leakage around the primer. Check the case head of rifle cartridges for a circular area of raised brass. This is a sure sign of excessive pressure in most guns, with large-frame (.308 Winchester) AR-10 types being a *possible* exception. Even moderate-pressure factory ammo will display some *slight* ejector flow due to the extreme violence of that rifle's action. If you see any of these signs, STOP shooting.

Compare your spent cartridge case with the cases from the factory ammo you fired as your baseline. If everything looks similar and none of the issues above are observed, proceed with your next handloaded rounds. Always begin with the lowest charges and work up, constantly watching for any signs of high pressure. If you see any signs of excessive pressure, STOP shooting. You will need to pull those loads apart and reduce your powder charge.

Brass gets harder each time it is stretched or "worked" during firing or resizing. Eventually, split necks will result. Throw this brass away (without exception) and check the rest of that lot.

When brass begins to flow into the ejector area of our boltface, we're already over the line. Back your charge down by at least 1% and re-evaluate.

It is common but undesirable for some semi-auto rifles to deform the rims of fired cases. Rather than relying solely on this evidence, use this as one of several signs of high pressure.

Points Of Impact

Don't be surprised if your handloads shoot to a different point of impact than the factory ammo you fired as a baseline. The use of different bullets or different powders can cause significantly different points of impact, even from the same rifle. Once you've developed a load that is reliable and accurate in your rifle, you should stick with that combination and configuration as much as your components inventory will allow.

Load Development

If you follow the recommendations in some reloading manuals for their "accuracy load" or "most accurate" powder and charge, it's possible you'll discover acceptable or good accuracy with your first load-development efforts. I'm usually not that fortunate, but I know some guys who are. Usually, a bit more work is required. Take a look at the target picture for the results of one round of load development. This workup was done with a DPMS semi-auto in .260 Remington firing a Hornady 123gr AMAX bullet. You'll notice the initial point of impact (POI) near the first dot on the left. We aren't trying to hit these dots, by the way. The dots are simply aiming points, and as long as we're close enough to put all of our hits on paper somewhere near the dot, we're ok. We're looking for consistency, then precision with our initial efforts. The POI starts out a few inches high and a couple of inches left of our dot. That's fine, but I go ahead and make an adjustment down 0.7 MILS and right 0.3, as noted on the target. Write those adjustments down! The center of the second group is just below the bottom of the dot. As you can see, groups three through seven all have a practically identical point of impact. That's the sort of consistency we're looking for when we start loading for precision. Remember, we want consistency *along with* precision. The consistent point of impact we see across those five groups suggests that we've found a very forgiving and consistent range of powder charges that perform similarly.

Different types of ammo and different handloads will produce significantly different points of impact (POI). Notice how the POI (group centers marked by red dots) changes among these three types of factory .30-06 hunting loads. These were all fired from the same rifle on the same day.

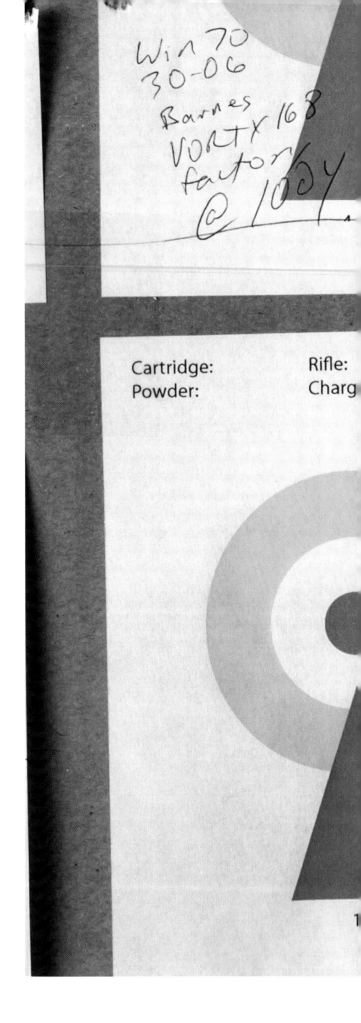

Cartridge: Rifle:
Powder: Charg

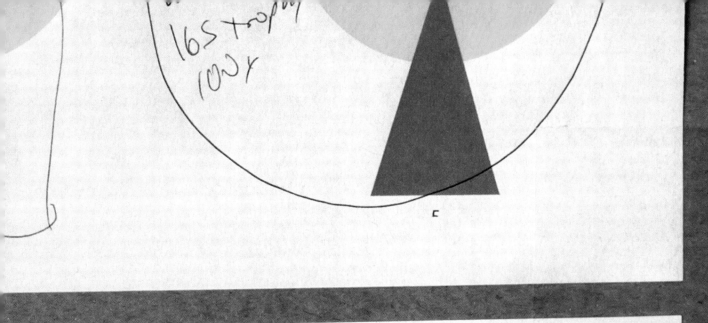

Bullet: Primer: Case:

Shooter: Temp: Date:

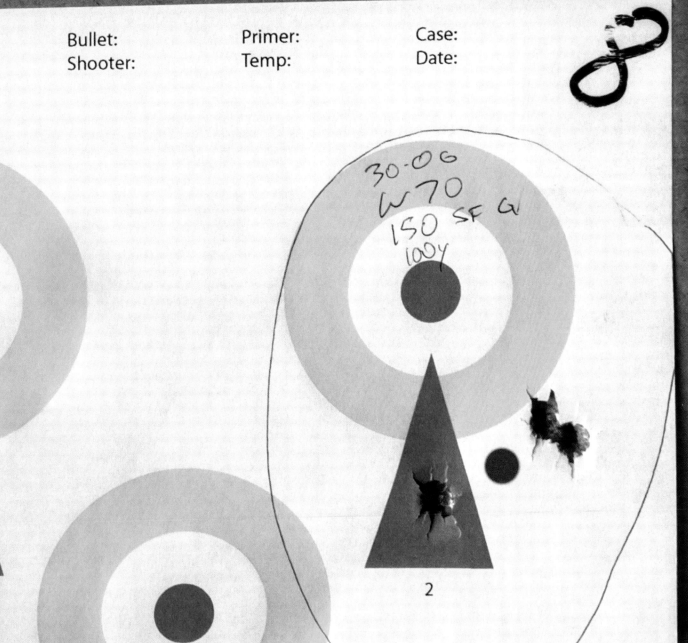

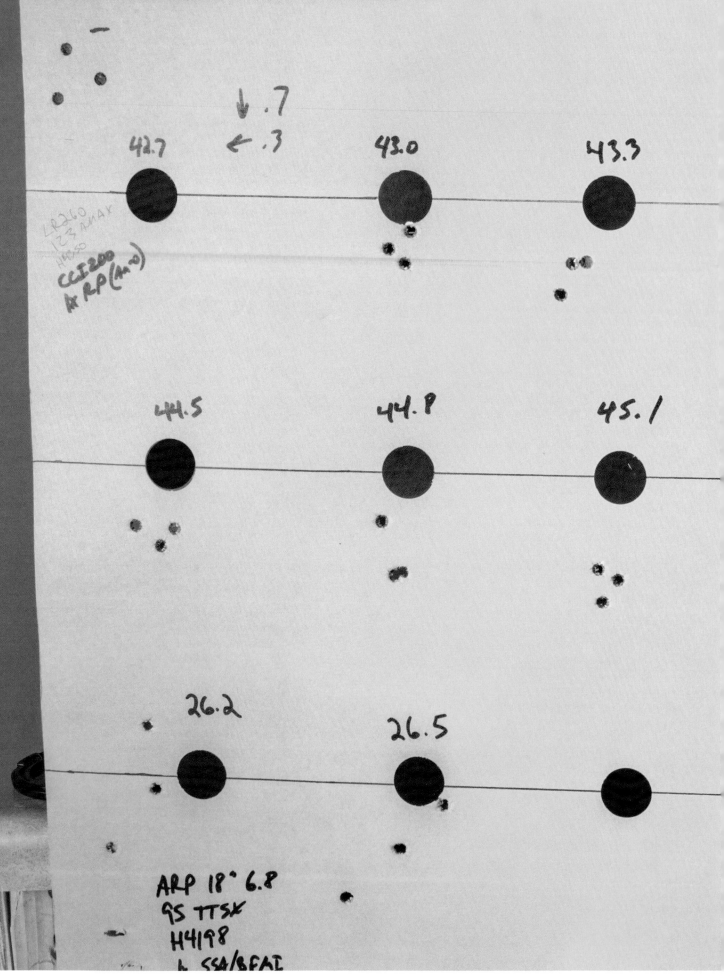

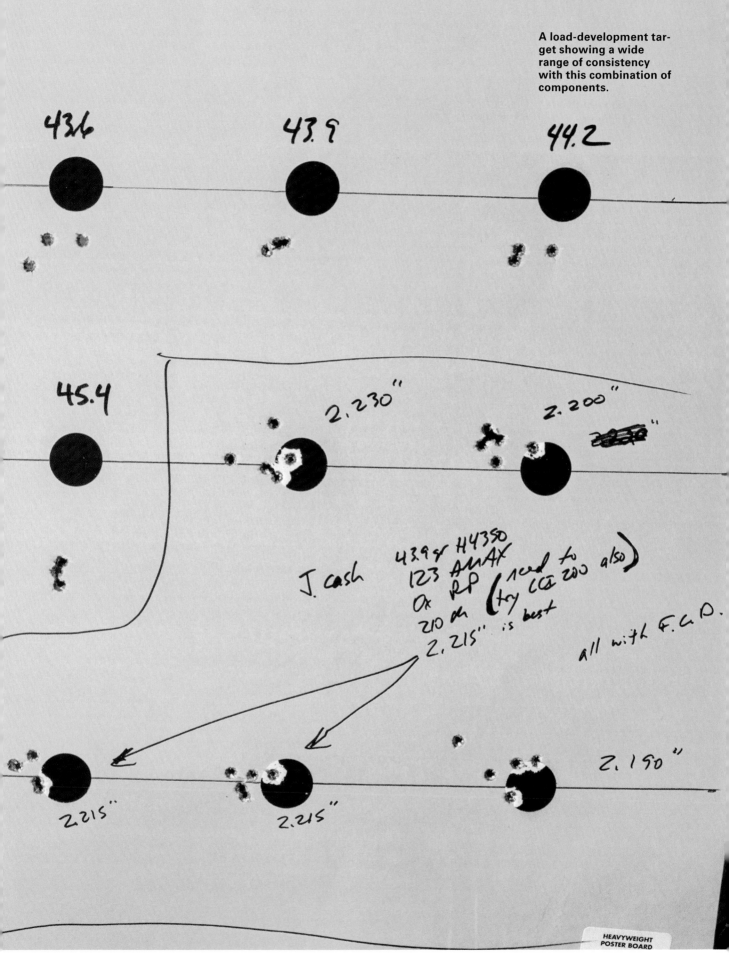

A load-development target showing a wide range of consistency with this combination of components.

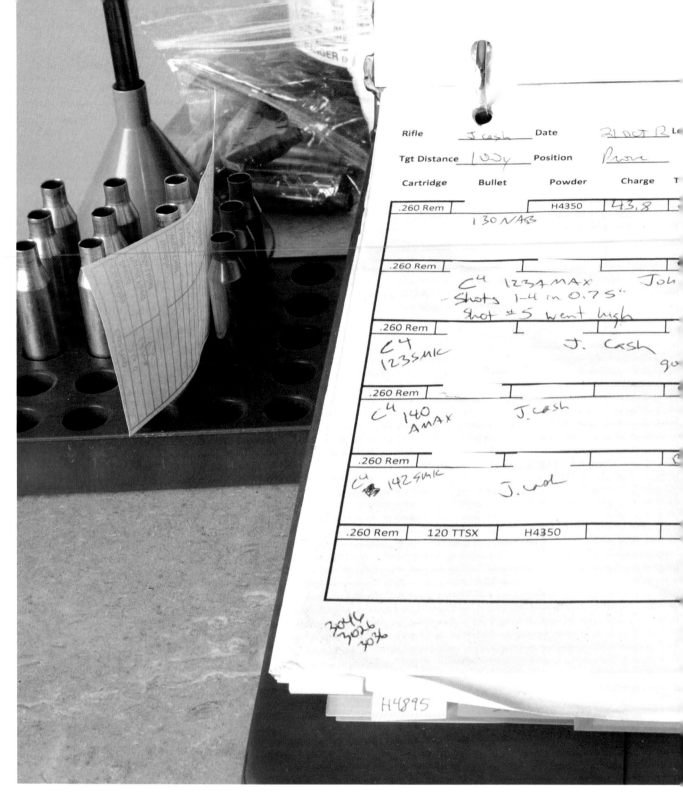

These five charges cover a range of 1.5 grains of powder. In this example, that's about a 3 percent variation. Even if we charge directly from the least consistent powder measure on the market, we can still achieve far better consistency than that! As we examine these groups further, we see that group number five is incredibly precise. That's the exact combination of components I'll explore further.

Consistent POI

While it's tempting to take results like we see on this target at face value and load up a couple hundred rounds of the load that performed so well here, it's important to understand that a single three-shot group, or a single five or ten-shot group, doesn't provide enough data to rely on. I suggest you assemble 15 rounds of the load that seems to be the "sweet spot" and fire three five-

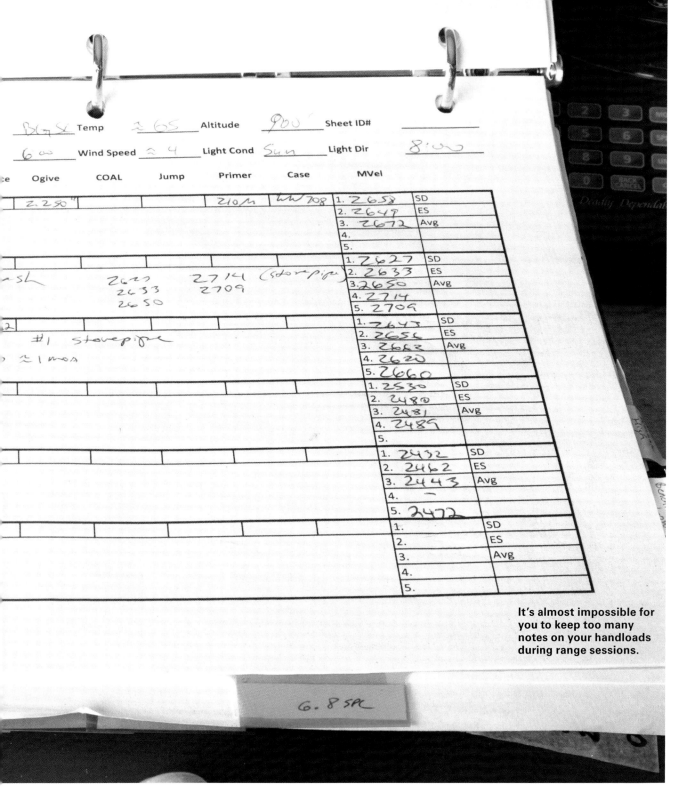

It's almost impossible for you to keep too many notes on your handloads during range sessions.

round groups for tactical or heavy barrelled rifles (or five three-round groups for lightweight or hunting rifles). This will give us enough data to at least *begin* to have confidence in our load. If the second round of testing (the batch of 15 rounds) confirms the results you saw during initial load development, load another batch of validation rounds, usually 30-50 cartridges, and repeat the process on a different day and, ideally, at extended range. My local gun club has a 300-yard rifle range, so I'll typically validate a new load at 100 and 300 yards for most hunting or match rifle cartridges. Limited capacity rounds like the 6.8 SPC may also be fired at 200 yards initially. If the results are satisfactory there, I'll then stretch out to 300 yards. This allows me to begin to see if the consistency I observed up close (100 yards) translates into consistency at longer distances.

Sometimes it does, and sometimes it does not. Velocity deviation is often the culprit here, assuming my 100-yard targets demonstrated a consistent POI. If the load is wildly inconsistent at extended ranges, a decision must be made to accept the results or find another load. I've been really disappointed more than once when a load shot beautifully at 100 yards but fell apart at 300 yards. I'm a long-range competitive shooter, and I like to have long-range capability in all of my rifles, even if they're just for hunting. Needless to say, I'd go back to the drawing board with a load that didn't perform at longer ranges.

If you can't find the sort of consistency of POI we see in the target image, my first suggestion would be to try a different powder or a different bullet, depending on what your component inventory looks like. It's especially hard to find some powders right now, and some bullets are unavailable as well. Cross-reference multiple data sources and look for combinations of components that you can find. As I type this, I've seen ONE box of Sierra Match King 6.5mm 142-grain bullets in the last 13 months. I bought it for a rifle I already know likes that bullet, but I wouldn't start load development from scratch with a bullet so scarce. The 140-grain Hornady AMAX 6.5mm bullet has been much more readily available.

Also, keep in mind the accuracy level baseline you established with the factory ammo. Your goal is to match or surpass the performance of factory ammo. When you fire your handloads at longer distances, you can also determine if you've matched or surpassed the velocity of your factory ammo baseline load by measuring the amount of drop from your point of aim, assuming similar bullets are used. If your goal is plinking or blasting, the steps outlined above are overkill, so keep your intended use in mind, too. When I load 9mm pistol ammo, my only real goal is 100 percent reliability in my handguns. I've yet to find a load that was too inaccurate for use in defensive training or action-pistol type competitions. It just needs to go bang every time I pull the trigger! The same applies when I load 55-grain FMJ training ammo for my AR15s. That's a mil spec FMJ bullet that I'll fire from a rack-grade rifle, and I don't expect or waste time trying to achieve match-grade accuracy.

If you own (or can borrow) a chronograph and you've recorded the average velocity of your factory baseline ammo from your gun, then you can compare the average velocity of the handloads you've developed to get an idea of how close you are to factory pressure levels. Be sure you're comparing apples to apples. If your baseline ammo launches a 168-grain bullet, you can't make a direct velocity comparison with your 150-grain or 180-grain handloads. If you see a significant increase in velocity when comparing your handloads to factory ammo, I would strongly suggest you retrace your steps and make sure you've followed the load data in your reloading manual(s). You may have found a more efficient combination of components, you may have a longer barrel than the one used in the reloading manual, or you may be in danger of damaging your gun and yourself with an overly "hot" load.

Watch for the signs of pressure we previously discussed like leaking, pierced, loose, or lost primers. Watch for ejector marks on the case head of rifle cartridges or pull marks on the rims of rounds fired in a semi-auto. If ammo is loaded too hot, it will likely sound and feel different, though you can't always rely on that to be true. Look at all factors: gun function, ammo function, sight, sound, feel, chronograph data, and common sense. These will help you evaluate the safety and performance of your handloads.

As you fire your handloads, be sure to make extensive notes about velocity, point of impact, precision, pressure signs, cycling, target information, felt recoil, observed damage to brass, wind conditions, light conditions, temperature, altitude, time of day, and anything else that occurs at the range. When in doubt, make more notes. You will learn what info is most important over time, but for now be extremely thorough in your data recording. Return fired brass to the container it arrived in, along with the load-planning and case preparation sheets. Be sure you've identified which target, dot, bullseye, or aiming point each group was fired on while that info is still fresh. Better yet, before you fire a group, write down which target you're aiming at. After firing the group, add any additional information generated from shooting those rounds.

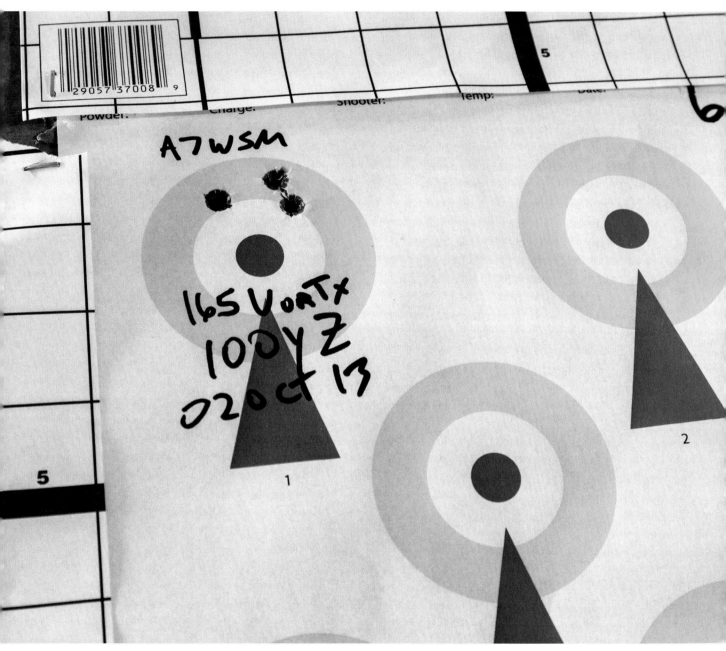

When you collect your targets at the end of the day, be sure to mark each target with the type of ammo or charge that corresponds to the shots on the page. You've already recorded this in your range data sheet, but it needs to be on the target, too. As I previously explained, it is incredibly frustrating to find a great-looking

Write down dates, ammo types, distance, and rifle info on your zero confirmation targets. I even put the same information on a piece of tape on the riflescope housing. This helps me know which rifles are field-ready and what ammo to grab with them.

load-development target in your stack of reloading records only to realize you have no idea what firearm or ammo was used to create it. Amend any notes you made on your range data sheet if the actual holes in the paper differ in your hands from what you thought you saw through your spotting scope 100 (or more) yards away. What you thought was a staple may turn out to be an errant bullet hole that ruins an otherwise great group.

Once you return home and get your gear packed away, it's time to sit down and evaluate what your targets are telling you. It's also crucial to make sure your notebooks are in order. My preference is to place load-development targets into my benchtop data binder for future reference. I'll add a zero confirmation target before hunting season every year and make notes about any adjustments that were required to my zero. When you have a good, consistent, reliable load, rifle, optic, and mounting system, you'll almost never have to make adjustments. Keeping notes on this info will help you paint a picture of the overall performance of your system over time. I've sold more than one rifle over the years after a thorough off-season review of my benchtop data book revealed a pattern of inconsistency or a broad lack of precision across many differ-

ent bullets, powders, or other variables that I'd rather not have to worry about.

You'll also need to start a new case-preparation card for your fired brass. What was once-fired brass is now twice-fired brass. Now is the time to update those notes. Once you've started a new case prep card with updated info, you can pitch the old one. You can begin planning your next loading session now, too. If you had promising results at the range, it's time to make those 15 validation cartridges. Make your load-planning card now while the relevant info is fresh in your mind and your range data book and today's targets are on top of the pile. Inspect your brass for any damage or pressure signs you may have missed and then throw them in the tumbler to clean, ensuring that case prep card stays with the brass. Check your components inventory. Do you have enough powder, projectiles, or primers to complete the load plan you just made based on today's results? The longer you wait to do these things, the more difficult it will be to do them and to do them correctly. Prepare for the next step now, and you will be a lot more efficient.

Ready for the field.

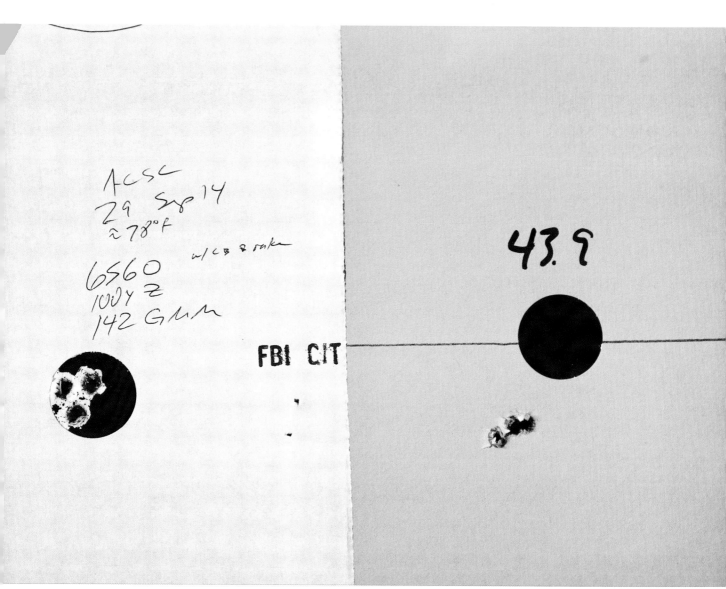

The group on the right is more precise, but the group on the left is more accurate. Our goal is to maximize both.

Conclusion

Loading your own ammunition is a fascinating and enjoyable activity that can quickly become a new hobby on its own, just like the shooting hobby it's intended to support. Once a new hand-loader gets a taste of success and experiences the thrill of solving the puzzle of accuracy, velocity, reliability, and in-field performance with his carefully selected components, the thrill of the chase may be too much to resist. Simply going to the range or field to shoot whatever ammo the factory has determined to be good enough will no longer be adequate. We can achieve better accuracy, increased consistency, and improved results in the field or in the winner's circle, all at a reduced cost, by loading our own ammunition. If you've purchased this book, followed along step by step, and made it this far, you already know that.

Thanks for reading *Making Ammo: A Be-*

ginner's Guide to Handloading. Check out my upcoming work *Book of the Long Range Rifle* for an in-depth look at state-of-the-art precision rifles. You'll learn about stocks, barrels, actions, optics, mounts, gunsmithing, ammunition, triggers, magazines, slings, and all sorts of rifle-shooting accessories. In addition to the industry's most comprehensive look at rifles and related equipment, you'll enjoy personal interviews and perspectives from the most well-known and respected experts in the long-range rifle world. Find the *Book of the Long Range Rifle* and many more firearms-related resources from 2nd Amendment Media at *www.RedBookofGuns.com*.

Appendix A

Terms and Definitions

Action – The part of a firearm that loads, fires, and ejects a cartridge. Includes: lever action, pump action, bolt action, and semi-automatic. The first three are found in weapons that fire a single shot. Firearms that can shoot multiple rounds ("repeaters") include all these types of actions, but only the semi-automatic does not require manual operation between rounds. A truly "automatic" action is found on a machine gun.

Anneal – to soften metal by a process of heating so that the metal becomes malleable for reshaping

Boat tail – the heel of a bullet that is tapered and used in military and match loads

Bullet jump – distance the bullet travels from where it's held in the case neck to the point where it first contacts the barrel's lands

Cannelure – a groove on a bullet used for identification, for crimping, or for lubricating

Chamfer – to bevel the interior edge of a cartridge case mouth to ensure proper bullet seating

Collet – a cone-shaped sleeve to secure loading components in a lathe or other machine

Concentricity – Merriam Webster defines the root word concentric as "having a common center" or "having a common axis." We use the term concentricity in reference to the centerline of the bore and our attempts to load ammunition that is concentric thereto or in reference to individual components or assemblies that have been put together having the same axis.

Decap – deprime, or remove the primer from the cartridge case

External Ballistics – The subject covering all effects of and on a bullet as it moves between muzzle and impact. Examples of external ballistics topics include: trajectory, drift, time of flight, stabilization, etc.

Headspace – The distance from the bolt face to the datum line on the shoulder portion of the chamber (bottleneck cases) or from the bolt face to a predetermined point in the chamber. This pre-determined point is based on the cartridge design and may be the result of case mouth position, case belt position, or case rim thickness.

Mandrel – a metal rod around which metal may be shaped

Meplat – the diameter of the blunt end of a projectile

Minute of Angle (MOA) – An angular measurement equal to one 60th of one degree. This nearly corresponds with a dispersion of one inch per 100 yards.

Muzzle Velocity – The bullet's speed the moment it leaves the muzzle

Ogive – the curve of a bullet's forward portion; between the meplat and where the curve begins

Plinking – informal target shooting

Runout – bullet's relationship to its seating depth in the cartridge case as it relates to how far the bullet's ogive sits off of the barrel's lands

Sabot – a lightweight cover containing a smaller bullet and used in larger bore barrel

Shell plate – in a multi-stage press, the shell plate holds the case rim secure during the reloading process

Shooter's MOA – An angular measurement equal to a dispersion of one inch per 100 yards. A shooter's MOA equals one inch at 100 yards, two inches at 200 yards, three inches at 300 yards, and so on.

Swage – to form metal by means of pressure

Work harden – to change the grain structure of metal by drawing or working

Appendix B

Range Equipment List

- Chronograph
  - Spare battery
  - Instruction manual
  - Tripod for chronograph (if shooting from a bench)
- Good rifle rests
  - Front
- Bipod
- Bags
- Adjustable rest
  - Rear
  - Field
  - Bench/table
- Shooting Mat
- Staple gun & Staples
- Cardboard backer
- Masking tape
- Permanent marker(s)
- Pen(s)
- Spotting scope with tripod (optional)
- Chamber flag, gun case, & other items required by range or club rules

Additional Equipment:

Acknowledgements:

Through my years in the firearms industry, I've been fortunate to develop relationships with many fine people who represent some of the best names and products in the business. Several of the photos, some of the information, and a few of the components used in this book were sent to me courtesy of the companies listed below. Others are simply listed to recognize industry friends who have offered support or guidance over the years, or as examples of fine companies every shooter or reloader should know.

I'd like to thank Neal Emery at Hornady for his help with a number of the images you'll see in this book and in *Book of the Long-Range Rifle*, also published by 2nd Amendment Media.

George Gardner of GA Precision builds tremendous rifles and has been a good friend and wonderful resource on the subject of accuracy with respect to rifles and ammunition.

Adam Braverman at Lapua has been a great resource and industry friend over the years and has gone out of his way to provide technical support when needed.

Robin Sharpless at Redding Reloading is one of the nicest guys in the business and a wealth of information on all things reloading.

Ryan Melancon at McMillan is a great friend and has been either visionary enough (or perhaps foolish enough) to listen to me a few times over the years when I've approached him with new projects.

Paul Heflin at EXOS Defense and Molding Solutions is a good man and a good friend. Paul is responsible for bringing many products to the shooting public and to our military from a manufacturing standpoint, and few folks know about him. You'll see the Ti-7 buttstock from EXOS Defense pictured in this book.

Tom Fuller of Armageddon Gear is one of my closest friends and a Ranger buddy that I served with 2/75. Tom has never met a stranger and knows just about everyone in the industry. You'll see several Armageddon Gear products pictured later in the book as well.

The whole crew at Thunder Beast Arms deserves mention for what they've done and continue to do for the long-range shooting community. Zak Smith, Ray Sanchez, and Shane Coppinger are friends and real ambassadors for the sport. Their suppressors have changed expectations and standards when it comes to suppressed precision rifles.

Every single person I've met at Vortex Optics is easily on the list of the industry's nicest folks. Paul Neess has been especially helpful with elk hunting advice over the years, not to mention technical information and support related to rifle optics. Vortex is still a relatively new name, but their focus on finding the right mix of performance and value has made them a booming success.

John Paul at JP Enterprises has been a great resource on all things gas-gun related and does a lot for competitive tactical long-range sports as well.

Melvin Forbes at New Ultralight Arms is an industry pioneer who has gone out of his way to help me with one of his rifles that I purchased used at no profit to him. If you're looking for a mind-blowing ultralight hunting rifle, check him out.

My good friends at Barrett have contributed photos you'll see on later pages, and Chris Barrett is always up for a chat about our favorite cartridges, and he has shared some good info and test results on reloading and shooting.

Finally, my company, Game Scout, is a resource for several shooting and reloading products as well as marketing services focused specifically on the firearms and hunting industry. We're small but growing, and I'm proud of the innovation we've brought to the market already.

Kyle Lynch, Author

Armageddon Gear
www.armageddongear.com

Barrett Firearms
www.barrett.net

EXOS Defense
www.exosdefense.com

GA Precision
www.gaprecision.net

Game Scout
www.gamescoutusa.com

Hornady
www.hornady.com

JP Enterprises
www.jprifles.com

Lapua
www.lapua.com

McMillan Riflestocks
www.mcmillanusa.com

New Ultralight Arms
www.newultralight.com

Redding Reloading
www.redding-reloading.com

Thunder Beast Arms Corporation
www.thunderbeastarms.com

Vortex Optics
www.vortexoptics.com

>>Index

FREE Test Taking Tips DVD Offer

To help us better serve you, we have developed a Test Taking Tips DVD that we would like to give you for FREE. **This DVD covers world-class test taking tips that you can use to be even more successful when you are taking your test.**

All that we ask is that you email us your feedback about your study guide. Please let us know what you thought about it – whether that is good, bad or indifferent.

To get your **FREE Test Taking Tips DVD**, email freedvd@studyguideteam.com with "FREE DVD" in the subject line and the following information in the body of the email:

    a. The title of your study guide.

    b. Your product rating on a scale of 1-5, with 5 being the highest rating.

    c. Your feedback about the study guide. What did you think of it?

    d. Your full name and shipping address to send your free DVD.

If you have any questions or concerns, please don't hesitate to contact us at freedvd@studyguideteam.com.

Thanks again!

OAR Test Prep

Officer Aptitude Rating Study Guide and Practice Exam Questions for the Navy [4th Edition]

Joshua Rueda

Interested in buying more than 10 copies of our product? Contact us about bulk discounts:
bulkorders@studyguideteam.com

ISBN 13: 9781637751084
ISBN 10: 1637751087

Table of Contents

Quick Overview

As you draw closer to taking your exam, effective preparation becomes more and more important. Thankfully, you have this study guide to help you get ready. Use this guide to help keep your studying on track and refer to it often.

This study guide contains several key sections that will help you be successful on your exam. The guide contains tips for what you should do the night before and the day of the test. Also included are test-taking tips. Knowing the right information is not always enough. Many well-prepared test takers struggle with exams. These tips will help equip you to accurately read, assess, and answer test questions.

A large part of the guide is devoted to showing you what content to expect on the exam and to helping you better understand that content. In this guide are practice test questions so that you can see how well you have grasped the content. Then, answer explanations are provided so that you can understand why you missed certain questions.

Don't try to cram the night before you take your exam. This is not a wise strategy for a few reasons. First, your retention of the information will be low. Your time would be better used by reviewing information you already know rather than trying to learn a lot of new information. Second, you will likely become stressed as you try to gain a large amount of knowledge in a short amount of time. Third, you will be depriving yourself of sleep. So be sure to go to bed at a reasonable time the night before. Being well-rested helps you focus and remain calm.

Be sure to eat a substantial breakfast the morning of the exam. If you are taking the exam in the afternoon, be sure to have a good lunch as well. Being hungry is distracting and can make it difficult to focus. You have hopefully spent lots of time preparing for the exam. Don't let an empty stomach get in the way of success!

When travelling to the testing center, leave earlier than needed. That way, you have a buffer in case you experience any delays. This will help you remain calm and will keep you from missing your appointment time at the testing center.

Be sure to pace yourself during the exam. Don't try to rush through the exam. There is no need to risk performing poorly on the exam just so you can leave the testing center early. Allow yourself to use all of the allotted time if needed.

Remain positive while taking the exam even if you feel like you are performing poorly. Thinking about the content you should have mastered will not help you perform better on the exam.

Once the exam is complete, take some time to relax. Even if you feel that you need to take the exam again, you will be well served by some down time before you begin studying again. It's often easier to convince yourself to study if you know that it will come with a reward!

Test-Taking Strategies

1. Predicting the Answer

When you feel confident in your preparation for a multiple-choice test, try predicting the answer before reading the answer choices. This is especially useful on questions that test objective factual knowledge. By predicting the answer before reading the available choices, you eliminate the possibility that you will be distracted or led astray by an incorrect answer choice. You will feel more confident in your selection if you read the question, predict the answer, and then find your prediction among the answer choices. After using this strategy, be sure to still read all of the answer choices carefully and completely. If you feel unprepared, you should not attempt to predict the answers. This would be a waste of time and an opportunity for your mind to wander in the wrong direction.

2. Reading the Whole Question

Too often, test takers scan a multiple-choice question, recognize a few familiar words, and immediately jump to the answer choices. Test authors are aware of this common impatience, and they will sometimes prey upon it. For instance, a test author might subtly turn the question into a negative, or he or she might redirect the focus of the question right at the end. The only way to avoid falling into these traps is to read the entirety of the question carefully before reading the answer choices.

3. Looking for Wrong Answers

Long and complicated multiple-choice questions can be intimidating. One way to simplify a difficult multiple-choice question is to eliminate all of the answer choices that are clearly wrong. In most sets of answers, there will be at least one selection that can be dismissed right away. If the test is administered on paper, the test taker could draw a line through it to indicate that it may be ignored; otherwise, the test taker will have to perform this operation mentally or on scratch paper. In either case, once the obviously incorrect answers have been eliminated, the remaining choices may be considered. Sometimes identifying the clearly wrong answers will give the test taker some information about the correct answer. For instance, if one of the remaining answer choices is a direct opposite of one of the eliminated answer choices, it may well be the correct answer. The opposite of obviously wrong is obviously right! Of course, this is not always the case. Some answers are obviously incorrect simply because they are irrelevant to the question being asked. Still, identifying and eliminating some incorrect answer choices is a good way to simplify a multiple-choice question.

4. Don't Overanalyze

Anxious test takers often overanalyze questions. When you are nervous, your brain will often run wild, causing you to make associations and discover clues that don't actually exist. If you feel that this may be a problem for you, do whatever you can to slow down during the test. Try taking a deep breath or counting to ten. As you read and consider the question, restrict yourself to the particular words used by the author. Avoid thought tangents about what the author *really* meant, or what he or she was *trying* to say. The only things that matter on a multiple-choice test are the words that are actually in the question. You must avoid reading too much into a multiple-choice question, or supposing that the writer meant something other than what he or she wrote.

5. No Need for Panic

It is wise to learn as many strategies as possible before taking a multiple-choice test, but it is likely that you will come across a few questions for which you simply don't know the answer. In this situation, avoid panicking. Because most multiple-choice tests include dozens of questions, the relative value of a single wrong answer is small. As much as possible, you should compartmentalize each question on a multiple-choice test. In other words, you should not allow your feelings about one question to affect your success on the others. When you find a question that you either don't understand or don't know how to answer, just take a deep breath and do your best. Read the entire question slowly and carefully. Try rephrasing the question a couple of different ways. Then, read all of the answer choices carefully. After eliminating obviously wrong answers, make a selection and move on to the next question.

6. Confusing Answer Choices

When working on a difficult multiple-choice question, there may be a tendency to focus on the answer choices that are the easiest to understand. Many people, whether consciously or not, gravitate to the answer choices that require the least concentration, knowledge, and memory. This is a mistake. When you come across an answer choice that is confusing, you should give it extra attention. A question might be confusing because you do not know the subject matter to which it refers. If this is the case, don't eliminate the answer before you have affirmatively settled on another. When you come across an answer choice of this type, set it aside as you look at the remaining choices. If you can confidently assert that one of the other choices is correct, you can leave the confusing answer aside. Otherwise, you will need to take a moment to try to better understand the confusing answer choice. Rephrasing is one way to tease out the sense of a confusing answer choice.

7. Your First Instinct

Many people struggle with multiple-choice tests because they overthink the questions. If you have studied sufficiently for the test, you should be prepared to trust your first instinct once you have carefully and completely read the question and all of the answer choices. There is a great deal of research suggesting that the mind can come to the correct conclusion very quickly once it has obtained all of the relevant information. At times, it may seem to you as if your intuition is working faster even than your reasoning mind. This may in fact be true. The knowledge you obtain while studying may be retrieved from your subconscious before you have a chance to work out the associations that support it. Verify your instinct by working out the reasons that it should be trusted.

8. Key Words

Many test takers struggle with multiple-choice questions because they have poor reading comprehension skills. Quickly reading and understanding a multiple-choice question requires a mixture of skill and experience. To help with this, try jotting down a few key words and phrases on a piece of scrap paper. Doing this concentrates the process of reading and forces the mind to weigh the relative importance of the question's parts. In selecting words and phrases to write down, the test taker thinks about the question more deeply and carefully. This is especially true for multiple-choice questions that are preceded by a long prompt.

9. Subtle Negatives

One of the oldest tricks in the multiple-choice test writer's book is to subtly reverse the meaning of a question with a word like *not* or *except*. If you are not paying attention to each word in the question, you can easily be led astray by this trick. For instance, a common question format is, "Which of the following is...?" Obviously, if the question instead is, "Which of the following is not...?," then the answer will be quite different. Even worse, the test makers are aware of the potential for this mistake and will include one answer choice that would be correct if the question were not negated or reversed. A test taker who misses the reversal will find what he or she believes to be a correct answer and will be so confident that he or she will fail to reread the question and discover the original error. The only way to avoid this is to practice a wide variety of multiple-choice questions and to pay close attention to each and every word.

10. Reading Every Answer Choice

It may seem obvious, but you should always read every one of the answer choices! Too many test takers fall into the habit of scanning the question and assuming that they understand the question because they recognize a few key words. From there, they pick the first answer choice that answers the question they believe they have read. Test takers who read all of the answer choices might discover that one of the latter answer choices is actually *more* correct. Moreover, reading all of the answer choices can remind you of facts related to the question that can help you arrive at the correct answer. Sometimes, a misstatement or incorrect detail in one of the latter answer choices will trigger your memory of the subject and will enable you to find the right answer. Failing to read all of the answer choices is like not reading all of the items on a restaurant menu: you might miss out on the perfect choice.

11. Spot the Hedges

One of the keys to success on multiple-choice tests is paying close attention to every word. This is never truer than with words like almost, most, some, and sometimes. These words are called "hedges" because they indicate that a statement is not totally true or not true in every place and time. An absolute statement will contain no hedges, but in many subjects, the answers are not always straightforward or absolute. There are always exceptions to the rules in these subjects. For this reason, you should favor those multiple-choice questions that contain hedging language. The presence of qualifying words indicates that the author is taking special care with their words, which is certainly important when composing the right answer. After all, there are many ways to be wrong, but there is only one way to be right! For this reason, it is wise to avoid answers that are absolute when taking a multiple-choice test. An absolute answer is one that says things are either all one way or all another. They often include words like *every*, *always*, *best*, and *never*. If you are taking a multiple-choice test in a subject that doesn't lend itself to absolute answers, be on your guard if you see any of these words.

12. Long Answers

In many subject areas, the answers are not simple. As already mentioned, the right answer often requires hedges. Another common feature of the answers to a complex or subjective question are qualifying clauses, which are groups of words that subtly modify the meaning of the sentence. If the question or answer choice describes a rule to which there are exceptions or the subject matter is complicated, ambiguous, or confusing, the correct answer will require many words in order to be expressed clearly and accurately. In essence, you should not be deterred by answer choices that seem excessively long. Oftentimes, the author of the text will not be able to write the correct answer without

offering some qualifications and modifications. Your job is to read the answer choices thoroughly and completely and to select the one that most accurately and precisely answers the question.

13. Restating to Understand

Sometimes, a question on a multiple-choice test is difficult not because of what it asks but because of how it is written. If this is the case, restate the question or answer choice in different words. This process serves a couple of important purposes. First, it forces you to concentrate on the core of the question. In order to rephrase the question accurately, you have to understand it well. Rephrasing the question will concentrate your mind on the key words and ideas. Second, it will present the information to your mind in a fresh way. This process may trigger your memory and render some useful scrap of information picked up while studying.

14. True Statements

Sometimes an answer choice will be true in itself, but it does not answer the question. This is one of the main reasons why it is essential to read the question carefully and completely before proceeding to the answer choices. Too often, test takers skip ahead to the answer choices and look for true statements. Having found one of these, they are content to select it without reference to the question above. Obviously, this provides an easy way for test makers to play tricks. The savvy test taker will always read the entire question before turning to the answer choices. Then, having settled on a correct answer choice, he or she will refer to the original question and ensure that the selected answer is relevant. The mistake of choosing a correct-but-irrelevant answer choice is especially common on questions related to specific pieces of objective knowledge. A prepared test taker will have a wealth of factual knowledge at their disposal, and should not be careless in its application.

15. No Patterns

One of the more dangerous ideas that circulates about multiple-choice tests is that the correct answers tend to fall into patterns. These erroneous ideas range from a belief that B and C are the most common right answers, to the idea that an unprepared test-taker should answer "A-B-A-C-A-D-A-B-A." It cannot be emphasized enough that pattern-seeking of this type is exactly the WRONG way to approach a multiple-choice test. To begin with, it is highly unlikely that the test maker will plot the correct answers according to some predetermined pattern. The questions are scrambled and delivered in a random order. Furthermore, even if the test maker was following a pattern in the assignation of correct answers, there is no reason why the test taker would know which pattern he or she was using. Any attempt to discern a pattern in the answer choices is a waste of time and a distraction from the real work of taking the test. A test taker would be much better served by extra preparation before the test than by reliance on a pattern in the answers.

FREE DVD OFFER

Don't forget that doing well on your exam includes both understanding the test content and understanding how to use what you know to do well on the test. We offer a completely FREE Test Taking Tips DVD that covers world class test taking tips that you can use to be even more successful when you are taking your test.

All that we ask is that you email us your feedback about your study guide. To get your **FREE Test Taking Tips DVD**, email freedvd@studyguideteam.com with "FREE DVD" in the subject line and the following information in the body of the email:

- The title of your study guide.
- Your product rating on a scale of 1-5, with 5 being the highest rating.
- Your feedback about the study guide. What did you think of it?
- Your full name and shipping address to send your free DVD.

Introduction to the OAR

Function of the Test

The Officer Aptitude Rating test is used by the United States Navy to determine how individuals will perform while in Naval Officer Candidate School (non-aviation). This test is part of the larger Aviation Selection Test Battery (ASTB), which is used to select individuals for flight and pilot officer training programs offered by the U.S. Coast Guard, the U.S. Marine Corps, and the U.S. Navy.

Test Administration

Candidates can take the OAR test at over 250 registered locations all over the world. These locations include naval officer recruiting stations, military institutes, and NROTC units at major universities. Officer recruiters schedule test dates for exam candidates once examinees prove that they meet the qualifications. The exam is offered in two different formats: a paper form and a computer adaptive (online) exam via the web-based APEX system. No two online tests are identical due to the fact that they are automatically generated from a library containing hundreds of potential questions for each of the three subtests.

Test takers who wish to retake the OAR test must wait thirty days in order to do so, and they can only take the OAR test a total of three times over the course of their lifetime. It is important to note that a test taker's most recent test score replaces any OAR score on record, even if he or she received a higher score on a previous test. Test scores are valid for life.

Test Format

The Math Skills Test (MST) utilizes both word problems and equations to test examinees on high school math concepts dealing with algebra and geometry. The Reading Comprehension Test (RCT) incorporates text passages to test examinees on their ability to extract information and make conclusions from what they have read. Finally, the Mechanical Comprehension Test (MCT) is comprised of questions involving high school physics concepts, as well simple machine mechanics.

The following table contains a breakdown of the number of questions and the corresponding time limits for the various subtests of the OAR exam for examinees that choose to take the paper version:

Sections of the OAR Test – Paper Version		
Subject Areas	Number of Questions (Multiple-Choice)	Time Limit
Math Skills Test	30	40 minutes
Reading Comprehension Test	20	30 minutes
Mechanical Comprehension Test	30	15 minutes
Total	80	85 minutes

For test takers who choose to take the online version of the exam, the test can take ninety minutes to two hours to complete. There are between twenty and thirty multiple-choice questions within each subtest.

Scoring

Individuals taking the paper version of the OAR test are penalized for leaving questions unanswered, and thus, they are best served by making educated guesses when time is running out or the answer is unknown. In contrast, individuals who are taking the online version of the OAR will not benefit from guessing. It is more important to work accurately and quickly on this version of the exam.

The OAR test is scored on a scale of 20-80 in increments of one point, and a minimum score of 35 is needed to pass. However, most individuals who take the exam receive scores between 40-60. There has been no real difference in passing rates between test takers who take the paper version of the exam versus those who take the online version.

Test takers who attempt the online version of the exam receive their scores immediately upon completion of the exam, while those who take the paper version must send their completed tests into Navy Medicine Operational Training Center (NMOTC) to be scored.

Recent/Future Developments

At the end of 2013, a new version of the larger Aviation Selection Test Battery (ASTB) was released, which is known as the ASTB-E. Three of the seven subtests still make up the OAR test. The overall exam is now better able to gauge how test takers think in different dimensions (hand-eye coordination, physical dexterity, and dividing attention between tasks) through what is known as the Performance Based Measures Battery.

Study Prep Plan for the OAR

 **Schedule** - Use one of our study schedules below or come up with one of your own.

 **Relax** - Test anxiety can hurt even the best students. There are many ways to reduce stress. Find the one that works best for you.

 **Execute** - Once you have a good plan in place, be sure to stick to it.

One Week Study Schedule

Day	Topic
Day 1	Arithmetic
Day 2	Solving Linear Equations and Inequalities
Day 3	Perimeters and Areas
Day 4	Reading Comprehension
Day 5	Mechanical Comprehension
Day 6	Practice Test
Day 7	Take Your Exam!

Two Week Study Schedule

Day	Topic	Day	Topic
Day 1	Math Skills	Day 8	Reading Comprehension
Day 2	Algebra	Day 9	Practice Quiz
Day 3	Solving Linear Equations and Inequalities	Day 10	Mechanical Comprehension
Day 4	Geometry	Day 11	Power
Day 5	Perimeters and Areas	Day 12	Practice Quiz
Day 6	Probability and Statistics	Day 13	Practice Test
Day 7	Practice Quiz	Day 14	Take Your Exam!

One Month Study Schedule					
Day 1	Math Skills	Day 11	Symmetry	Day 21	Mechanical Comprehension
Day 2	Fractions	Day 12	Volumes and Surface Areas	Day 22	Newton's Laws
Day 3	Ratios and Proportions	Day 13	Probability and Statistics	Day 23	Potential and Kinetic Energy
Day 4	Algebra	Day 14	Describing Distributions	Day 24	Power
Day 5	Evaluating Simple Formulas and Expressions	Day 15	Practice Quiz	Day 25	Machines
Day 6	Simplifying Algebraic Fractions	Day 16	Reading Comprehension	Day 26	Pulleys
Day 7	Matrices	Day 17	Paragraph Comprehension	Day 27	Practice Quiz
Day 8	Solving Verbal Problems Presented in an Algebraic Context	Day 18	Inferences in a Text	Day 28	Practice Test
Day 9	Geometry	Day 19	Figurative Language	Day 29	Answer Explanations
Day 10	Polygons	Day 20	Practice Quiz	Day 30	Take Your Exam!

Math Skills

Arithmetic

How to Prepare

These problems involve basic arithmetic skills as well as the ability to break down a word problem to see where to apply these skills in order to get the correct answer. The basics of arithmetic and the approach to solving word problems are discussed here.

Note that math requires practice in order to become proficient. Make sure to not just read through the material here, but also try out the practice questions, as well as check the answers provided. Just reading through examples does not necessarily mean that someone can do the problems themselves. Note that sometimes there can be multiple approaches to getting a solution when doing the problems. What matters is getting the correct answer, so it is okay if the approach to a problem is different than the solution method provided.

Definitions

A few definitions:

Whole numbers: describes a set of numbers that does not contain any fractions or decimals. The set of whole numbers includes zero.

>   Example: 0, 1, 2, 3, 4, 189, 293 are all whole numbers.

Integers: describes whole numbers and their negative counterparts. (Zero does not have a negative counterpart here. Instead, zero is its own negative.)

>   Example: -1, -2, -3, -4, -5, 0, 1, 2, 3, 4, 5 are all integers.

-1, -2, -3, -4, -5 are considered negative integers, and 1, 2, 3, 4, 5 are considered positive integers.

Absolute value: describes the value of a number regardless of its sign. The symbol for absolute value is $|x|$.

>   Example: The absolute value of 24 is 24 or $|24| = 24$.

The absolute value of -693 is 693 or $|-693| = 693$.

Even numbers: describes any number that can be divided by 2 evenly, meaning the answer has no decimal or remainder portion.

>   Example: 2, 4, 9082, -2, -16, -504 are all considered even numbers, because they can be divided by 2, without leaving a remainder or forming a decimal. It does not matter whether the number is positive or negative.

Odd numbers: describes any number that does not divide evenly by 2.

Example: 1, 21, 541, 3003, -9, -63, -1257 are all considered odd numbers, because they cannot be divided by 2 without a remainder or a decimal.

Prime numbers: describes a number that is only evenly divisible, resulting in no remainder or decimal, by 1 and itself.

Example: 2, 3, 7, 13, 113 are all considered prime numbers because each can only be evenly divided by 1 and itself.

Composite numbers: describes a positive integer that is formed by multiplying two smaller integers together. Composite numbers can be divided evenly by numbers other than 1 or itself.

Example: 10, 24, 66, 2348, 10002 are all considered composite numbers because they are the result of multiplying two smaller integers together. In particular, these are all divisible by 2.

Decimals: designated by a decimal point which indicates that what follows the point is a value that is less than 1 and is added to the integer number preceding the decimal point. The digit immediately following the decimal point is in the tenths place, the digit following the tenths place is in the hundredths place, and so on.

Example: the decimal number 1.735 has a value greater than 1 but less than 2. The 7 represents seven tenths of the unit 1 (0.7 or $\frac{7}{10}$); the 3 represents three hundredths of 1 (0.03 or $\frac{3}{100}$); and the 5 represents five thousandths of 1 (0.005 or $\frac{5}{1000}$).

Real numbers: describes rational numbers and irrational numbers.

Rational numbers: describes any number that can be expressed as a fraction, with a non-zero denominator. Since any integer can be written with 1 in the denominator without changing its value, all integers are considered rational numbers. Every rational number has a decimal expression that terminates or repeats. That is, any rational number either will have a countable number of nonzero digits or will end with an ellipsis or a bar (3.6666... or $3.\overline{6}$) to depict repeating decimal digits. Some examples of rational numbers include 12, -3.54, $110.\overline{256}$, $\frac{-35}{10}$, and $4.\overline{7}$.

Irrational numbers: describes numbers that cannot be written as a finite decimal. Pi (π) is considered to be an irrational number because its decimal portion is unending or a non-repeating decimal. The most common irrational number is π, which has an endless and non-repeating decimal, but there are other well-known irrational numbers like e and $\sqrt{2}$.

Basic Operations of Arithmetic

There are four different basic operations used with numbers: addition, subtraction, multiplication, and division.

- Addition takes two numbers and combines them into a total called the sum. The sum is the total when combining two collections into one. If there are 5 things in one collection and 3 in another, then after combining them, there is a total of $5 + 3 = 8$. Note the order does not matter when adding numbers. For example, $3 + 5 = 8$.

- Subtraction is the opposite (or "inverse") operation to addition. Whereas addition combines two quantities together, subtraction takes one quantity away from another. For example, if there are 20 gallons of fuel and 5 are removed, that gives $20 - 5 = 15$ gallons remaining. Note that for subtraction, the order does matter because it makes a difference which quantity is being removed from which.

- Multiplication is repeated addition. 3×4 can be thought of as putting together 3 sets of items, each set containing 4 items. The total is 12 items. Another way to think of this is to think of each number as the length of one side of a rectangle. If a rectangle is covered in tiles with 3 columns of 4 tiles each, then there are 12 tiles in total. From this, one can see that the answer is the same if the rectangle has 4 rows of 3 tiles each: $4 \times 3 = 12$. By expanding this reasoning, the order the numbers are multiplied does not matter.

- Division is the opposite of multiplication. It means taking one quantity and dividing it into sets the size of the second quantity. If there are 16 sandwiches to be distributed to 4 people, then each person gets $16 \div 4 = 4$ sandwiches. As with subtraction, the order in which the numbers appear does matter for division.

Addition

Addition is the combination of two numbers so their quantities are added together cumulatively. The sign for an addition operation is the + symbol. For example, $9 + 6 = 15$. The 9 and 6 combine to achieve a cumulative value, called a **sum**.

Addition holds the commutative property, which means that the order of the numbers in an addition equation can be switched without altering the result. The formula for the commutative property is $a + b = b + a$. Let's look at a few examples to see how the commutative property works:

$$7 = 3 + 4 = 4 + 3 = 7$$

$$20 = 12 + 8 = 8 + 12 = 20$$

Addition also holds the associative property, which means that the grouping of numbers doesn't matter in an addition problem. In other words, the presence or absence of parentheses is irrelevant. The formula for the associative property is $(a + b) + c = a + (b + c)$. Here are some examples of the associative property at work:

$$30 = (6 + 14) + 10 = 6 + (14 + 10) = 30$$

$$35 = 8 + (2 + 25) = (8 + 2) + 25 = 35$$

Subtraction

Subtraction is taking away one number from another, so their quantities are reduced. The sign designating a subtraction operation is the − symbol, and the result is called the difference. For example, $9 - 6 = 3$. The number 6 detracts from the number 9 to reach the difference 3.

Unlike addition, subtraction follows neither the commutative nor associative properties. The order and grouping in subtraction impact the result.

$$15 = 22 - 7 \neq 7 - 22 = -15$$

$$3 = (10 - 5) - 2 \neq 10 - (5 - 2) = 7$$

When working through subtraction problems involving larger numbers, it's necessary to regroup the numbers. Let's work through a practice problem using regrouping:

$$\begin{array}{r} 3\ 2\ 5 \\ -\ 7\ 7 \\ \hline \end{array}$$

Here, it is clear that the ones and tens columns for 77 are greater than the ones and tens columns for 325. To subtract this number, borrow from the tens and hundreds columns. When borrowing from a column, subtracting 1 from the lender column will add 10 to the borrower column:

$$\begin{array}{ccc} 3\text{-}1 & 10+2\text{-}1 & 10+5 \\ - & 7 & 7 \end{array} = \begin{array}{r} 2\ \ 11\ \ 15 \\ -\ \ \ \ \ 7\ \ 7 \\ \hline 2\ \ 4\ \ 8 \end{array}$$

After ensuring that each digit in the top row is greater than the digit in the corresponding bottom row, subtraction can proceed as normal, and the answer is found to be 248.

Multiplication

Multiplication involves adding together multiple copies of a number. It is indicated by an $\times$ symbol or a number immediately outside of a parenthesis. For example:

$$5(8 - 2)$$

The two numbers being multiplied together are called factors, and their result is called a **product**. For example, $9 \times 6 = 54$. This can be shown alternatively by expansion of either the 9 or the 6:

$$9 \times 6 = 9 + 9 + 9 + 9 + 9 + 9 = 54$$

$$9 \times 6 = 6 + 6 + 6 + 6 + 6 + 6 + 6 + 6 + 6 = 54$$

Like addition, multiplication holds the commutative and associative properties:

$$115 = 23 \times 5 = 5 \times 23 = 115$$

$$84 = 3 \times (7 \times 4) = (3 \times 7) \times 4 = 84$$

Multiplication also follows the distributive property, which allows the multiplication to be distributed through parentheses. The formula for distribution is $a \times (b + c) = ab + ac$. This is clear after the examples:

$$45 = 5 \times 9 = 5(3 + 6) = (5 \times 3) + (5 \times 6) = 15 + 30 = 45$$

$$20 = 4 \times 5 = 4(10 - 5) = (4 \times 10) - (4 \times 5) = 40 - 20 = 20$$

Multiplication becomes slightly more complicated when multiplying numbers with decimals. The easiest way to answer these problems is to ignore the decimals and multiply as if they were whole numbers. After multiplying the factors, place a decimal in the product. The placement of the decimal is determined by taking the cumulative number of decimal places in the factors.

For example:

$$
\begin{array}{r}
0.7 \\
\times\ 3 \\
\hline
2.1
\end{array}
\qquad
\begin{array}{r}
2.6 \\
\times\ 4.2 \\
\hline
10.92
\end{array}
\qquad
\begin{array}{r}
1.5 \\
\times\ 6.4 \\
\hline
9.60
\end{array}
$$

Let's tackle the first example. First, ignore the decimal and multiply the numbers as though they were whole numbers to arrive at a product: 21. Second, count the number of digits that follow a decimal (one). Finally, move the decimal place that many positions to the left, as the factors have only one decimal place. The second example works the same way, except that there are two total decimal places in the factors, so the product's decimal is moved two places over. In the third example, the decimal should be moved over two digits, but the digit zero is no longer needed, so it is erased, and the final answer is 9.6.

Division

Division and multiplication are inverses of each other in the same way that addition and subtraction are opposites. The signs designating a division operation are the ÷ and / symbols. In division, the second number divides into the first.

The number before the division sign is called the dividend or, if expressed as a fraction, the numerator. For example, in $a \div b$, a is the dividend, while in $\frac{a}{b}$, a is the numerator.

The number after the division sign is called the divisor or, if expressed as a fraction, the denominator. For example, in $a \div b$, b is the divisor, while in $\frac{a}{b}$, b is the denominator.

Like subtraction, division doesn't follow the commutative property, as it matters which number comes before the division sign, and division doesn't follow the associative or distributive properties for the same reason.

For example:

$$\frac{3}{2} = 9 \div 6 \neq 6 \div 9 = \frac{2}{3}$$

$$2 = 10 \div 5 = (30 \div 3) \div 5 \neq 30 \div (3 \div 5) = 30 \div \frac{3}{5} = 50$$

$$25 = 20 + 5 = (40 \div 2) + (40 \div 8) \neq 40 \div (2 + 8) = 40 \div 10 = 4$$

If a divisor doesn't divide into a dividend an integer number of times, whatever is left over is termed the remainder. The remainder can be further divided out into decimal form by using long division; however, this doesn't always give a quotient with a finite number of decimal places, so the remainder can also be expressed as a fraction over the original divisor.

Division with decimals is similar to multiplication with decimals in that when dividing a decimal by a whole number, ignore the decimal and divide as if it were a whole number.

Upon finding the answer, or quotient, place the decimal at the decimal place equal to that in the dividend.

$$15.75 \div 3 = 5.25$$

When the divisor is a decimal number, multiply both the divisor and dividend by 10. Repeat this until the divisor is a whole number, then complete the division operation as described above.

$$17.5 \div 2.5 = 175 \div 25 = 7$$

Fractions

A **fraction** is a number used to express a ratio. It is written as a number x over a line with another number y underneath: $\frac{x}{y}$, and can be thought of as x out of y equal parts. The number on top (x) is called the **numerator**, and the number on the bottom is called the **denominator** (y). It is important to remember the only restriction is that the denominator is not allowed to be 0.

Imagine that an apple pie has been baked for a holiday party, and the full pie has eight slices. After the party, there are five slices left. How could the amount of the pie that remains be expressed as a fraction? The numerator is 5 since there are 5 pieces left, and the denominator is 8 since there were eight total slices in the whole pie. Thus, expressed as a fraction, the leftover pie totals $\frac{5}{8}$ of the original amount.

Another way of thinking about fractions is like this: $\frac{x}{y} = x \div y$.

Two fractions can sometimes equal the same number even when they look different. The value of a fraction will remain equal when multiplying both the numerator and the denominator by the same number. The value of the fraction does not change when dividing both the numerator and the denominator by the same number. For example, $\frac{4}{8} = \frac{2}{4} = \frac{1}{2}$. If two fractions look different, but are actually the same number, these are **equivalent fractions**.

A number that can divide evenly into a second number is called a **divisor** or **factor** of that second number; 3 is a divisor of 6, for example. If the numerator and denominator in a fraction have no common factors other than 1, the fraction is said to be **simplified**. $\frac{2}{4}$ is not simplified (since the numerator and denominator have a factor of 2 in common), but $\frac{1}{2}$ is simplified. Often, when solving a problem, the final answer generally requires us to simplify the fraction.

It is often useful when working with fractions to rewrite them so they have the same denominator. This process is called finding a **common denominator**. The common denominator of two fractions needs to be a number that is a multiple of both denominators. For example, given $\frac{1}{6}$ and $\frac{5}{8}$, a common denominator is $6 \times 8 = 48$. However, there are often smaller choices for the common denominator. The smallest number that is a multiple of two numbers is called the **least common multiple** of those numbers. For this example, use the numbers 6 and 8. The multiples of 6 are 6, 12, 18, 24… and the multiples of 8 are 8, 16, 24…, so the least common multiple is 24. The two fractions are rewritten as $\frac{4}{24}, \frac{15}{24}$.

If two fractions have a common denominator, then the numerators can be added or subtracted. For example:

$$\frac{4}{5} - \frac{3}{5} = \frac{4-3}{5} = \frac{1}{5}$$

If the fractions are not given with the same denominator, a common denominator needs to be found before adding or subtracting them.

It is always possible to find a common denominator by multiplying the denominators by each other. However, when the denominators are large numbers, this method is unwieldy, especially if the answer must be provided in its simplest form. Thus, it's beneficial to find the least common denominator of the fractions—the least common denominator is incidentally also the least common multiple.

Once equivalent fractions have been found with common denominators, simply add or subtract the numerators to arrive at the answer:

$$1) \; \frac{1}{2} + \frac{3}{4} = \frac{2}{4} + \frac{3}{4} = \frac{5}{4}$$

$$2) \; \frac{3}{12} + \frac{11}{20} = \frac{15}{60} + \frac{33}{60} = \frac{48}{60} = \frac{4}{5}$$

$$3) \; \frac{7}{9} - \frac{4}{15} = \frac{35}{45} - \frac{12}{45} = \frac{23}{45}$$

$$4) \; \frac{5}{6} - \frac{7}{18} = \frac{15}{18} - \frac{7}{18} = \frac{8}{18} = \frac{4}{9}$$

One of the most fundamental concepts of fractions is their ability to be manipulated by multiplication or division. This is possible since $\frac{n}{n} = 1$ for any non-zero integer. As a result, multiplying or dividing by $\frac{n}{n}$ will not alter the original fraction since any number multiplied or divided by 1 doesn't change the value of that number. Fractions of the same value are known as equivalent fractions. For example, $\frac{2}{4}, \frac{4}{8}, \frac{50}{100}$, and $\frac{75}{150}$ are equivalent, as they all equal $\frac{1}{2}$.

To multiply two fractions, multiply the numerators to get the new numerator as well as multiply the denominators to get the new denominator. For example:

$$\frac{3}{5} \times \frac{2}{7} = \frac{3 \times 2}{5 \times 7} = \frac{6}{35}$$

Switching the numerator and denominator is called taking the **reciprocal** of a fraction. So, the reciprocal of $\frac{4}{5}$ is $\frac{5}{4}$.

To divide one fraction by another, multiply the first fraction by the reciprocal of the second. So:

$$\frac{3}{4} \div \frac{2}{5} = \frac{3}{4} \times \frac{5}{2} = \frac{15}{8}$$

If the numerator is smaller than the denominator, the fraction is a **proper fraction**. Otherwise, the fraction is said to be **improper**.

A **mixed number** is a number that is an integer plus some proper fraction, and is written with the integer first and the proper fraction to the right of it. Any mixed number can be written as an improper fraction by multiplying the integer by the denominator, adding the product to the value of the numerator, and dividing the sum by the original denominator. For example:

$$3\frac{1}{2} = \frac{3 \times 2 + 1}{2} = \frac{7}{2}$$

Whole numbers can also be converted into fractions by placing the whole number as the numerator and making the denominator 1. For example, $3 = \frac{3}{1}$.

Distribution of a Quantity into its Fractional Parts

A quantity may be broken into its fractional parts. For example, a toy box holds three types of toys for kids. $\frac{1}{3}$ of the toys are Type A and $\frac{1}{4}$ of the toys are Type B. With that information, how many Type C toys are there?

First, the sum of Type A and Type B must be determined by finding a common denominator to add the fractions. The lowest common multiple is 12, so that is what will be used. The sum is:

$$\frac{1}{3} + \frac{1}{4} = \frac{4}{12} + \frac{3}{12} = \frac{7}{12}$$

This value is subtracted from 1 to find the number of Type C toys. The value is subtracted from 1 because 1 represents a whole. The calculation is:

$$1 - \frac{7}{12}$$

$$\frac{12}{12} - \frac{7}{12} = \frac{5}{12}$$

This means that $\frac{5}{12}$ of the toys are Type C. To check the answer, add all fractions together, and the result should be 1.

Percentages

Think of percentages as fractions with a denominator of 100. In fact, percentage means "per hundred." Problems often require converting numbers from percentages, fractions, and decimals. The following explains how to work through those conversions.

- Converting Fractions to Percentages: Convert the fraction by using an equivalent fraction with a denominator of 100. For example:

$$\frac{3}{4} = \frac{3}{4} \times \frac{25}{25}$$

$$\frac{75}{100} = 75\%$$

- Converting Percentages to Fractions: Percentages can be converted to fractions by turning the percentage into a fraction with a denominator of 100. Be wary of questions asking the converted fraction to be written in the simplest form. For example, $35\% = \frac{35}{100}$ which, although correctly written, has a numerator and denominator with a greatest common factor of 5 and can be simplified to $\frac{7}{20}$.

- Converting Percentages to Decimals: As a percentage is based on "per hundred," decimals and percentages can be converted by multiplying or dividing by 100. Practically speaking, this always amounts to moving the decimal point two places to the right or left, depending on the conversion. To convert a percentage to a decimal, move the decimal point two places to the left and remove the % sign. To convert a decimal to a percentage, move the decimal point two places to the right and add a "%" sign. Here are some examples:

$$65\% = 0.65$$
$$0.33 = 33\%$$
$$0.215 = 21.5\%$$
$$99.99\% = 0.9999$$
$$500\% = 5.00$$
$$7.55 = 755\%$$

Questions dealing with percentages can be difficult when they are phrased as word problems. These word problems almost always come in three varieties. The first type will ask to find what percentage of some number will equal another number. The second asks to determine what number is some percentage of another given number. The third will ask what number another number is a given percentage of.

One of the most important parts of correctly answering percentage word problems is to identify the numerator and the denominator. This fraction can then be converted into a percentage, as described above.

The following word problem shows how to make this conversion:

A department store carries several different types of footwear. The store is currently selling 8 athletic shoes, 7 dress shoes, and 5 sandals. What percentage of the store's footwear are sandals?

First, calculate what serves as the "whole," as this will be the denominator. How many total pieces of footwear does the store sell? The store sells 20 different types (8 athletic + 7 dress + 5 sandals).

Second, what footwear type is the question specifically asking about? Sandals. Thus, 5 is the numerator.

Third, the resultant fraction must be expressed as a percentage. The first two steps indicate that $\frac{5}{20}$ of the footwear pieces are sandals. This fraction must now be converted into a percentage:

$$\frac{5}{20} \times \frac{5}{5} = \frac{25}{100} = 25\%$$

Ratios and Proportions

Ratios are used to show the relationship between two quantities. The ratio of oranges to apples in the grocery store may be 3 to 2. That means that for every 3 oranges, there are 2 apples. This comparison can be expanded to represent the actual number of oranges and apples, such as 36 oranges to 24 apples. Another example may be the number of boys to girls in a math class. If the ratio of boys to girls is given as 2 to 5, that means there are 2 boys to every 5 girls in the class. Ratios can also be compared if the units in each ratio are the same. The ratio of boys to girls in the math class can be compared to the ratio of boys to girls in a science class by stating which ratio is higher and which is lower.

Rates are used to compare two quantities with different units. **Unit rates** are the simplest form of rate. With unit rates, the denominator in the comparison of two units is one. For example, if someone can type at a rate of 1000 words in 5 minutes, then their unit rate for typing is $\frac{1000}{5} = 200$ words in one minute or 200 words per minute. Any rate can be converted into a unit rate by dividing to make the denominator one. 1000 words in 5 minutes has been converted into the unit rate of 200 words per minute.

Ratios and rates can be used together to convert rates into different units. For example, if someone is driving 50 kilometers per hour, that rate can be converted into miles per hour by using a ratio known as the **conversion factor**. Since the given value contains kilometers and the final answer needs to be in miles, the ratio relating miles to kilometers needs to be used. There are 0.62 miles in 1 kilometer. This, written as a ratio and in fraction form, is:

$$\frac{0.62 \text{ miles}}{1 \text{ km}}$$

To convert 50km/hour into miles per hour, the following conversion needs to be set up:

$$\frac{50 \text{ km}}{\text{hour}} \times \frac{0.62 \text{ miles}}{1 \text{ km}} = 31 \text{ miles per hour}$$

The ratio between two similar geometric figures is called the **scale factor**. For example, a problem may depict two similar triangles, A and B. The scale factor from the smaller triangle A to the larger triangle B is given as 2 because the length of the corresponding side of the larger triangle, 16, is twice the corresponding side on the smaller triangle, 8. This scale factor can also be used to find the value of a missing side, x, in triangle A. Since the scale factor from the smaller triangle (A) to larger one (B) is 2, the larger corresponding side in triangle B (given as 25) can be divided by 2 to find the missing side in A ($x = 12.5$). The scale factor can also be represented in the equation $2A = B$ because two times the lengths of A gives the corresponding lengths of B. This is the idea behind similar triangles.

Much like a scale factor can be written using an equation like $2A = B$, a **relationship** is represented by the equation $Y = kX$. X and Y are proportional because as values of X increase, the values of Y also increase. A relationship that is inversely proportional can be represented by the equation $Y = \frac{k}{X}$, where the value of Y decreases as the value of x increases and vice versa.

Proportional reasoning can be used to solve problems involving ratios, percentages, and averages. Ratios can be used in setting up proportions and solving them to find unknowns. For example, if a student completes an average of 10 pages of math homework in 3 nights, how long would it take the student to complete 22 pages? Both ratios can be written as fractions. The second ratio would contain the unknown.

The following proportion represents this problem, where x is the unknown number of nights:

$$\frac{10 \text{ pages}}{3 \text{ nights}} = \frac{22 \text{ pages}}{x \text{ nights}}$$

Solving this proportion entails cross-multiplying and results in the following equation:

$$10x = 22 \times 3$$

Simplifying and solving for x results in the exact solution: $x = 6.6$ nights. The result would be rounded up to 7 because the homework would actually be completed on the 7th night.

The following problem uses ratios involving percentages:

If 20% of the class is girls and 30 students are in the class, how many girls are in the class?

To set up this problem, it is helpful to use the common proportion:

$$\frac{\%}{100} = \frac{\text{is}}{\text{of}}$$

Within the proportion, % is the percentage of girls, 100 is the total percentage of the class, *is* is the number of girls, and *of* is the total number of students in the class. Most percentage problems can be written using this language. To solve this problem, the proportion should be set up as $\frac{20}{100} = \frac{x}{30}$, and then solved for x. Cross-multiplying results in the equation $20 \times 30 = 100x$, which results in the solution $x = 6$. There are 6 girls in the class.

Problems involving volume, length, and other units can also be solved using ratios. A problem may ask for the volume of a cone that has a radius, $r = 7$ m and a height, $h = 16$ m. Referring to the formulas provided on the test, the volume of a cone is given as:

$$V = \pi r^2 \frac{h}{3}$$

r is the radius, and h is the height. Plugging $r = 7$ and $h = 16$ into the formula, the following is obtained:

$$V = \pi (7^2) \frac{16}{3}$$

Therefore, the volume of the cone is found to be approximately 821 m³. Sometimes, answers in different units are sought. If this problem wanted the answer in liters, 821 m³ would need to be converted.

Using the equivalence statement 1 m³ = 1000 L, the following ratio would be used to solve for liters:

$$821 \text{ m}^3 \times \frac{1000 \text{ L}}{1 \text{ m}^3}$$

Cubic meters in the numerator and denominator cancel each other out, and the answer is converted to 821,000 liters, or 8.21×10^5 L.

Other conversions can also be made between different given and final units. If the temperature in a pool is 30°C, what is the temperature of the pool in degrees Fahrenheit? To convert these units, an equation is used relating Celsius to Fahrenheit. The following equation is used:

$$T_{°F} = 1.8T_{°C} + 32$$

Plugging in the given temperature and solving the equation for T yields the result:

$$T_{°F} = 1.8(30) + 32 = 86°F$$

Units in both the metric system and U.S. customary system are widely used.

Measurement Conversions

The United States customary system and the metric system each consist of distinct units to measure lengths and volume of liquids. The U.S. customary units for length, from smallest to largest, are: inch (in), foot (ft), yard (yd), and mile (mi). The metric units for length, from smallest to largest, are: millimeter (mm), centimeter (cm), decimeter (dm), meter (m), and kilometer (km). The relative size of each unit of length is shown below.

U.S. Customary	Metric	Conversion
12 in = 1 ft	10 mm = 1 cm	1 in = 2.54 cm
36 in = 3 ft = 1 yd	10 cm = 1 dm (decimeter)	1 m ≈ 3.28 ft ≈ 1.09 yd
5,280 ft = 1,760 yd = 1mi	100 cm = 10 dm = 1 m	1 mi ≈ 1.6 km
	1000 m = 1 km	

The U.S. customary units for volume of liquids, from smallest to largest, are: fluid ounces (fl oz), cup (c), pint (pt), quart (qt), and gallon (gal). The metric units for volume of liquids, from smallest to largest, are: milliliter (mL), centiliter (cL), deciliter (dL), liter (L), and kiloliter (kL). The relative size of each unit of liquid volume is shown below.

U.S. Customary	Metric	Conversion
8 fl oz = 1 c	10 mL = 1 cL	1 pt ≈ 0.473 L
2 c = 1 pt	10 cL = 1 dL	1 L ≈ 1.057 qt
4 c = 2 pt = 1 qt	1,000 mL = 100 cL = 10 dL = 1 L	1 gal ≈ 3.785 L
4 qt = 1 gal	1,000 L = 1 kL	

The U.S. customary system measures weight (how strongly Earth is pulling on an object) in the following units, from least to greatest: ounce (oz), pound (lb), and ton. The metric system measures mass (the

quantity of matter within an object) in the following units, from least to greatest: milligram (mg), centigram (cg), gram (g), kilogram (kg), and metric ton (MT). The relative sizes of each unit of weight and mass are shown below.

U.S. Measures of Weight	Metric Measures of Mass
$16\text{ oz} = 1\text{ lb}$	$10\text{ mg} = 1\text{ cg}$
$2{,}000\text{ lb} = 1\text{ ton}$	$100\text{ cg} = 1\text{ g}$
	$1{,}000\text{ g} = 1\text{ kg}$
	$1{,}000\text{ kg} = 1\text{ MT}$

Note that weight and mass DO NOT measure the same thing.

Word Problems

Word problems can appear daunting, but don't let the verbiage psych you out. No matter the scenario or specifics, the key to answering them is to translate the words into a math problem. Always keep in mind what the question is asking and what operations could lead to that answer.

Translating Words into Math

When asked to rewrite a mathematical expression as a situation or translated from a word problem into an expression, look for a series of key words indicating addition, subtraction, multiplication, or division:

Addition: add, altogether, together, plus, increased by, more than, in all, sum, and total

Subtraction: minus, less than, difference, decreased by, fewer than, remain, and take away

Multiplication: times, twice, of, double, and triple

Division: divided by, cut up, half, quotient of, split, and shared equally

Identifying and utilizing the proper units for the scenario requires knowing how to apply the conversion rates for money, length, volume, and mass. For example, given a scenario that requires subtracting 8 inches from $2\frac{1}{2}$ feet, both values should first be expressed in the same unit (they could be expressed $\frac{2}{3}$ft & $2\frac{1}{2}$ ft, or 8in and 30in). The desired unit for the answer may also require converting back to another unit.

Consider the following scenario: A parking area along the river is only wide enough to fit one row of cars and is $\frac{1}{2}$ kilometers long. The average space needed per car is 5 meters. How many cars can be parked along the river? First, all measurements should be converted to similar units: $\frac{1}{2}$ km $= 500$ m. The operation(s) needed should be identified. Because the problem asks for the number of cars, the total space should be divided by the space per car. 500 meters divided by 5 meters per car yields a total of 100 cars. Written as an expression, the meters unit cancels and the cars unit is left: $\frac{500\text{ m}}{5\text{ m/car}}$ is the same **as**:

$$500\text{ m} \times \frac{1\text{ car}}{5\text{ m}} = 100\text{ cars}$$

When dealing with problems involving elapsed time, breaking the problem down into workable parts is helpful. For example, suppose the length of time between 1:15pm and 3:45pm must be determined. From 1:15pm to 2:00pm is 45 minutes (knowing there are 60 minutes in an hour). From 2:00pm to 3:00pm is 1 hour. From 3:00pm to 3:45pm is 45 minutes. The total elapsed time is 45 minutes plus 1 hour plus 45 minutes. This sum produces 1 hour and 90 minutes. 90 minutes is over an hour, so this is converted to 1 hour (60 minutes) and 30 minutes. The total elapsed time can now be expressed as 2 hours and 30 minutes.

Example 1
Alexandra made $96 during the first 3 hours of her shift as a temporary worker at a law office. She will continue to earn money at this rate until she finishes in 5 more hours. How much does Alexandra make per hour? How much will Alexandra have made at the end of the day?

The hourly rate can be figured by dividing $96 by 3 hours to get $32 per hour. Now her total pay can be figured by multiplying $32 per hour by 8 hours, which comes out to $256.

Example 2
Bernard wishes to paint a wall that measures 20 feet wide by 8 feet high. It costs $0.10 to paint 1 square foot. How much money will Bernard need for paint?

The final quantity to compute is the *cost* to paint the wall. This will be ten cents ($0.10) for each square foot of area needed to paint. The area to be painted is unknown, but the dimensions of the wall are given; thus, it can be calculated.

The dimensions of the wall are 20 feet wide and 8 feet high. Since the area of a rectangle is length multiplied by width, the area of the wall is $8 \times 20 = 160$ square feet. Multiplying 0.1 x 160 yields $16 as the cost of the paint.

Algebra

Computing with Integers and Negative Rational Numbers

Integers are the whole numbers together with their negatives. They include numbers like 5, 24, 0, -6, and 15. They do not include fractions or numbers that have digits after the decimal point.

Rational numbers are all numbers that can be written as a fraction using integers. A **fraction** is written as $\frac{x}{y}$ and represents the quotient of x being divided by y. More practically, it means dividing the whole into y equal parts, then taking x of those parts.

Examples of rational numbers include $\frac{1}{2}$ and $\frac{5}{4}$. The number on the top is called the **numerator**, and the number on the bottom is called the **denominator**. Because every integer can be written as a fraction with a denominator of 1, (e.g., $\frac{3}{1} = 3$), every integer is also a rational number.

When adding integers and negative rational numbers, there are some basic rules to determine if the solution is negative or positive:

Adding two positive numbers results in a positive number: $3.3 + 4.8 = 8.1$

Adding two negative numbers results in a negative number: $(-8) + (-6) = -14$

Adding one positive and one negative number requires taking the absolute values and finding the difference between them. Then, the sign of the number that has the higher absolute value for the final solution is used.

For example, (-9) + 11, has a difference of absolute values of 2. The final solution is 2 because 11 has the higher absolute value. Another example is 9 + (-11), which has a difference of absolute values of 2. The final solution is -2 because 11 has the higher absolute value.

When subtracting integers and negative rational numbers, one has to change the problem to adding the opposite and then apply the rules of addition.

Subtracting two positive numbers is the same as adding one positive and one negative number.

For example, $4.9 - 7.1$ is the same as $4.9 + (-7.1)$. The solution is -2.2 since the absolute value of -7.1 is greater. Another example is $8.5 - 6.4$ which is the same as $8.5 + (-6.4)$. The solution is 2.1 since the absolute value of 8.5 is greater.

Subtracting a positive number from a negative number results in negative value.

For example, $(-12) - 7$ is the same as $(-12) + (-7)$ with a solution of -19.

Subtracting a negative number from a positive number results in a positive value.

For example, $12 - (-7)$ is the same as $12 + 7$ with a solution of 19.

For multiplication and division of integers and rational numbers, if both numbers are positive or both numbers are negative, the result is a positive value.

For example, $(-1.7)(-4)$ has a solution of 6.8 since both numbers are negative values.

If one number is positive and another number is negative, the result is a negative value.

For example, $\frac{-15}{5}$ has a solution of -3 since there is one negative number.

Using Absolute Values

The **absolute value** represents the distance a number is from 0. The **absolute value symbol** is | | with a number between the bars. The |10| = 10 and the |-10| = 10.

When simplifying an algebraic expression, the value of the absolute value expression is determined first, much like parenthesis in the order of operations. See the example below:

$$|8 - 12| + 5 = |-4| + 5 = 4 + 5 = 9$$

Radicals and Exponents

Exponents are used in mathematics to express a number or variable multiplied by itself a certain number of times. For example, x^3 means x is multiplied by itself three times. In this expression, x is called the **base**, and 3 is the **exponent**. Exponents can be used in more complex problems when they contain fractions and negative numbers.

Fractional exponents can be explained by looking first at the inverse of exponents, which are **roots**. Given the expression x^2, the square root can be taken, $\sqrt{x^2}$, cancelling out the 2 and leaving x by itself, if x is positive. Cancellation occurs because $\sqrt{x}$ can be written with exponents, instead of roots, as $x^{\frac{1}{2}}$. The numerator of 1 is the exponent, and the denominator of 2 is called the root (which is why it's referred to as **square root**). Taking the square root of x^2 is the same as raising it to the $\frac{1}{2}$ power. Written out in mathematical form, it takes the following progression:

$$\sqrt{x^2} = (x^2)^{\frac{1}{2}} = x$$

From properties of exponents,

$$2 \times \frac{1}{2} = 1$$

1 is the actual exponent of x. Another example can be seen with $x^{\frac{4}{7}}$. The variable x, raised to four-sevenths, is equal to the seventh root of x to the fourth power: $\sqrt[7]{x^4}$. In general,

$$x^{\frac{1}{n}} = \sqrt[n]{x} \ and \ x^{\frac{m}{n}} = \sqrt[n]{x^m}$$

Negative exponents also involve fractions. Whereas y^3 can also be rewritten as $\frac{y^3}{1}$, y^{-3} can be rewritten as $\frac{1}{y^3}$. A negative exponent means the exponential expression must be moved to the opposite spot in a fraction to make the exponent positive. If the negative appears in the numerator, it moves to the denominator. If the negative appears in the denominator, it is moved to the numerator. In general, $a^{-n} = \frac{1}{a^n}$, and a^{-n} and a^n are reciprocals.

Take, for example, the following expression:

$$\frac{a^{-4}b^2}{c^{-5}}$$

Since a is raised to the negative fourth power, it can be moved to the denominator. Since c is raised to the negative fifth power, it can be moved to the numerator. The b variable is raised to the positive second power, so it does not move.

The simplified expression is as follows:

$$\frac{b^2c^5}{a^4}$$

There are a few rules for working with exponents. For any numbers a, b, m, n, the following hold true:

$$a^1 = a$$

$$1^a = 1$$

$$a^0 = 1$$

$$a^m \times a^n = a^{m+n}$$

$$a^m \div a^n = a^{m-n}$$

$$(a^m)^n = a^{m \times n}$$

$$(a \times b)^m = a^m \times b^m$$

$$(a \div b)^m = a^m \div b^m$$

$$\left(\frac{a}{b}\right)^m = \frac{a^m}{b^m}$$

Any number, including a fraction, can be an exponent. The same rules apply.

A **root** is a different way to write an exponent when the exponent is the reciprocal of a whole number. We use the **radical** symbol to write this in the following way:

$$\sqrt[n]{a} = a^{\frac{1}{n}}$$

This quantity is called the **n-th root** of a. The n is called the **index** of the radical.

Note that if the n-th root of a is multiplied by itself n times, the result will just be a. If no number n is written by the radical, it is assumed that n is 2:

$$\sqrt{5} = 5^{\frac{1}{2}}$$

The special case of the 2nd root is called the **square root**, and the third root is called the **cube root**.

A **perfect square** is a whole number that is the square of another whole number. For example, sixteen and 64 are perfect squares because 16 is the square of 4, and 64 is the square of 8.

Ordering

When working with longer expressions, parentheses are used to show the order in which the operations should be performed. Operations inside the parentheses should be completed first. Thus, $(3 - 1) \div 2$ means one should first subtract 1 from 3, and then divide that result by 2.

The **order of operations** gives an order for how a mathematical expression is to be simplified:

- Parentheses
- Exponents
- Multiplication
- Division

- Addition
- Subtraction

To help remember this, many people like to use the mnemonic PEMDAS. Some associate this word with a phrase to help them, such as "Pirates Eat Many Donuts at Sea." Here is a quick example:

$$\text{Evaluate } 2^2 \times (3 - 1) \div 2 + 3.$$

$$\text{Parenthesis: } 2^2 \times 2 \div 2 + 3.$$

$$\text{Exponents: } 4 \times 2 \div 2 + 3$$

$$\text{Multiply: } 8 \div 2 + 3.$$

$$\text{Divide: } 4 + 3.$$

$$\text{Addition: } 7$$

Evaluating Simple Formulas and Expressions

To evaluate simple formulas and expressions, the first step is to substitute the given values in for the variable(s). Then, the order of operations is used to simplify.

Example 1

Evaluate $\frac{1}{2}x^2 - 3, x = 4$.

The first step is to substitute in 4 for *x* in the expression:

$$\frac{1}{2}(4)^2 - 3$$

Then, the order of operations is used to simplify.

The exponent comes first, $\frac{1}{2}(16) - 3$, then the multiplication $8 - 3$, and then, after subtraction, the solution is 5.

Example 2

Evaluate $4|5 - x| + 2y, x = 4, y = -3$.

The first step is to substitute 4 in for *x* and -3 in for *y* in the expression:

$$4|5 - 4| + 2(-3)$$

Then, the absolute value expression is simplified, which is:

$$|5 - 4| = |1| = 1$$

The expression is $4(1) + 2(-3)$ which can be simplified using the order of operations.

First is the multiplication, $4 + (-6)$; then addition yields an answer of -2.

Example 3

Find the perimeter of a rectangle with a length of 6 inches and a width of 9 inches.

The first step is substituting in 6 for the length and 9 for the width in the perimeter of a rectangle formula:

$$P = 2(6) + 2(9)$$

Then, the order of operations is used to simplify.

First is multiplication (resulting in $12 + 18$) and then addition for a solution of 30 inches.

Performing Operations with Monomials and Polynomials

To add or subtract polynomials, add the coefficients of terms with the same exponent. For instance:

$$(-2x^2 + 3x + 1) + (4x^2 - x)$$

$$(-2 + 4)x^2 + (3 - 1)x + 1$$

$$2x^2 + 2x + 1$$

To multiply polynomials, each term of the first polynomial multiplies each term of the second polynomial and adds up the results. Here's an example:

$$(3x^4 + 2x^2)(2x^2 + 3)$$

$$3x^4 \times 2x^2 + 3x^4 \times 3 + 2x^2 \times 2x^2 + 2x^2 \times 3$$

Then, add like terms with a result of:

$$6x^6 + 9x^4 + 4x^4 + 6x^2 = 6x^6 + 13x^4 + 6x^2$$

A polynomial with two terms is called a **binomial**. Another way to remember the rule for multiplying two binomials is to use the acronym **FOIL**: multiply the *First* terms together, then the *Outside* terms (terms on the far left and far right), then the *Inner* terms, and finally the *Last* two terms. For longer polynomials, there is no such convenient mnemonic, so remember to multiply each term of the first polynomial by each term of the second, and add the results.

To divide one polynomial by another, the procedure is similar to long division. At each step, one needs to figure out how to get the term of the dividend with the highest exponent as a multiple of the divisor. The divisor is multiplied by the multiple to get that term, which goes in the quotient. Then, the product of this term is subtracted with the dividend from the divisor and repeat the process. An example of polynomial long division is shown below.

Example

$$(4x^3 + x^2 - x + 4) \div (2x - 1)$$

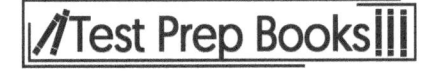

The first step is to set up the problem for long division:

$$2x - 1 \,\overline{\big)\, 4x^3 + x^2 - x + 4}$$

Divide the first term of the divisor on the left by the first term of the numerator under the bar and place the result over the bar:

$$\begin{array}{r} 2x^2 \\ 2x - 1 \,\overline{\big)\, 4x^3 + x^2 - x + 4} \end{array}$$

Next, multiply that new term by the divisor and place it beneath the numerator:

$$\begin{array}{r} 2x^2 \\ 2x - 1 \,\overline{\big)\, 4x^3 + x^2 - x + 4} \\ 4x^3 - 2x^2 \end{array}$$

Subtract it from the numerator to find the difference, a new polynomial:

$$\begin{array}{r} 2x^2 \\ 2x - 1 \,\overline{\big)\, 4x^3 + x^2 - x + 4} \\ - 4x^3 - 2x^2 \\ \hline 3x^2 - x + 4 \end{array}$$

The same steps are repeated until there are no more x terms in the quotient above the bar or the subtraction step leaves no remainder.

Divide the first term of the new polynomial by the first term of the divisor, multiply that by divisor, and subtract from the new polynomial:

$$2x - 1 \overline{\smash{\big)}\ 4x^3 + x^2 - x + 4}$$

with quotient $2x^2 + \frac{3}{2}x + \frac{1}{4}$

$$
\begin{array}{r}
2x^2 + \frac{3}{2}x + \frac{1}{4} \\
2x - 1 \overline{\smash{\big)}\ 4x^3 + x^2 - x + 4} \\
-\ \underline{4x^3 - 2x^2} \\
3x^2 - x + 4 \\
-\ \underline{3x^2 - \frac{3}{2}x} \\
\frac{1}{2}x + 4 \\
-\ \underline{\frac{1}{2}x - \frac{1}{4}} \\
\frac{17}{4}
\end{array}
$$

In this case, the problem doesn't divide evenly, so the answer is the quotient, plus the remainder divided by the divisor $(2x - 1)$:

$$2x^2 + \frac{3}{2}x + \frac{1}{4} + \frac{17}{4(2x - 1)}$$

Simplifying Algebraic Fractions

A **rational expression** is a fraction with a polynomial in the numerator and denominator. The denominator polynomial cannot be zero. An example of a rational expression is:

$$\frac{3x^4 - 2}{-x + 1}$$

The same rules for working with addition, subtraction, multiplication, and division with rational expressions apply as when working with regular fractions.

The first step is to find a common denominator when adding or subtracting. This can be done just as with regular fractions. For example, if $\frac{a}{b} + \frac{c}{d}$, then a common denominator can be found by multiplying to find the following fractions: $\frac{ad}{bd}, \frac{cb}{db}$.

A **complex fraction** is a fraction in which the numerator and denominator are themselves fractions, of the form $\frac{\left(\frac{a}{b}\right)}{\left(\frac{c}{d}\right)}$. These can be simplified by following the usual rules for the order of operations, or by

remembering that dividing one fraction by another is the same as multiplying by the reciprocal of the divisor. This means that any complex fraction can be rewritten using the following form:

$$\frac{\left(\frac{a}{b}\right)}{\left(\frac{c}{d}\right)} = \frac{a}{b} \times \frac{d}{c}$$

The following problem is an example of solving a complex fraction:

$$\frac{\left(\frac{5}{4}\right)}{\left(\frac{3}{8}\right)} = \frac{5}{4} \times \frac{8}{3} = \frac{40}{12} = \frac{10}{3}$$

Factoring

Factors for polynomials are similar to factors for integers—they are numbers, variables, or polynomials that, when multiplied together, give a product equal to the polynomial in question. One polynomial is a factor of a second polynomial if the second polynomial can be obtained from the first by multiplying by a third polynomial.

$6x^6 + 13x^4 + 6x^2$ can be obtained by multiplying together:

$$(3x^4 + 2x^2)(2x^2 + 3)$$

This means $2x^2 + 3$ and $3x^4 + 2x^2$ are factors of:

$$6x^6 + 13x^4 + 6x^2$$

In general, finding the factors of a polynomial can be tricky. However, there are a few types of polynomials that can be factored in a straightforward way.

If a certain monomial is in each term of a polynomial, it can be factored out. There are several common forms polynomials take, which if you recognize, you can solve. The first example is a perfect square trinomial. To factor this polynomial, first expand the middle term of the expression:

$$x^2 + 2xy + y^2$$

$$x^2 + xy + xy + y^2$$

Factor out a common term in each half of the expression (in this case x from the left and y from the right):

$$x(x + y) + y(x + y)$$

Then the same can be done again, treating $(x + y)$ as the common factor:

$$(x + y)(x + y) = (x + y)^2$$

Therefore, the formula for this polynomial is:

$$x^2 + 2xy + y^2 = (x + y)^2$$

Next is another example of a perfect square trinomial. The process is the similar, but notice the difference in sign:

$$x^2 - 2xy + y^2$$

$$x^2 - xy - xy + y^2$$

Factor out the common term on each side:

$$x(x - y) - y(x - y)$$

Factoring out the common term again:

$$(x - y)(x - y) = (x - y)^2$$

Thus:

$$x^2 - 2xy + y^2 = (x - y)^2$$

The next is known as a difference of squares. This process is effectively the reverse of binomial multiplication:

$$x^2 - y^2$$

$$x^2 - xy + xy - y^2$$

$$x(x - y) + y(x - y)$$

$$(x + y)(x - y)$$

Therefore:

$$x^2 - y^2 = (x + y)(x - y)$$

The following two polynomials are known as the sum or difference of cubes. These are special polynomials that take the form of $x^3 + y^3$ or $x^3 - y^3$. The following formula factors the sum of cubes:

$$x^3 + y^3 = (x + y)(x^2 - xy + y^2)$$

Next is the difference of cubes, but note the change in sign. The formulas for both are similar, but the order of signs for factoring the sum or difference of cubes can be remembered by using the acronym SOAP, which stands for "same, opposite, always positive." The first sign is the same as the sign in the first expression, the second is opposite, and the third is always positive. The next formula factors the difference of cubes:

$$x^3 - y^3 = (x - y)(x^2 + xy + y^2)$$

The following two examples are expansions of cubed binomials. Similarly, these polynomials always follow a pattern:

$$x^3 + 3x^2y + 3xy^2 + y^3 = (x + y)^3$$

$$x^3 - 3x^2y + 3xy^2 - y^3 = (x - y)^3$$

These rules can be used in many combinations with one another. For example, the expression $3x^3 - 24$ has a common factor of 3, which becomes:

$$3(x^3 - 8)$$

A difference of cubes still remains which can then be factored out:

$$3(x - 2)(x^2 + 2x + 4)$$

There are no other terms to be pulled out, so this expression is completely factored.

When factoring polynomials, a good strategy is to multiply the factors to check the result. Let's try another example:

$$4x^3 + 16x^2$$

Both sides of the expression can be divided by 4, and both contain x^2, because $4x^3$ can be thought of as $4x^2(x)$, so the common term can simply be factored out:

$$4x^2(x + 4)$$

It sometimes can be necessary to rewrite the polynomial in some clever way before applying the above rules. Consider the problem of factoring $x^4 - 1$. This does not immediately look like any of the previous polynomials. However, it's possible to think of this polynomial as $x^4 - 1 = (x^2)^2 - (1^2)^2$, and now it can be treated as a difference of squares to simplify this:

$$(x^2)^2 - (1^2)^2$$

$$(x^2)^2 - x^2 1^2 + x^2 1^2 - (1^2)^2$$

$$x^2(x^2 - 1^2) + 1^2(x^2 - 1^2)$$

$$(x^2 + 1^2)(x^2 - 1^2)$$

$$(x^2 + 1)(x^2 - 1)$$

Solving Linear Equations and Inequalities

The simplest equations to solve are **linear equations**, which have the form $ax + b = 0$. These have the solution $x = -\frac{b}{a}$.

For instance, in the equation $\frac{1}{3}x - 4 = 0$, it can be determined that $\frac{1}{3}x = 4$ by adding 4 on each side. Next, both sides of the equation are multiplied by 3 to get $x = 12$.

Solving an inequality is very similar to solving equations, with one important issue to keep track of: if multiplying or dividing both sides of an inequality by a negative number, the direction of the inequality *flips*.

For example, consider the inequality $-4x < 12$. Solving this inequality requires the division of -4. When the negative four is divided, the less-than sign changes to a greater-than sign. The solution becomes $x > -3$.

<u>Example</u>

$-4x - 3 \leq -2x + 1$

2x is added to both sides, and 3 is added to both sides, leaving $-2x \leq 4$.

$-2x \leq 4$ is multiplied by $-\frac{1}{2}$, which means flipping the direction of the inequality.

This gives $x \geq -2$.

An **absolute inequality** is an inequality that is true for all real numbers. An inequality that is only true for some real numbers is called **conditional**.

In addition to the inequalities above, there are also **double inequalities** where three quantities are compared to one another, such as $3 \leq x + 4 < 5$. The rules for double inequalities include always performing any operations to every part of the inequality and reversing the direction of the inequality when multiplying or dividing by a negative number.

When solving equations and inequalities, the solutions can be checked by plugging the answer back in to the original problem. If the solution makes a true statement, the solution is correct.

Systems of Equations

A **system of equations** is a group of equations that have the same variables or unknowns. These equations can be linear, but they are not always so. Finding a solution to a system of equations means finding the values of the variables that satisfy each equation. For a linear system of two equations and two variables, there could be a single solution, no solution, or infinitely many solutions.

A single solution occurs when there is one value for *x* and *y* that satisfies the system. This would be shown on the graph where the lines cross at exactly one point. When there is no solution, the lines are parallel and do not ever cross. With infinitely many solutions, the equations may look different, but they are the same line. One equation will be a multiple of the other, and on the graph, they lie on top of each other.

The **process of elimination** can be used to solve a system of equations. For example, the following equations make up a system: $x + 3y = 10$ and $2x - 5y = 9$. Immediately adding these equations does not eliminate a variable, but it is possible to change the first equation by multiplying the whole equation by -2. This changes the first equation to $-2x - 6y = -20$. The equations can be then added to obtain $-11y = -11$. Solving for *y* yields $y = 1$. To find the rest of the solution, 1 can be substituted in for *y* in either original equation to find the value of $x = 7$. The solution to the system is (7, 1) because it makes both equations true, and it is the point in which the lines intersect. If the system is **dependent**—having infinitely many solutions—then both variables will cancel out when the elimination method is used, resulting in an equation that is true for many values of *x* and *y*. Since the system is dependent, both equations can be simplified to the same equation or line.

A system can also be solved using **substitution**. This involves solving one equation for a variable and then plugging that solved equation into the other equation in the system. For example, $x - y = -2$ and $3x + 2y = 9$ can be solved using substitution. The first equation can be solved for *x*, where $x = -2 + y$. Then it can be plugged into the other equation:

$$3(-2 + y) + 2y = 9$$

Solving for y yields $-6 + 3y + 2y = 9$, where $y = 3$. If $y = 3$, then $x = 1$. This solution can be checked by plugging in these values for the variables in each equation to see if it makes a true statement.

Finally, a solution to a system of equations can be found graphically. The solution to a linear system is the point or points where the lines cross. The values of x and y represent the coordinates (x, y) where the lines intersect. Using the same system of equations as above, they can be solved for y to put them in slope-intercept form, $y = mx + b$. These equations become $y = x + 2$ and:

$$y = -\frac{3}{2}x + 4.5$$

This system with the solution is shown below:

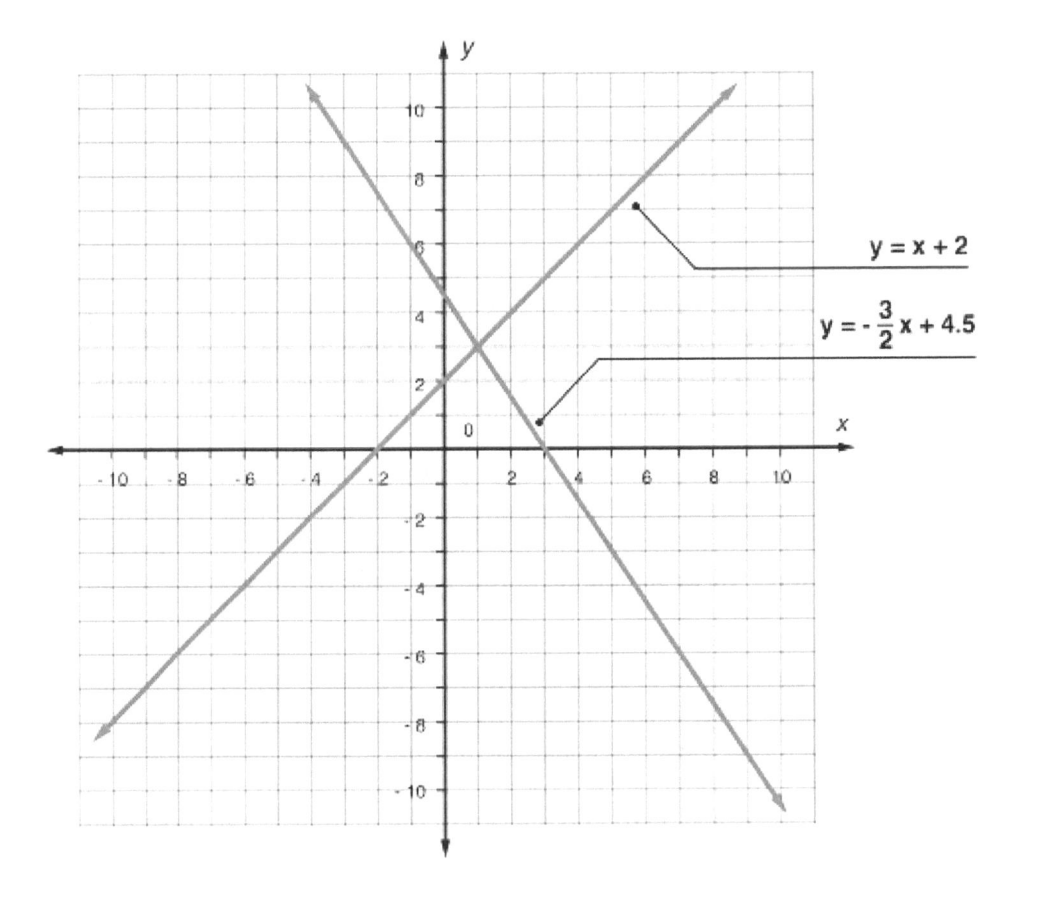

Matrices

Matrices can be used to represent linear equations, solve systems of equations, and manipulate data to simulate change. Matrices consist of numerical entries in both rows and columns. The following matrix A is a 3×4 matrix because it has three rows and four columns:

$$A = \begin{bmatrix} 3 & 2 & -5 & 3 \\ 3 & 6 & 2 & -5 \\ -1 & 3 & 7 & 0 \end{bmatrix}$$

Matrices can be added or subtracted only if they have the same dimensions. For example, the following matrices can be added by adding corresponding matrix entries:

$$\begin{bmatrix} 3 & 4 \\ 2 & -6 \end{bmatrix} + \begin{bmatrix} -1 & 4 \\ 4 & 2 \end{bmatrix} = \begin{bmatrix} 2 & 8 \\ 6 & -4 \end{bmatrix}$$

Multiplication can also be used to manipulate matrices. **Scalar multiplication** involves multiplying a matrix by a constant. Each matrix entry needs to be multiplied by the constant. The following example shows a 3×2 matrix being multiplied by the constant 6:

$$6 \times \begin{bmatrix} 3 & 4 \\ 2 & -6 \\ 1 & 0 \end{bmatrix} = \begin{bmatrix} 18 & 24 \\ 12 & -36 \\ 6 & 0 \end{bmatrix}$$

Matrix multiplication of two matrices involves finding multiple dot products. The **dot product** of a row and column is the sum of the products of each corresponding row and column entry. In the following example, a 2×2 matrix is multiplied by a 2×2 matrix. The dot product of the first row and column is:

$$(2 \times 1) + (1 \times 2) = (2) + (2) = 4$$

$$\begin{bmatrix} 2 & 1 \\ 3 & 5 \end{bmatrix} \times \begin{bmatrix} 1 & 4 \\ 2 & 0 \end{bmatrix} = \begin{bmatrix} 4 & 8 \\ 13 & 12 \end{bmatrix}$$

The same process is followed to find the other three values in the solution matrix. Matrices can only be multiplied if the number of columns in the first matrix equals the number of rows in the second matrix. The previous example is also an example of square matrix multiplication because they are both square matrices. A **square matrix** has the same number of rows and columns. For square matrices, the order in which they are multiplied does matter. Therefore, matrix multiplication does not satisfy the commutative property. It does, however, satisfy the associative and distributive properties.

Another transformation of matrices can be found by using the **identity matrix**—also referred to as the **"I" matrix**. The identity matrix is similar to the number one in normal multiplication. The identity matrix is a square matrix with ones in the diagonal spots and zeros everywhere else. The identity matrix is also the result of multiplying a matrix by its inverse. This process is similar to multiplying a number by its reciprocal.

The **zero matrix** is also a matrix acting as an additive identity. The zero matrix consists of zeros in every entry. It does not change the values of a matrix when using addition.

The **inverse of a matrix** is useful for solving complex systems of equations. Not all matrices have an inverse, but this can be checked by finding the **determinant** of the matrix. If the determinant of the matrix is 0, it is not inversible. Additionally, only square matrices are inversible. To find the determinant of any matrix, each value of the first row is multiplied by the determinant of submatrix consisting of all except the row and column for that value. The results of multiplication are alternatingly subtracted and added for 3×3 or larger matrices. The determinant of a matrix can be represented with straight bars (such as $|A|$) or the function det (A), where A is a matrix.

Using the *square 2 x 2 matrix*, the determinant is: $|A| = \begin{vmatrix} a & b \\ c & d \end{vmatrix} = ad - bc$

The absolute value of the determinant of matrix A is equal to the area of a parallelogram with vertices $(0, 0)$, (a, b), (c, d), and $(a + b, c + d)$.

For example, the determinant of the matrix $\begin{bmatrix} -5 & 1 \\ 3 & 4 \end{bmatrix}$ is:

$$-5(4) - 1(3) = -20 - 3 = -23$$

Solving Quadratic Equations by Factoring

Solving quadratic equations is a little trickier. If they take the form $ax^2 - b = 0$, then the equation can be solved by adding b to both sides and dividing by a to get $x^2 = \frac{b}{a}$.

Using the sixth rule above, the solution is $x = \pm\sqrt{\frac{b}{a}}$. Note that this is actually two separate solutions, unless b happens to be 0.

If a quadratic equation has no constant—so that it takes the form $ax^2 + bx = 0$—then the x can be factored out to get $x(ax + b) = 0$. Then, the solutions are $x = 0$, together with the solutions to $ax + b = 0$. Both factors x and $(ax + b)$ can be set equal to zero to solve for x because one of those values must be zero for their product to equal zero. For an equation $ab = 0$ to be true, either $a = 0$, or $b = 0$.

A given quadratic equation $x^2 + bx + c$ can be factored into $(x + A)(x + B)$, where $A + B = b$, and $AB = c$. Finding the values of A and B can take time, but such a pair of numbers can be found by guessing and checking. Looking at the positive and negative factors for c offers a good starting point.

For example, in $x^2 - 5x + 6$, the factors of 6 are 1, 2, and 3. Now, $(-2)(-3) = 6$, and $-2 - 3 = -5$. In general, however, this may not work, in which case another approach may need to be used.

A quadratic equation of the form $x^2 + 2xb + b^2 = 0$ can be factored into $(x + b)^2 = 0$. Similarly, $x^2 - 2xy + y^2 = 0$ factors into $(x - y)^2 = 0$.

In general, the constant term may not be the right value to be factored this way. A more general method for solving these quadratic equations must then be found. The following two methods will work in any situation.

Completing the Square

The first method is called **completing the square**. The idea here is that in any equation:

$$x^2 + 2xb + c = 0$$

something could be added to both sides of the equation to get the left side to look like $x^2 + 2xb + b^2$, meaning it could be factored into $(x + b)^2 = 0$.

Example
$x^2 + 6x - 1 = 0$

The left-hand side could be factored if the constant were equal to 9, since:

$$x^2 + 6x + 9 = (x + 3)^2$$

To get a constant of 9 on the left, 10 needs to be added to both sides.

That changes the equation to:

$$x^2 + 6x + 9 = 10$$

Factoring the left gives $(x + 3)^2 = 10$.

Then, the square root of both sides can be taken (remembering that this introduces a $\pm$):

$$x + 3 = \pm\sqrt{10}$$

Finally, 3 is subtracted from both sides to get two solutions: $x = -3 \pm \sqrt{10}$.

The Quadratic Formula

The first method of completing the square can be used in finding the second method, the quadratic formula. It can be used to solve any quadratic equation. This formula may be the longest method for solving quadratic equations and is commonly used as a last resort after other methods are ruled out.

It can be helpful in memorizing the formula to see where it comes from, so here are the steps involved.

The most general form for a quadratic equation is:

$$ax^2 + bx + c = 0$$

First, dividing both sides by a leaves us with:

$$x^2 + \frac{b}{a}x + \frac{c}{a} = 0$$

To complete the square on the left-hand side, $\frac{c}{a}$ can be subtracted on both sides to get:

$$x^2 + \frac{b}{a}x = -\frac{c}{a}$$

$(\frac{b}{2a})^2$ is then added to both sides.

This gives:

$$x^2 + \frac{b}{a}x + (\frac{b}{2a})^2 = (\frac{b}{2a})^2 - \frac{c}{a}$$

The left can now be factored and the right-hand side simplified to give:

$$(x + \frac{b}{2a})^2 = \frac{b^2 - 4ac}{4a}$$

Taking the square roots gives:

$$x + \frac{b}{2a} = \pm\frac{\sqrt{b^2 - 4ac}}{2a}$$

Solving for x yields the quadratic formula:

$$x = \frac{-b \pm \sqrt{b^2 - 4ac}}{2a}$$

It isn't necessary to remember how to get this formula but memorizing the formula itself is the goal.

If an equation involves taking a root, then the first step is to move the root to one side of the equation and everything else to the other side. That way, both sides can be raised to the index of the radical in order to remove it, and solving the equation can continue.

Exponential Functions

An **exponential function** is a function of the form $f(x) = b^x$, where b is a positive real number other than 1. In such a function, b is called the **base**.

The **domain** of an exponential function is all real numbers, and the **range** is all positive real numbers. There will always be a horizontal asymptote of $y = 0$ on one side. If b is greater than 1, then the graph will be increasing when moving to the right. If b is less than 1, then the graph will be decreasing when moving to the right. Exponential functions are one-to-one. The basic exponential function graph will go through the point (0, 1).

The following example demonstartes this more clearly:

Solve $5^{x+1} = 25$.

The first step is to get the x out of the exponent by rewriting the equation $5^{x+1} = 5^2$ so that both sides have a base of 5. Since the bases are the same, the exponents must be equal to each other. This leaves $x + 1 = 2$ or $x = 1$. To check the answer, the x-value of 1 can be substituted back into the original equation.

Exponential growth and decay can be found in real-world situations. For example, if a piece of notebook paper is folded 25 times, the thickness of the paper can be found. To model this situation, a table can be used. The initial point is one-fold, which yields a thickness of 2 papers. For the second fold, the thickness is 4. Since the thickness doubles each time, the table below shows the thickness for the next few folds. Notice the thickness changes by the same factor each time. Since this change for a constant interval of folds is a factor of 2, the function is exponential. The equation for this is $y = 2^x$. For twenty-five folds, the thickness would be 33,554,432 papers.

x (folds)	y (paper thickness)
0	1
1	2
2	4
3	8
4	16
5	32

One exponential formula that is commonly used is the **interest formula**: $A = Pe^{rt}$. In this formula, interest is compounded continuously. A is the value of the investment after the time, t, in years. P is the initial amount of the investment, r is the interest rate, and e is the constant equal to approximately

2.718. Given an initial amount of $200 and a time of 3 years, if interest is compounded continuously at a rate of 6%, the total investment value can be found by plugging each value into the formula. The invested value at the end is $239.44. In more complex problems, the final investment may be given, and the rate may be the unknown. In this case, the formula becomes $239.44 = 200e^{r3}$. Solving for r requires isolating the exponential expression on one side by dividing by 200, yielding the equation $1.20 = e^{r3}$. Taking the natural log of both sides results in $\ln(1.2) = r3$. Using a calculator to evaluate the logarithmic expression, $r = 0.06 = 6\%$.

When working with logarithms and exponential expressions, it is important to remember the relationship between the two. In general, the logarithmic form is $y = \log_b x$ for an exponential form $b^y = x$. Logarithms and exponential functions are inverses of each other.

Logarithmic Functions

A **logarithmic function** is an inverse for an exponential function. The inverse of the base b exponential function is written as $\log_b(x)$, and is called the **base b logarithm**. The domain of a logarithm is all positive real numbers. It has the properties that $\log_b(b^x) = x$. For positive real values of x, $b^{\log_b(x)} = x$.

When there is no chance of confusion, the parentheses are sometimes skipped for logarithmic functions: $\log_b(x)$ may be written as $\log_b x$. For the special number e, the base e logarithm is called the **natural logarithm** and is written as $\ln x$. Logarithms are one-to-one.

When working with logarithmic functions, it is important to remember the following properties. Each one can be derived from the definition of the logarithm as the inverse to an exponential function:

- $\log_b 1 = 0$
- $\log_b b = 1$
- $\log_b b^p = p$
- $\log_b MN = \log_b M + \log_b N$
- $\log_b \frac{M}{N} = \log_b M - \log_b N$
- $\log_b M^p = p \log_b M$

When solving equations involving exponentials and logarithms, the following fact should be used:

If f is a one-to-one function, $a = b$ is equivalent to $f(a) = f(b)$.

Using this, together with the fact that logarithms and exponentials are inverses, allows for manipulations of the equations to isolate the variable as is demonstrated in the following example:

Solve $4 = \ln(x - 4)$.

Using the definition of a logarithm, the equation can be changed to $e^4 = e^{\ln(x-4)}$. The functions on the right side cancel with a result of $e^4 = x - 4$. This then gives:

$$x = 4 + e^4$$

Solving Verbal Problems Presented in an Algebraic Context

There is a four-step process in problem-solving that can be used as a guide:

- Understand the problem and determine the unknown information.
- Translate the verbal problem into an algebraic equation.
- Solve the equation by using inverse operations.
- Check the work and answer the given question.

Example

Three times the sum of a number plus 4 equals the number plus 8. What is the number?

The first step is to determine the unknown, which is the number, or x.

The second step is to translate the problem into the equation, which is $3(x + 4) = x + 8$.

The equation can be solved as follows:

Apply the distributive property	$3x + 12 = x + 8$
Subtract 12 from both sides of the equation	$3x = x - 4$
Subtract x from both sides of the equation	$2x = -4$
Divide both sides of the equation by 2	$x = -2$

The final step is checking the solution. Plugging the value for x back into the equation yields the following problem:

$$3(-2) + 12 = -2 + 8$$

Using the order of operations shows that a true statement is made: $6 = 6$.

Determining Distance, Rate, and Time

In word problems, multiple quantities are often provided with a request to find some kind of relation between them. This often will mean that one variable (the dependent variable whose value needs to be found) can be written as a function of another variable (the independent variable whose value can be figured from the given information). The usual procedure for solving these problems is to start by giving each quantity in the problem a variable, and then figuring the relationship between these variables.

For example, suppose a car gets 25 miles per gallon. How far will the car travel if it uses 2.4 gallons of fuel? In this case, y would be the distance the car has traveled in miles, and x would be the amount of fuel burned in gallons (2.4). Then the relationship between these variables can be written as an algebraic equation, $y = 25x$. In this case, the equation is $y = 25 \times 2.4 = 60$, so the car has traveled 60 miles.

Some word problems require more than just one simple equation to be written and solved. Consider the following situations and the linear equations used to model them.

Suppose Margaret is 2 miles to the east of John at noon. Margaret walks to the east at 3 miles per hour. How far apart will they be at 3 p.m.? To solve this, x would represent the time in hours past noon, and y would represent the distance between Margaret and John. Now, noon corresponds to the equation

where *x* is 0, so the *y* intercept is going to be 2. It's also known that the slope will be the rate at which the distance is changing, which is 3 miles per hour.

This means that the slope will be 3 (be careful at this point: if units were used, other than miles and hours, for *x* and *y* variables, a conversion of the given information to the appropriate units would be required first). The simplest way to write an equation given the *y*-intercept, and the slope is the Slope-Intercept form, is $y = mx + b$. Recall that *m* here is the slope and *b* is the *y* intercept. So, $m = 3$ and $b = 2$. Therefore, the equation will be $y = 3x + 2$. The word problem asks how far to the east Margaret will be from John at 3 p.m., which means when *x* is 3. So, substitute $x = 3$ into this equation to obtain:

$$y = 3 \times 3 + 2 = 9 + 2 = 11$$

Therefore, she will be 11 miles to the east of him at 3 p.m.

For another example, suppose that a box with 4 cans in it weighs 6 lbs., while a box with 8 cans in it weighs 12 lbs. Find out how much a single can weighs. To do this, let *x* denote the number of cans in the box, and *y* denote the weight of the box with the cans in lbs. This line touches two pairs: $(4, 6)$ and $(8, 12)$. A formula for this relation could be written using the two-point form, with $x_1 = 4, y_1 = 6, x_2 = 8, y_2 = 12$. This would yield:

$$\frac{y - 6}{x - 4} = \frac{12 - 6}{8 - 4} \text{ or } \frac{y - 6}{x - 4} = \frac{6}{4} = \frac{3}{2}$$

However, only the slope is needed to solve this problem, since the slope will be the weight of a single can. From the computation, the slope is $\frac{3}{2}$. Therefore, each can weighs $\frac{3}{2}$ lb.

Determining Tax, Tip, and Discounts

Determining the tax and tip on a bill involves calculating a proportion. For example, a standard tip on a restaurant bill is 20%. This can be calculated using mental math by shifting the decimal place for the bill amount one place to the left (to calculate 10%) and then doubling that amount. Consider the following:

A bill for dinner is $68.24. What will be a 20% tip on the bill?

Step one: Take $68.24 and shift the decimal point one place to the left: $6.82

This is 10% of the bill.

Step two: Double the amount in step one: $6.82 × 2 = $13.64.

This is the 20% tip.

Computations can be taken one step further by adding the tip to the bill to find the total:

$$\$68.24 + \$13.64 = \$81.88$$

Discounts and the amount something depreciates in value also involve calculating percentages and then subtracting the calculated amounts from the initial value. For example, consider the following problem and the two proposed methods to arrive at the answer:

A store is having a spring sale, where everything is 70% off. You have $45.00 to spend. A jacket is regularly priced at $80.00. Do you have enough to buy the jacket and a pair of gloves, regularly priced at $20.00?

There are two ways to approach this.

Method 1:

Set up the equations to find the sale prices: the original price minus the amount discounted.
$80.00 - ($80.00 (0.70)) = sale cost of the jacket.
$20.00 – ($20.00 (0.70)) = sale cost of the gloves.
Solve for the sale cost.
$24.00 = sale cost of the jacket.
$6.00 = sale cost of the gloves.
Determine if you have enough money for both.
$24.00 + $6.00 = total sale cost.
$30.00 is less than $45.00, so you can afford to purchase both.

Method 2:

Determine the percent of the original price that you will pay.
100% – 70% = 30%
Set up the equations to find the sale prices.
$80.00 (0.30) = cost of the jacket.
$20.00 (0.30) = cost of the gloves.
Solve.
$24.00 = cost of the jacket.
$6.00 = cost of the gloves.
Determine if you have enough money for both.
$24.00 + $6.00 = total sale cost.
$30.00 is less than $45.00, so you can afford to purchase both.

Physics Applications

Description of Motion in One and Two Dimensions

The description of motion is known as **kinetics**, and the causes of motion are known as **dynamics**. Motion in one dimension is a **scalar** quantity. It consists of one measurement such as length (length/distance is also known as **displacement**), speed, or time. Motion in two dimensions is a **vector** quantity. This is a speed with a direction, or **velocity**.

Velocity is the measure of the change in distance over the change in time. All vector quantities have a direction that can be relayed through the sign of an answer, such as -5.0 m/s or +5.0 m/s. The objects

registering these velocities would be traveling in opposite directions, where the change in distance is denoted by Δx and the change in time is denoted by Δt:

$$v = \frac{\Delta x}{\Delta t}$$

Acceleration is the measure of the change in an object's velocity over a change in time, where the change in velocity, $v_2 - v_1$, is denoted by Δv and the change in time, $t_1 - t_2$, is denoted by Δt:

$$a = \frac{\Delta v}{\Delta t}$$

The **linear momentum,** p, of an object is the result of the object's mass, m, multiplied by its velocity, v, and is described by the equation:

$$p = mv$$

This aspect becomes important when one object hits another object. For example, the linear momentum of a small sports car will be much smaller than the linear momentum of a large semi-truck. Thus, the semi-truck will cause more damage to the car than the car to the truck.

Mechanical Advantage

Using a simple machine employs an advantage to the user. This is referred to as the **mechanical advantage**. It can be calculated by comparing the force input by the user to the simple machine with the force output from the use of the machine (also displayed as a ratio).

$$\text{Mechanical Advantage} = \frac{\text{output force}}{\text{input force}}$$

$$MA = \frac{F_{out}}{F_{in}}$$

When using a lever, it can be helpful to calculate the torque, or circular force, necessary to move something. This is also employed when using a wrench to loosen a bolt.

$$\text{Torque} = F \times \text{distance of lever arm from the axis of rotation}$$

$$T = F \times d$$

Geometry

Perimeter

The **perimeter** is the distance around a figure or the sum of all sides of a polygon.

The *formula for the perimeter of a square* is four times the length of a side.

For example, the following square has side lengths of 5 meters:

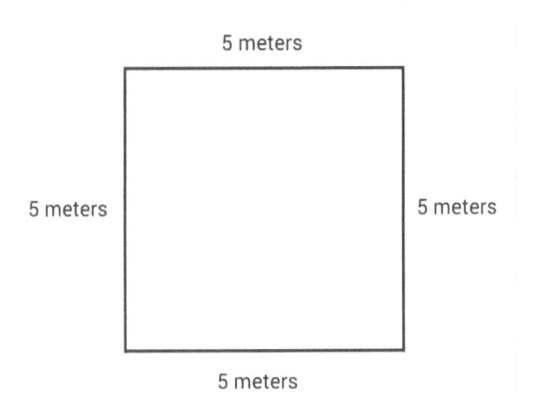

The perimeter is 20 meters because 4 times 5 is 20.

The *formula for a perimeter of a rectangle* is the sum of twice the length and twice the width. For example, if the length of a rectangle is 10 inches and the width 8 inches, then the perimeter is 36 inches because:

$$P = 2l + 2w$$

$$2(10) + 2(8)$$

$$20 + 16 = 36 \text{ inches}$$

Area

The area is the amount of space inside of a figure, and there are formulas associated with area.

The area of a triangle is the product of one-half the base and height. For example, if the base of the triangle is 2 feet and the height 4 feet, then the area is 4 square feet. The following equation shows the formula used to calculate the area of the triangle:

$$A = \frac{1}{2}bh = \frac{1}{2}(2)(4) = 4 \text{ square feet}$$

The area of a square is the length of a side squared. For example, if a side of a square is 7 centimeters, then the area is 49 square centimeters. The formula for this example is:

$$A = s^2 = 7^2 = 49 \text{ square centimeters}$$

The area of a rectangle is the length times the width. An example is if the rectangle has a length of 6 inches and a width of 7 inches, then the area is 42 square inches:

$$A = lw = 6(7) = 42 \text{ square inches}$$

The area of a trapezoid is one-half the height times the sum of the bases. For example, if the length of the bases are 2.5 and 3 feet and the height 3.5 feet, then the area is 9.625 square feet. The following formula shows how the area is calculated:

$$A = \frac{1}{2}h(b_1 + b_2)$$

$$\frac{1}{2}(3.5)(2.5 + 3)$$

$$\frac{1}{2}(3.5)(5.5) = 9.625 \text{ square feet}$$

The perimeter of a figure is measured in single units, while the area is measured in square units.

Plane Geometry

Points and Lines

The basic unit of geometry is a point. A point represents an exact location on a plane, or flat surface. The position of a point is indicated with a dot and usually named with a single uppercase letter, such as point *A* or point *T*. A point is a place, not a thing, and therefore has no dimensions or size. A set of points that lies on the same line is called collinear. A set of points that lies on the same plane is called coplanar.

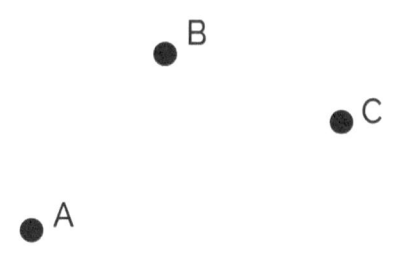

The image above displays point *A*, point *B*, and point *C*.

A line is a series of points that extends in both directions without ending. It consists of an infinite number of points and is drawn with arrows on both ends to indicate it extends infinitely. Lines can be named by two points on the line or with a single, cursive, lower case letter. The line below could be named line *AB* or line *BA* or $\overleftrightarrow{AB}$ or $\overleftrightarrow{BA}$.

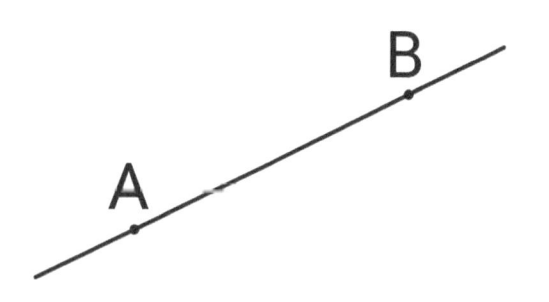

If the Cartesian coordinates for A and B are known, then the distance $d(A, B)$ along the line between them can be measured using the **Pythagorean formula**, which states that if $A = (x_1, y_1)$ and $B = (x_2, y_2)$, then the distance between them is:

$$d(A, B) = \sqrt{(x_2 - x_1)^2 + (y_2 - y_1)^2}$$

A ray has a specific starting point and extends in one direction without ending. The endpoint of a ray is its starting point. Rays are named using the endpoint first, and any other point on the ray. The following ray can be named ray AB and written $\overrightarrow{AB}$.

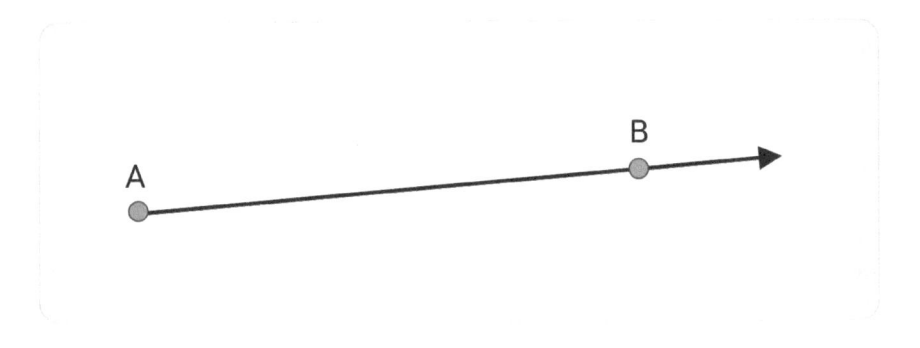

A line segment has specific starting and ending points. A line segment consists of two endpoints and all the points in between. Line segments are named by the two endpoints. The example below is named segment KL or segment LK, written $\overline{KL}$ or $\overline{LK}$.

Two lines are considered parallel to each other if, while extending infinitely, they will never intersect (or meet). Parallel lines point in the same direction and are always the same distance apart. Two lines are considered perpendicular if they intersect to form right angles. Right angles are 90°. Typically, a small box is drawn at the intersection point to indicate the right angle.

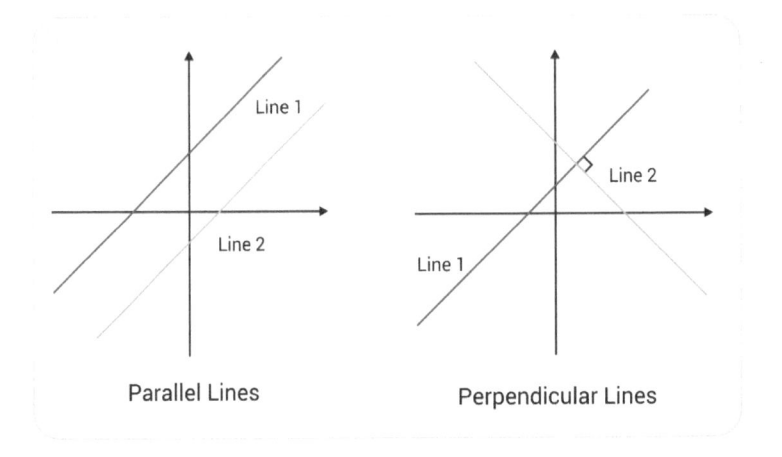

Parallel Lines　　　　　Perpendicular Lines

Line 1 is parallel to line 2 in the left image and is written as line 1 ∥ line 2. Line 1 is perpendicular to line 2 in the right image and is written as line 1 ⊥ line 2.

Angles

When two lines cross, they form an **angle**. The point where the lines cross is called the **vertex** of the angle. The angle below has a vertex at point *B* and the sides consist of ray *BA* and ray *BC*. An angle can be named in three ways:

- Using the vertex and a point from each side, with the vertex letter in the middle.
- Using only the vertex. This can only be used if it is the only angle with that vertex.
- Using a number that is written inside the angle.

The angle below can be written ∠*ABC* (read angle *ABC*), ∠*CBA*, ∠*B*, or ∠1.

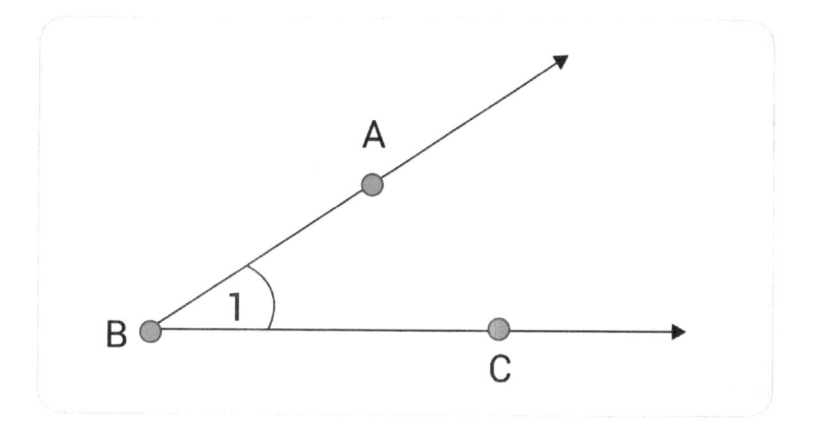

An angle divides a plane, or flat surface, into three parts: the angle itself, the interior (inside) of the angle, and the exterior (outside) of the angle. The figure below shows point *M* on the interior of the angle and point *N* on the exterior of the angle.

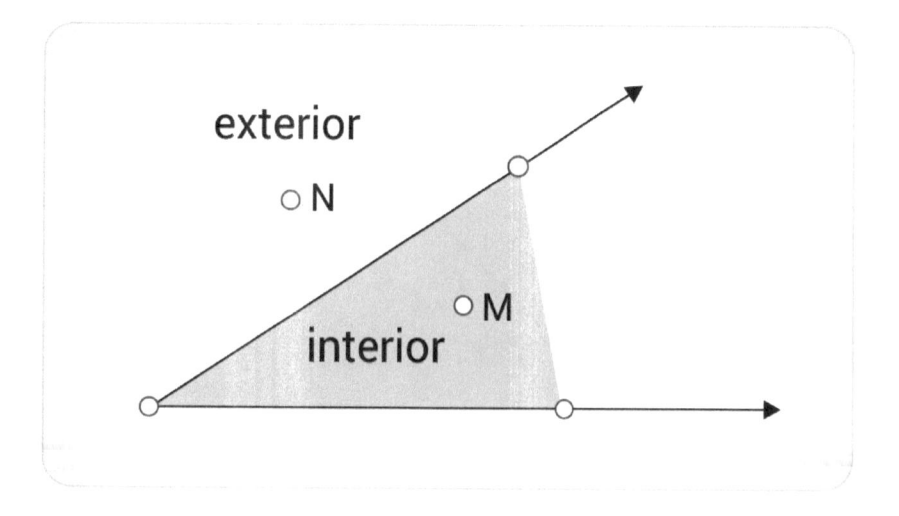

The measurement of an angle can be given in degrees or in radians. In degrees, a full circle is 360 degrees, written 360°. In radians, a full circle is 2π radians. The degree measure of an angle is between 0° and 180° and can be obtained by using a protractor.

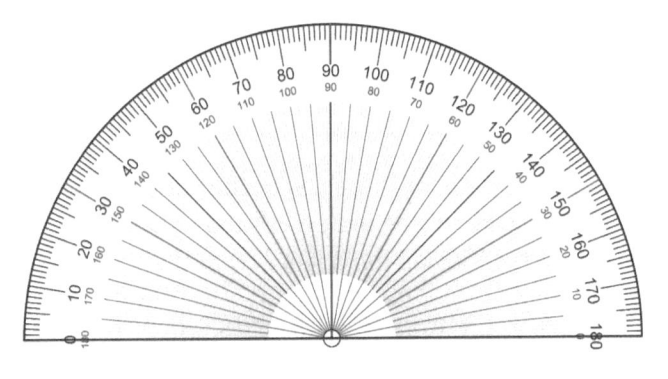

A straight angle (or simply a line) measures exactly 180°. A right angle's sides meet at the vertex to create a square corner. A right angle measures exactly 90° and is typically indicated by a box drawn in the interior of the angle. An acute angle has an interior that is narrower than a right angle. The measure of an acute angle is any value less than 90° and greater than 0°. For example, 89.9°, 47°, 12°, and 1°. An obtuse angle has an interior that is wider than a right angle. The measure of an obtuse angle is any value greater than 90° but less than 180°. For example, 90.1°, 110°, 150°, and 179.9°.

- Acute angles: Less than 90°
- Obtuse angles: Greater than 90°
- Right angles: 90°
- Straight angles: 180°

If two angles add together to give 90°, the angles are **complementary**.

If two angles add together to give 180°, the angles are **supplementary**.

When two lines intersect, the pairs of angles they form are always supplementary. The two angles marked here are supplementary:

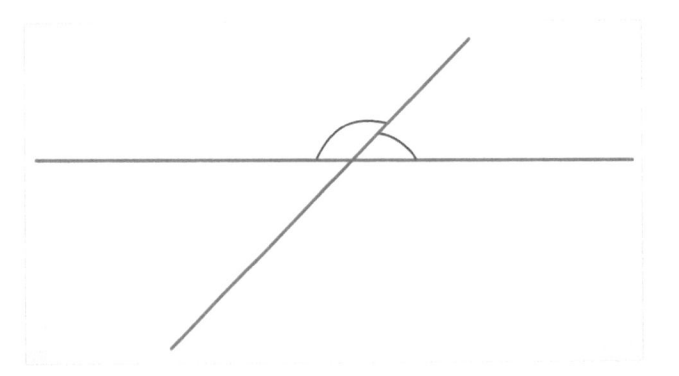

When two supplementary angles are next to one another or "adjacent" in this way, they always give rise to a straight line.

Polygons

A **polygon** is a closed figure (meaning it divides the plane into an inside and an outside) consisting of a collection of at least three line segments between points. These points are called the **vertices** of the

polygon. These line segments must not overlap one another. Note that the number of sides is equal to the number of angles, or vertices of the polygon. The angles between line segments meeting one another in the polygon are called **interior angles**.

A **regular polygon** is a polygon whose edges are all the same length and whose interior angles are all of equal measure. A *triangle* is a polygon with three sides. A **quadrilateral** is a polygon with four sides. Polygons can be classified by the number of sides (also equal to the number of angles) they have.

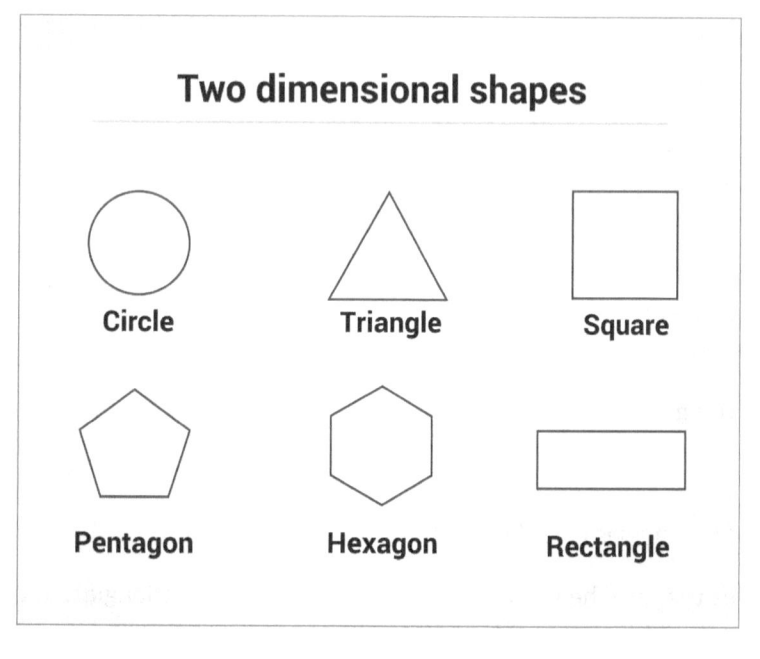

The following are the names of polygons with a given number of sides or angles:

Name of Polygon	# of Sides	Sum of Interior Angles
Triangle	3	180°
Quadrilateral	4	360°
Pentagon	5	540°
Hexagon	6	720°
Septagon (or heptagon)	7	900°
Octagon	8	1080°
Nonagon	9	1260°
Decagon	10	1440°
n-gon	n	(n - 2) * 180°

Triangles can be further classified by their sides and angles. A triangle with its largest angle measuring 90° is a right triangle. A triangle with the largest angle less than 90° is an acute triangle. A triangle with the largest angle greater than 90° is an obtuse triangle. Below is an example of a right triangle.

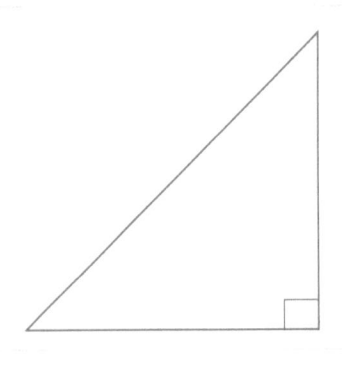

A triangle consisting of two equal sides and two equal angles is an isosceles triangle. A triangle with three equal sides and three equal angles is an equilateral triangle. A triangle with no equal sides or angles is a scalene triangle.

For any triangle, the **Triangle Inequality Theorem** says that the following holds true:

$$A + B > C, A + C > B, B + C > A$$

In addition, the sum of two angles must be less than 180°.

Two triangles with sides that are the same length must also be similar triangles. In this case, such triangles are called **congruent**. Congruent triangles have the same angles and lengths, even if they are rotated from one another.

Quadrilaterals can be further classified according to their sides and angles. A quadrilateral with exactly one pair of parallel sides is called a trapezoid. A **parallelogram** is a quadrilateral in which the opposite sides are parallel and equivalent to each other. Parallelograms include rhombuses, rectangles, and squares. A rhombus has four equal sides. A rectangle has four equal angles (90° each) and two sets of sides that are equal to one another. A square has four 90° angles and four equal sides. Therefore, a square is a regular polygon that is both a rhombus and a rectangle.

Transformations of a Plane

Given a figure drawn on a plane, many changes can be made to that figure, including **rotation**, **translation**, and **reflection**. Rotations turn the figure about a point, translations slide the figure, and reflections flip the figure over a specified line. When performing these transformations, the original figure is called the **pre-image**, and the figure after transformation is called the **image**.

More specifically, **translation** means that all points in the figure are moved in the same direction by the same distance. In other words, the figure is slid in some fixed direction. Of course, while the entire figure is slid by the same distance, this does not change any of the measurements of the figures involved. The result will have the same distances and angles as the original figure.

In terms of Cartesian coordinates, a translation means a shift of each of the original points (x, y) by a fixed amount in the x and y directions, to become $(x + a, y + b)$.

Another procedure that can be performed is called **reflection**. To do this, a line in the plane is specified, called the **line of reflection**. Then, take each point and flip it over the line so that it is the same distance from the line but on the opposite side of it. This does not change any of the distances or angles involved, but it does reverse the order in which everything appears.

To reflect something over the x-axis, the points (x, y) are sent to $(x, -y)$. To reflect something over the y-axis, the points (x, y) are sent to the points $(-x, y)$. Flipping over other lines is not something easy to express in Cartesian coordinates. However, by drawing the figure and the line of reflection, the distance to the line and the original points can be used to find the reflected figure.

Example: Reflect this triangle with vertices (-1, 0), (2, 1), and (2, 0) over the y-axis. The pre-image is shown below.

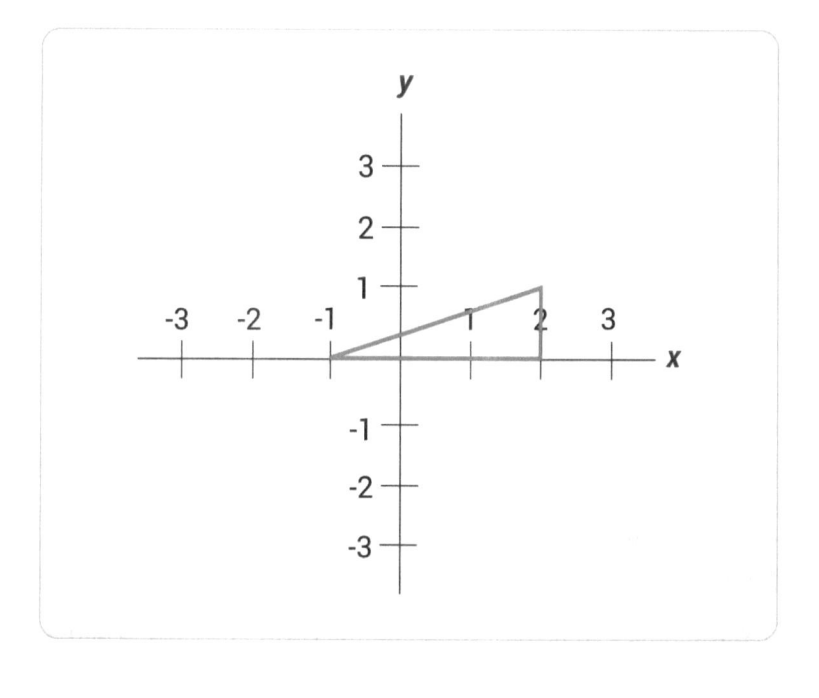

To do this, flip the x values of the points involved to the negatives of themselves, while keeping the y values the same. The image is shown here.

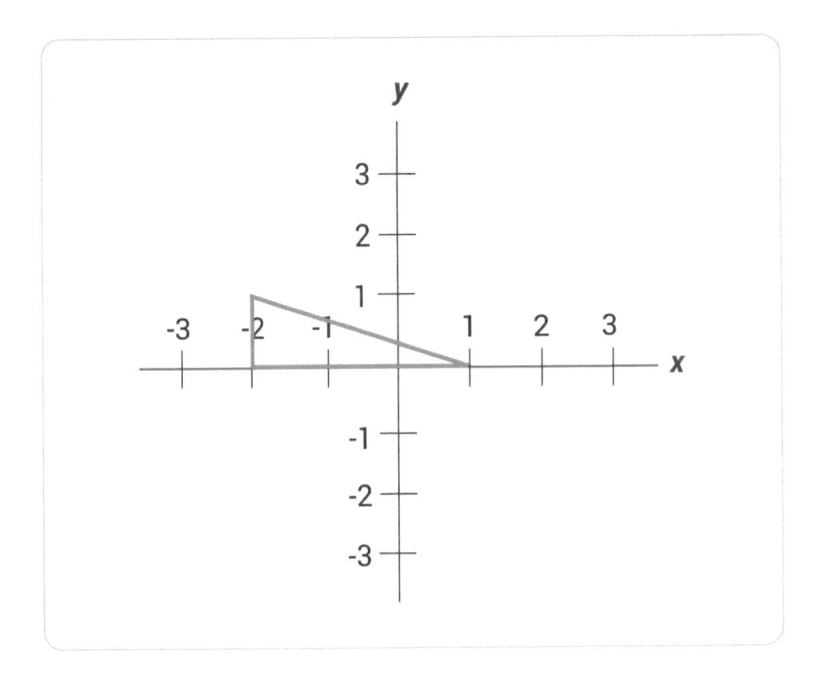

The new vertices will be (1, 0), (-2, 1), and (-2, 0).

Another procedure that does not change the distances and angles in a figure is **rotation**. In this procedure, pick a center point, then rotate every vertex along a circle around that point by the same angle. This procedure is also not easy to express in Cartesian coordinates, and this is not a requirement on this test. However, as with reflections, it's helpful to draw the figures and see what the result of the rotation would look like. This transformation can be performed using a compass and protractor.

Each one of these transformations can be performed on the coordinate plane without changes to the original dimensions or angles.

If two figures in the plane involve the same distances and angles, they are called **congruent figures**. In other words, two figures are congruent when they go from one form to another through reflection, rotation, and translation, or a combination of these.

Remember that rotation and translation will give back a new figure that is identical to the original figure, but reflection will give back a mirror image of it.

To recognize that a figure has undergone a rotation, check to see that the figure has not been changed into a mirror image, but that its orientation has changed (that is, whether the parts of the figure now form different angles with the x and y axes).

To recognize that a figure has undergone a translation, check to see that the figure has not been changed into a mirror image, and that the orientation remains the same.

To recognize that a figure has undergone a reflection, check to see that the new figure is a mirror image of the old figure.

Keep in mind that sometimes a combination of translations, reflections, and rotations may be performed on a figure.

Dilation

A **dilation** is a transformation that preserves angles, but not distances. This can be thought of as stretching or shrinking a figure. If a dilation makes figures larger, it is called an **enlargement**. If a dilation makes figures smaller, it is called a **reduction**. The easiest example is to dilate around the origin. In this case, multiply the x and y coordinates by a **scale factor**, k, sending points (x, y) to (kx, ky).

As an example, draw a dilation of the following triangle, whose vertices will be the points (-1, 0), (1, 0), and (1, 1).

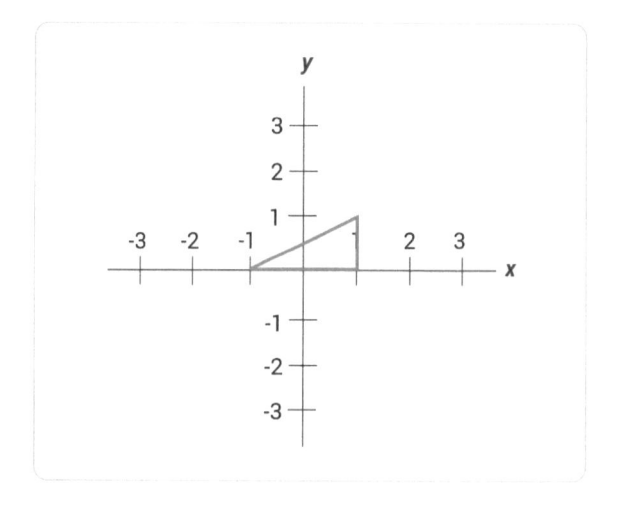

For this problem, dilate by a scale factor of 2, so the new vertices will be (-2, 0), (2, 0), and (2, 2).

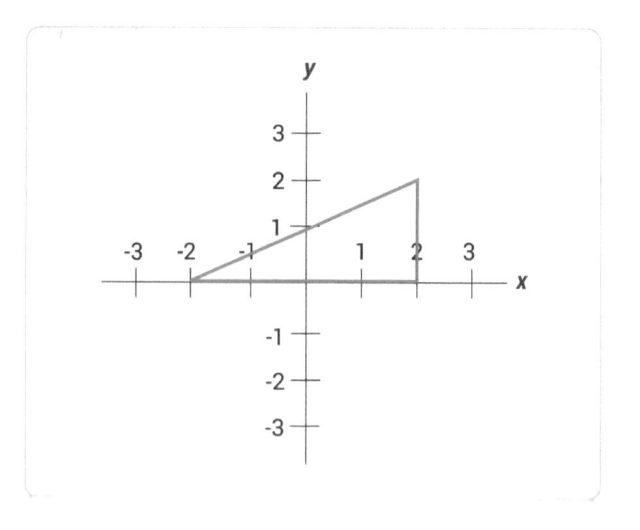

Note that after a dilation, the distances between the vertices of the figure will have changed, but the angles will remain the same. The two figures that are obtained by dilation, along with possibly translation, rotation, and reflection, are all **similar** to one another. Another way to think of this is that

similar figures have the same number of vertices and edges, and their angles are all the same. Similar figures have the same basic shape but are different in size.

Symmetry

Using the types of transformations above, if an object can undergo these changes and not appear to have changed, then the figure is symmetrical. If an object can be split in half by a line and flipped over that line to lie directly on top of itself, it is said to have **line symmetry**. An example of both types of figures is seen below.

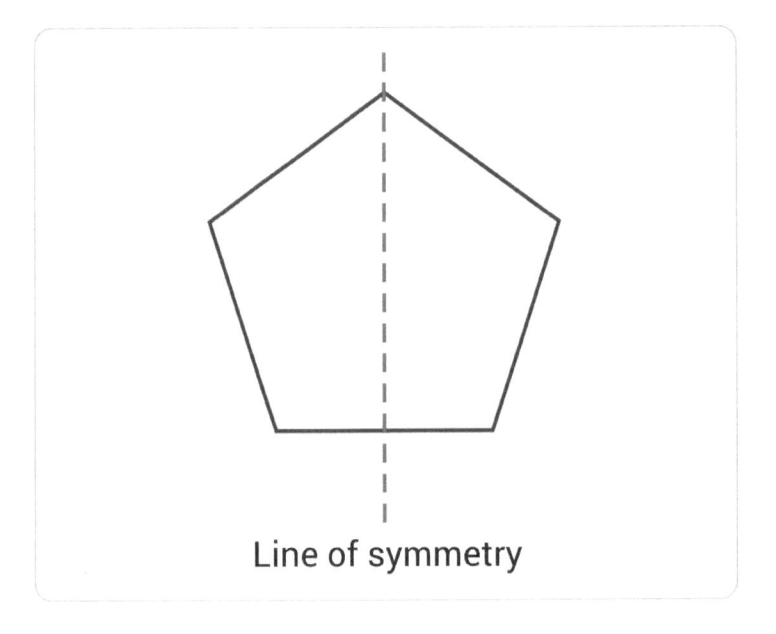

Line of symmetry

If an object can be rotated about its center to any degree smaller than 360, and it lies directly on top of itself, the object is said to have **rotational symmetry**. An example of this type of symmetry is shown below. The pentagon has an order of 5.

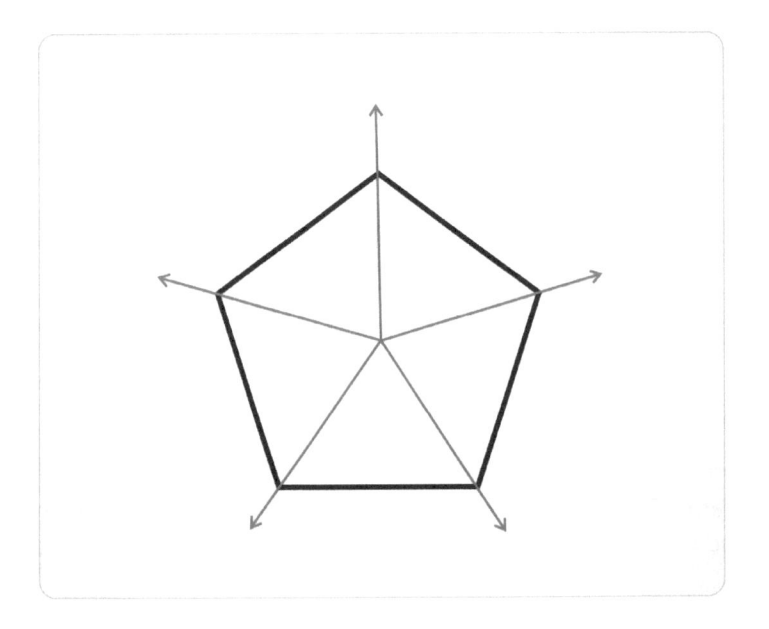

The rotational symmetry lines in the figure above can be used to find the angles formed at the center of the pentagon. Knowing that all of the angles together form a full circle, at 360 degrees, the figure can be split into 5 angles equally. By dividing the 360° by 5, each angle is 72°.

Given the length of one side of the figure, the perimeter of the pentagon can also be found using rotational symmetry. If one side length was 3 cm, that side length can be rotated onto each other side length four times. This would give a total of 5 side lengths equal to 3 cm. To find the perimeter, or distance around the figure, multiply 3 by 5. The perimeter of the figure would be 15 cm.

If a line cannot be drawn anywhere on the object to flip the figure onto itself or rotated less than or equal to 180 degrees to lay on top of itself, the object is asymmetrical. Examples of these types of figures are shown below.

No line of symmetry

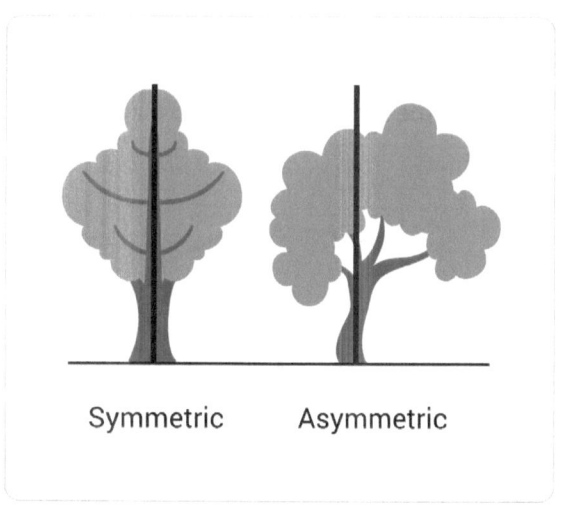

Symmetric   Asymmetric

Perimeters and Areas

The **perimeter** of a polygon is the total length of a trip around the whole polygon, starting and ending at the same point. It is found by adding up the lengths of each line segment in the polygon. Since a square has four equal sides, its perimeter can be calculated by multiplying the length of one side by 4. Thus, the formula is $P = 4 \times s$, where s equals one side. Like a square, a rectangle's perimeter is measured by adding together all of the sides. But as the sides are unequal, the formula is different. A rectangle has equal values for its lengths (long sides) and equal values for its widths (short sides), so the perimeter formula for a rectangle is $P = l + l + w + w = 2l + 2w$, where l is length and w is width. Perimeter is measured in simple units such as inches, feet, yards, centimeters, meters, miles, etc.

Given a pair of points A and B, a circle centered at A and passing through B can be formed. This is the set of points whose distance from A is exactly $d(A, B)$. The radius of this circle will be $d(A, B)$.

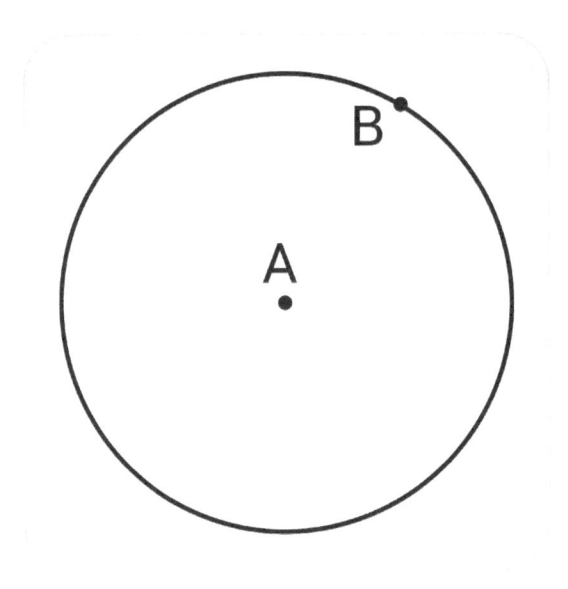

The **circumference** of a circle is the distance traveled by following the edge of the circle for one complete revolution, and the length of the circumference is given by $2\pi r$, where r is the radius of the circle. The circumference can also be considered the perimeter of a circle. The formula for circumference is $C = 2\pi r$.

Given two points on the circumference of a circle, the path along the circle between those points is called an **arc** of the circle. For example, the arc between B and C is denoted by a thinner line:

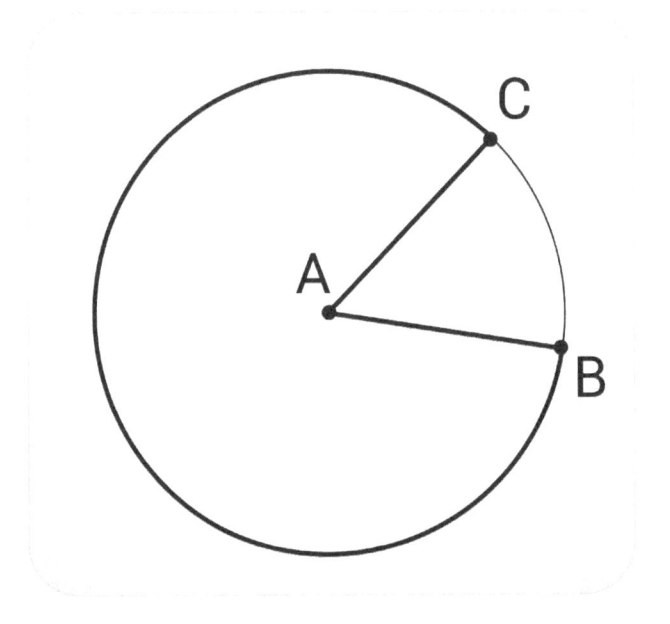

The length of the path along an arc is called the **arc length**. If the circle has radius r, then the arc length is given by multiplying the measure of the angle in radians by the radius of the circle.

The area of a polygon is the area of the region that it encloses. Regarding the area of a rectangle with sides of length x and y, the area is given by xy. For a triangle with a base of length b and a height of length h, the area is $\frac{1}{2}bh$.

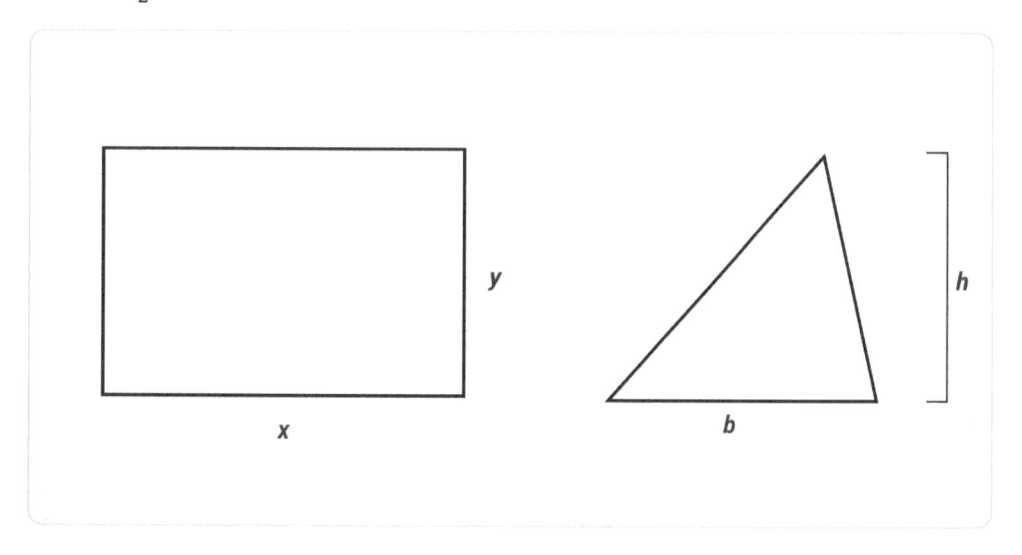

To find the areas of more general polygons, it is usually easiest to break up the polygon into rectangles and triangles. For example, find the area of the following figure whose vertices are (-1, 0), (-1, 2), (1, 3), and (1, 0).

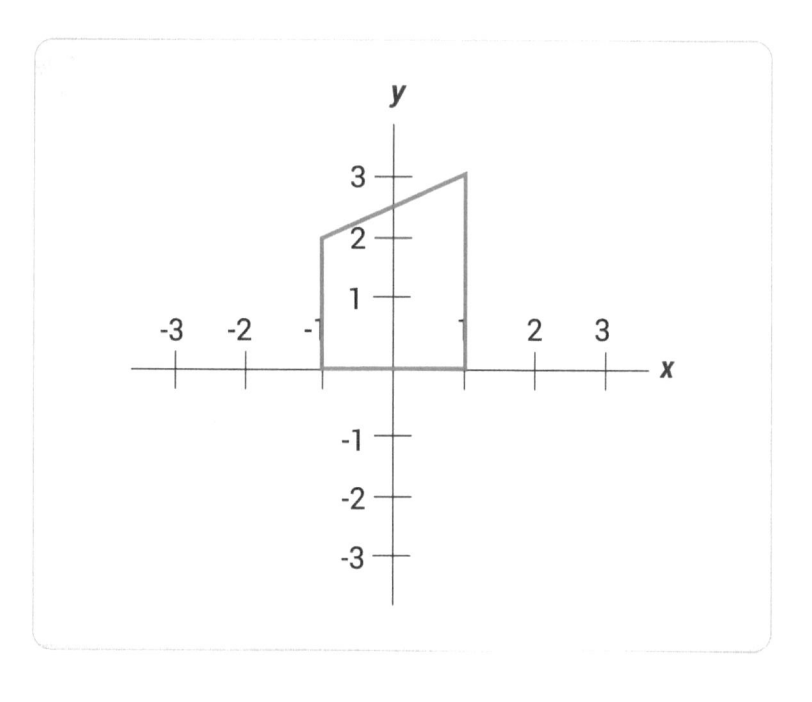

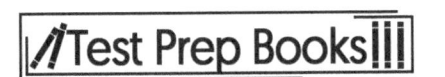

Separate this into a rectangle and a triangle as shown:

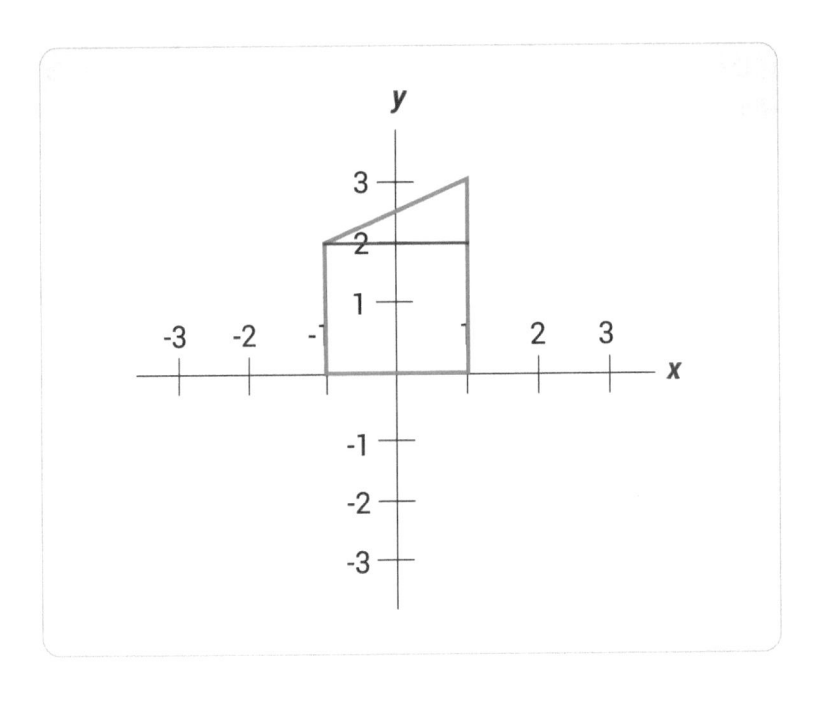

The rectangle has a height of 2 and a width of 2, so it has a total area of $2 \times 2 = 4$. The triangle has a width of 2 and a height of 1, so it has an area of $\frac{1}{2} 2 \times 1 = 1$. Therefore, the entire quadrilateral has an area of $4 + 1 = 5$.

As another example, suppose someone wants to tile a rectangular room that is 10 feet by 6 feet using triangular tiles that are 12 inches by 6 inches. How many tiles would be needed? To figure this, first find the area of the room, which will be $10 \times 6 = 60$ square feet. The dimensions of the triangle are 1 foot by ½ foot, so the area of each triangle is:

$$\frac{1}{2} \times 1 \times \frac{1}{2} = \frac{1}{4} \text{ square feet}$$

Notice that the dimensions of the triangle had to be converted to the same units as the rectangle. Now, take the total area divided by the area of one tile to find the answer:

$$\frac{60}{\frac{1}{4}} = 60 \times 4 = 240 \text{ tiles required}$$

Here are some formulas for the areas of basic planar shapes:

- The area of a rectangle is $l \times w$, where w is the width and l is the length
- The area of a square is s^2, where s is the length of one side (this follows from the formula for rectangles)
- The area of a triangle with base b and height h is $\frac{1}{2} bh$
- The area of a circle with radius r is πr^2

Volumes and Surface Areas

A solid figure, or simply solid, is a figure that encloses a part of space. Some solids consist of flat surfaces only while others include curved surfaces. Solid figures are often defined as three-dimensional shapes. Common three-dimensional shapes include spheres, prisms, cubes, pyramids, cylinders, and cones.

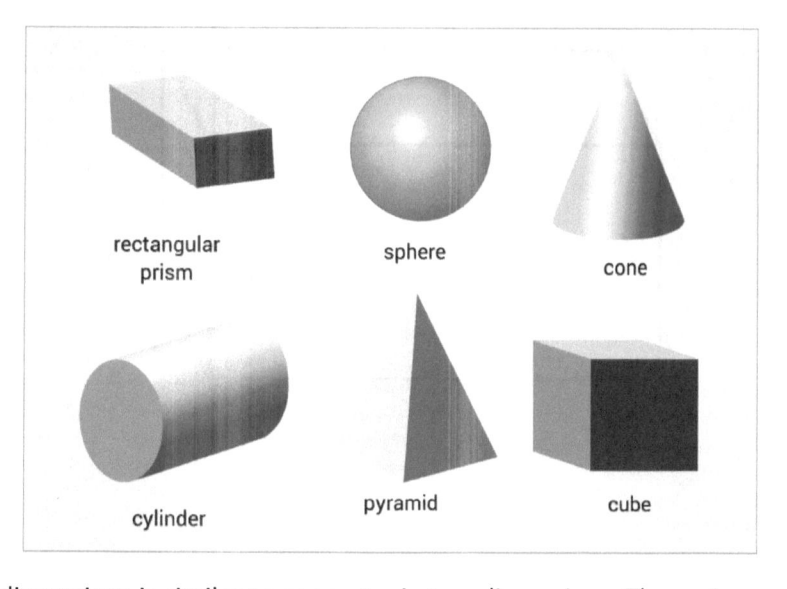

Geometry in three dimensions is similar to geometry in two dimensions. The main new feature is that three points now define a unique **plane** that passes through each of them. Three dimensional objects can be made by putting together two-dimensional figures in different surfaces.

Surface area and volume are two- and three-dimensional measurements. Surface area measures the total surface space of an object, like the six sides of a cube. Questions about surface area will ask how much of something is needed to cover a three-dimensional object, like wrapping a present. **Volume** is the measurement of how much space an object occupies, like how much space is in the cube. Volume questions will ask how much of something is needed to completely fill the object. It is measured in cubic units, such as cubic inches.

A rectangular prism is a box whose sides are all rectangles meeting at 90° angles. Such a box has three dimensions: length, width, and height. If the length is x, the width is y, and the height is z, then the volume is given by $V = xyz$.

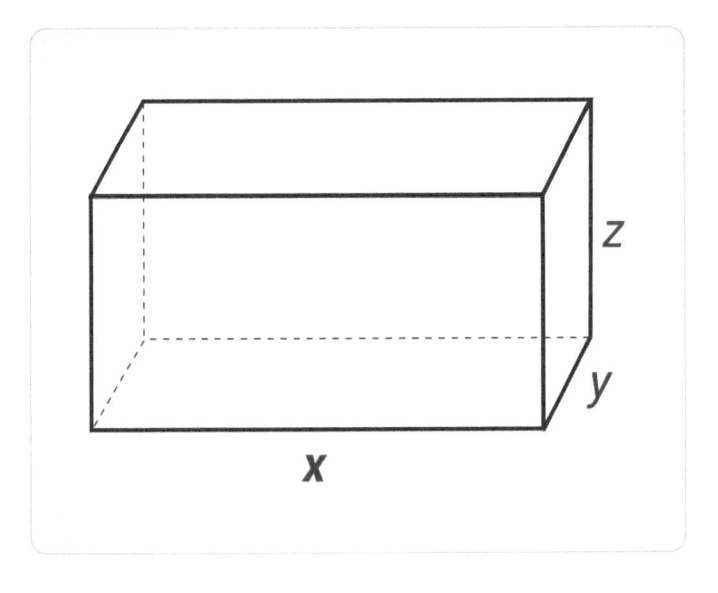

The surface area will be given by computing the surface area of each rectangle and adding them together. There are a total of six rectangles. Two of them have sides of length x and y, two have sides of length y and z, and two have sides of length x and z. Therefore, the total surface area will be given by:

$$SA = 2xy + 2yz + 2xz$$

A cube is similar to a rectangular prism except all six sides are congruent squares. The formula for a cube's surface area is $SA = 6 \times s^2$, where s is the length of a side. A cube has 6 equal sides, so the formula expresses the area of all the sides. Volume is simply measured by taking the cube of the length, so the formula is $V = s^3$.

A **rectangular pyramid** is a figure with a rectangular base and four triangular sides that meet at a single vertex. If the rectangle has sides of length x and y, then the volume will be given by $V = \frac{1}{3}xyh$.

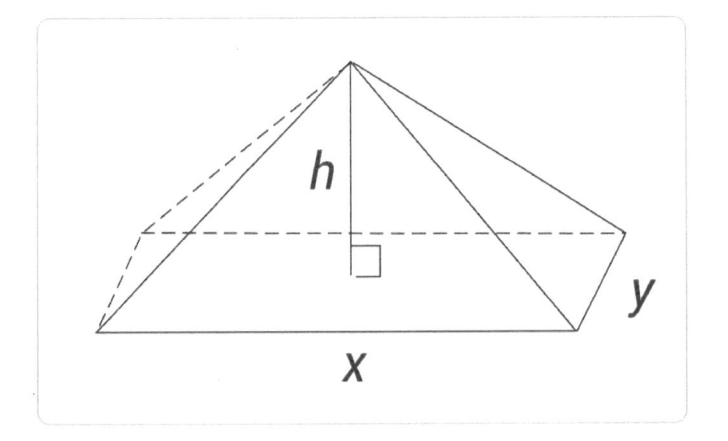

To find the surface area, the dimensions of each triangle need to be known. However, these dimensions can differ depending on the problem in question. Therefore, there is no general formula for calculating total surface area.

A **sphere** is a set of points all of which are equidistant from some central point. It is like a circle, but in three dimensions. The volume of a sphere of radius r is given by $V = \frac{4}{3}\pi r^3$. The surface area is given by $A = 4\pi r^2$.

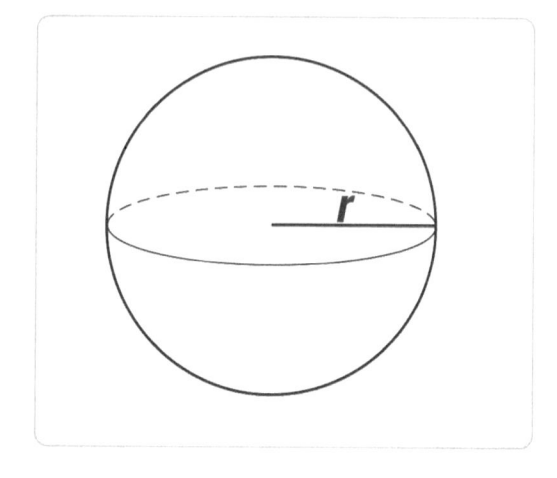

A cylinder consists of two parallel, congruent (same size) circles and a lateral curved surface. The formula to find the surface area of a cylinder is $2\pi rh + 2\pi r^2$. The formula to find the volume of a cylinder is $\pi r^2 h$. These formulas contain the formula for the area of a circle (πr^2) because the base of a cylinder is a circle. To calculate the volume of a cylinder, the slices of circles needed to build the entire height of the cylinder are added together. For example, if the radius is 5 feet and the height of the cylinder is 10 feet, the cylinder's volume is calculated by using the following equation: $\pi 5^2 \times 10$. Substituting 3.14 for π, the volume is 785.4 ft³.

The Pythagorean Theorem

The Pythagorean theorem is an important concept in geometry. It states that for right triangles, the sum of the squares of the two shorter sides will be equal to the square of the longest side (also called the **hypotenuse**). The longest side will always be the side opposite to the 90° angle. If this side is called c, and the other two sides are a and b, then the Pythagorean theorem states that $c^2 = a^2 + b^2$. Since lengths are always positive, this also can be written as:

$$c = \sqrt{a^2 + b^2}$$

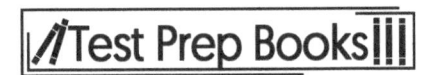

A diagram to show the parts of a triangle using the Pythagorean theorem is below.

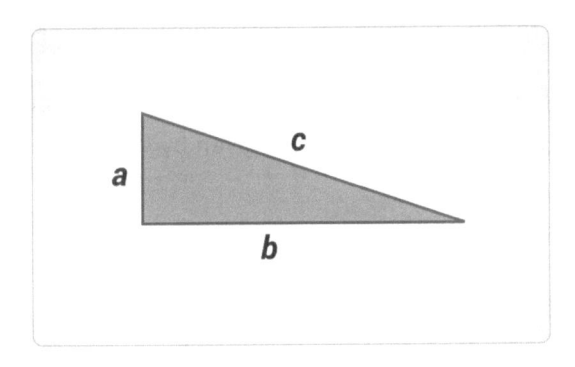

As an example of the theorem, suppose that Shirley has a rectangular field that is 5 feet wide and 12 feet long, and she wants to split it in half using a fence that goes from one corner to the opposite corner. How long will this fence need to be? To figure this out, note that this makes the field into two right triangles, whose hypotenuse will be the fence dividing it in half. Therefore, the fence length will be given by:

$$\sqrt{5^2 + 12^2} = \sqrt{169} = 13 \text{ feet long}$$

Similar Figures and Proportions

Sometimes, two figures are similar, meaning they have the same basic shape and the same interior angles, but they have different dimensions. If the ratio of two corresponding sides is known, then that ratio, or scale factor, holds true for all of the dimensions of the new figure.

Here is an example of applying this principle. Suppose that Lara is 5 feet tall and is standing 30 feet from the base of a light pole, and her shadow is 6 feet long. How high is the light on the pole? To figure this, it helps to make a sketch of the situation:

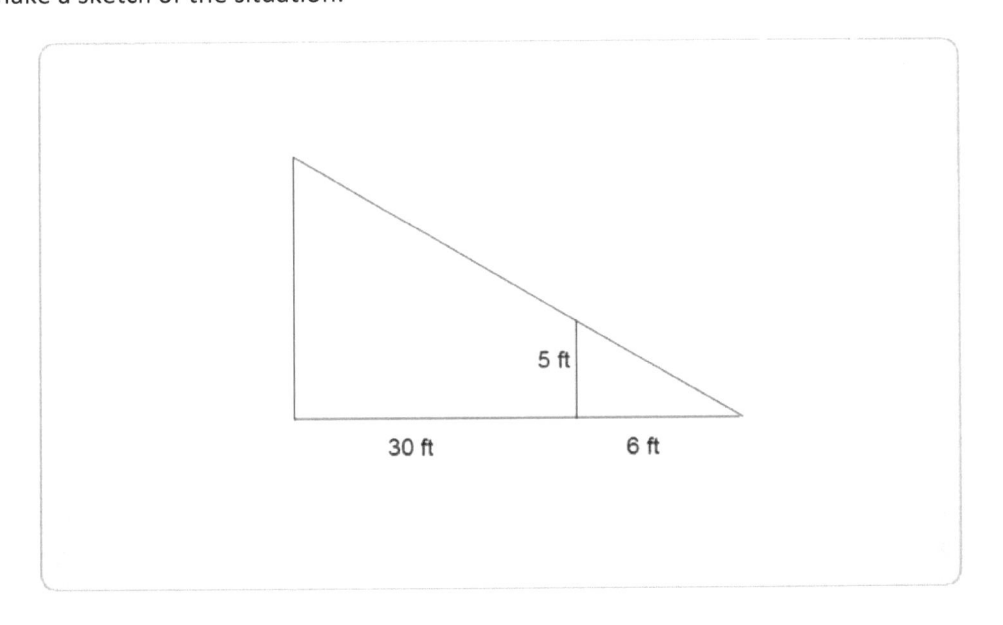

The light pole is the left side of the triangle. Lara is the 5-foot vertical line. Notice that there are two right triangles here, and that they have all the same angles as one another. Therefore, they form similar triangles. So, the ratio of proportionality between them must be determined.

The bases of these triangles are known. The small triangle, formed by Lara and her shadow, has a base of 6 feet. The large triangle formed by the light pole along with the line from the base of the pole out to the end of Lara's shadow is $30 + 6 = 36$ feet long. So, the ratio of the big triangle to the little triangle will be $\frac{36}{6} = 6$. The height of the little triangle is 5 feet. Therefore, the height of the big triangle will be $6 \times 5 = 30$ feet, meaning that the light is 30 feet up the pole.

Notice that the perimeter of a figure changes by the ratio of proportionality between two similar figures, but the area changes by the **square** of the ratio. This is because if the length of one side is doubled, the area is quadrupled.

As an example, suppose two rectangles are similar, but the edges of the second rectangle are three times longer than the edges of the first rectangle. The area of the first rectangle is 10 square inches. How much more area does the second rectangle have than the first?

To answer this, note that the area of the second rectangle is $3^2 = 9$ times the area of the first rectangle, which is 10 square inches. Therefore, the area of the second rectangle is going to be $9 \times 10 = 90$ square inches. This means it has $90 - 10 = 80$ square inches more area than the first rectangle.

As a second example, suppose X and Y are similar right triangles. The hypotenuse of X is 4 inches. The area of Y is $\frac{1}{4}$ the area of X. What is the hypotenuse of Y?

First, realize the area has changed by a factor of $\frac{1}{4}$. The area changes by a factor that is the **square** of the ratio of changes in lengths, so the ratio of the lengths is the square root of the ratio of areas. That means that the ratio of lengths must be $\sqrt{\frac{1}{4}} = \frac{1}{2}$, and the hypotenuse of Y must be:

$$\frac{1}{2} \times 4 = 2 \text{ inches}$$

Volumes between similar solids change like the cube of the change in the lengths of their edges. Likewise, if the ratio of the volumes between similar solids is known, the ratio between their lengths is known by finding the cube root of the ratio of their volumes.

For example, suppose there are two similar rectangular pyramids X and Y. The base of X is 1 inch by 2 inches, and the volume of X is 8 inches. The volume of Y is 64 inches. What are the dimensions of the base of Y?

To answer this, first find the ratio of the volume of Y to the volume of X. This will be given by $\frac{64}{8} = 8$. Now the ratio of lengths is the cube root of the ratio of volumes, or $\sqrt[3]{8} = 2$. So, the dimensions of the base of Y must be 2 inches by 4 inches.

Probability and Statistics

Computing Probabilities

Simple and Compound Events

A **simple event** consists of only one outcome. The most popular simple event is flipping a coin, which results in either heads or tails. A **compound event** results in more than one outcome and consists of more than one simple event. An example of a compound event is flipping a coin while tossing a die. The result is either heads or tails on the coin and a number from one to six on the die. The probability of a simple event is calculated by dividing the number of possible outcomes by the total number of outcomes. Therefore, the probability of obtaining heads on a coin is $\frac{1}{2}$, and the probability of rolling a 6 on a die is $\frac{1}{6}$. The probability of compound events is calculated using the basic idea of the probability of simple events. If the two events are independent, the probability of one outcome is equal to the product of the probabilities of each simple event. For example, the probability of obtaining heads on a coin and rolling a 6 is equal to:

$$\frac{1}{2} \times \frac{1}{6} = \frac{1}{12}$$

The probability of either A or B occurring is equal to the sum of the probabilities minus the probability that both A and B will occur. Therefore, the probability of obtaining either heads on a coin or rolling a 6 on a die is:

$$\frac{1}{2} + \frac{1}{6} - \frac{1}{12} = \frac{7}{12}$$

The two events aren't mutually exclusive because they can happen at the same time. If two events are mutually exclusive, and the probability of both events occurring at the same time is zero, the probability of event A or B occurring equals the sum of both probabilities. An example of calculating the probability of two mutually exclusive events is determining the probability of pulling a king or a queen from a deck of cards. The two events cannot occur at the same time.

Uniform and Non-Uniform Probability Models

A **uniform probability model** is one where each outcome has an equal chance of occurring, such as the probabilities of rolling each side of a die. A **non-uniform probability model** is one where each outcome has an unequal chance of occurring. In a uniform probability model, the conditional probability formulas for $P(B|A)$ and $P(A|B)$ can be multiplied by their respective denominators to obtain two formulas for $P(A \text{ and } B)$. Therefore, the multiplication rule is derived as:

$$P(A \text{ and } B) = P(A)P(B|A) = P(B)P(A|B)$$

In a model, if the probability of either individual event is known and the corresponding conditional probability is known, the multiplication rule allows the probability of the joint occurrence of A and B to be calculated.

Chance Processes and Probability Models

Counting Techniques

There are many counting techniques that can help solve problems involving counting possibilities. For example, the **Addition Principle** states that if there are m choices from Group 1 and n choices from Group 2, then $n + m$ is the total number of choices possible from Groups 1 and 2. For this to be true, the groups can't have any choices in common. The **Multiplication Principle** states that if Process 1 can be completed n ways and Process 2 can be completed m ways, the total number of ways to complete both Process 1 and Process 2 is $n \times m$. For this rule to be used, both processes must be independent of each other. Counting techniques also involve permutations. A **permutation** is an arrangement of elements in a set for which order must be considered. For example, if three letters from the alphabet are chosen, ABC and BAC are two different permutations. The multiplication rule can be used to determine the total number of possibilities. If each letter can't be selected twice, the total number of possibilities is:

$$26 \times 25 \times 24 = 15{,}600$$

A formula can also be used to calculate this total. In general, the notation $P(n, r)$ represents the number of ways to arrange r objects from a set of n and, the formula is:

$$P(n, r) = \frac{n!}{(n - r)!}$$

In the previous example:

$$P(26, 3) = \frac{26!}{23!} = 15{,}600$$

Contrasting permutations, a **combination** is an arrangement of elements in which order doesn't matter. In this case, ABC and BAC are the same combination. In the previous scenario, there are six permutations that represent each single combination. Therefore, the total number of possible combinations is:

$$15{,}600 \div 6 = 2{,}600$$

In general, $C(n, r)$ represents the total number of combinations of n items selected r at a time where order doesn't matter. Another way to represent the combinations of r items selected out of a set of n items is $\binom{n}{r}$. The formula for select combinations of items is:

$$\binom{n}{r} = C(n, r) = \frac{n!}{(n - r)! \, r!}$$

Therefore, the following relationship exists between permutations and combinations:

$$C(n, r) = \frac{P(n, r)}{n!} = \frac{P(n, r)}{P(r, r)}$$

Fundamental Counting Principle

The **fundamental counting principle** states that if there are m possible ways for an event to occur, and n possible ways for a second event to occur, there are $m \cdot n$ possible ways for both events to occur. For example, there are two events that can occur after flipping a coin and six events that can occur after rolling a die, so there are $2 \times 6 = 12$ total possible event scenarios if both are done simultaneously. This

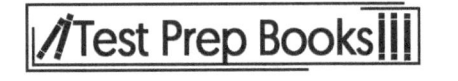

principle can be used to find probabilities involving finite sample spaces and independent trials because it calculates the total number of possible outcomes. For this principle to work, the events must be independent of each other.

Independence and Conditional Probability

Sample Subsets

A sample can be broken up into subsets that are smaller parts of the whole. For example, consider a sample population of females. The sample can be divided into smaller subsets based on the characteristics of each female. There can be a group of females with brown hair and a group of females that wear glasses. There also can be a group of females that have brown hair *and* wear glasses. This "and" relates to the **intersection** of the two separate groups of brunettes and those with glasses. Every female in that intersection group has both characteristics. Similarly, there also can be a group of females that either have brown hair *or* wear glasses. The "or" relates to the union of the two separate groups of brunettes and glasses. Every female in this group has at least one of the characteristics. Finally, the group of females who do not wear glasses can be discussed. This "not" relates to the **complement** of the glass-wearing group. No one in the complement has glasses. **Venn diagrams** are useful in highlighting these ideas. When discussing statistical experiments, this idea can also relate to events instead of characteristics.

Verifying Independent Events

Two events aren't always independent. For example, females with glasses and brown hair aren't independent characteristics. There definitely can be overlap because people with brown hair can wear glasses. Also, two events that exist at the same time don't have to have a relationship. For example, even if everyone in a given sample is wearing glasses, the characteristics aren't related. In this case, the probability of a brunette wearing glasses is equal to the probability of a person being a brunette multiplied by the probability of a person wearing glasses. This mathematical test of $P(A \cap B) = P(A)P(B)$ verifies that two events are independent.

Conditional Probability

Conditional probability is the probability that event A will happen given that event B has already occurred. An example of this is calculating the probability that a person will eat dessert once they have eaten dinner. This is different than calculating the probability of a person just eating dessert. The formula for the conditional probability of event A occurring given B is:

$$P(A|B) = \frac{P(A \text{ and } B)}{P(B)}$$

It's defined as the probability of both *A* and *B* occurring divided by the probability of event B occurring. If A and B are independent, then the probability of both A and B occurring is equal to $P(A)P(B)$, so $P(A|B)$ reduces to just $P(A)$. This means that A and B have no relationship, and the probability of A occurring is the same as the conditional probability of A occurring given B. Similarly, if A and B are independent:

$$P(B|A) = \frac{P(B \text{ and } A)}{P(A)} = P(B)$$

Independent Versus Related Events

To summarize, conditional probability is the probability that an event occurs given that another event has happened. If the two events are related, the probability that the second event will occur changes if

the other event has happened. However, if the two events aren't related and are therefore independent, the first event to occur won't impact the probability of the second event occurring.

Measuring Probabilities with Two-Way Frequency Tables

When measuring event probabilities, two-way frequency tables can be used to report the raw data and then used to calculate probabilities. If the frequency tables are translated into relative frequency tables, the probabilities presented in the table can be plugged directly into the formulas for conditional probabilities. By plugging in the correct frequencies, the data from the table can be used to determine if events are independent or dependent.

Differing Probabilities

The probability that event A occurs differs from the probability that event A occurs given B. When working within a given model, it's important to note the difference. $P(A|B)$ is determined using the formula:

$$P(A|B) = \frac{P(A \text{ and } B)}{P(B)}$$

This represents the total number of A's outcomes left that could occur after B occurs. $P(A)$ can be calculated without any regard for B. For example, the probability of a student finding a parking spot on a busy campus is different once class is in session.

The Addition Rule

The probability of event A or B occurring isn't equal to the sum of each individual probability. The probability that both events can occur at the same time must be subtracted from this total. This idea is shown in the **addition rule**:

$$P(A \text{ or } B) = P(A) + P(B) - P(A \text{ and } B)$$

The addition rule is another way to determine the probability of compound events that aren't mutually exclusive. If the events are mutually exclusive, the probability of both A and B occurring at the same time is 0.

Basic Statistics

The field of statistics describes relationships between quantities that are related, but not necessarily in a deterministic manner. For example, a graduating student's salary will often be higher when the student graduates with a higher GPA, but this is not always the case. Likewise, people who smoke tobacco are more likely to develop lung cancer, but, in fact, it is possible for non-smokers to develop the disease as well. **Statistics** describes these kinds of situations, where the likelihood of some outcome depends on the starting data.

Descriptive statistics involves analyzing a collection of data to describe its broad properties such as average (or mean), what percent of the data falls within a given range, and other such properties. An example of this would be taking all of the test scores from a given class and calculating the average test score. **Inferential statistics** attempts to use data about a subset of some population to make inferences about the rest of the population. An example of this would be taking a collection of students who received tutoring and comparing their results to a collection of students who did not receive tutoring, then using that comparison to try to predict whether the tutoring program in question is beneficial.

To be sure that inferences have a high probability of being true for the whole population, the subset that is analyzed needs to resemble a miniature version of the population as closely as possible. For this reason, statisticians like to choose random samples from the population to study, rather than picking a specific group of people based on some similarity. For example, studying the incomes of people who live in Portland does not reveal anything useful about the incomes of people who live in Tallahassee.

A statistical question is answered by collecting data with variability. Data consists of facts and/or statistics (numbers), and variability refers to a tendency to shift or change. Data is a broad term, inclusive of things like height, favorite color, name, salary, temperature, gas mileage, and language. Questions requiring data as an answer are not necessarily statistical questions. If there is no variability in the data, then the question is not statistical in nature. Consider the following examples: what is Mary's favorite color? How much money does your mother make? What was the highest temperature last week? How many miles did your car get on its last tank of gas? How much taller than Bob is Ed?

None of the above are statistical questions because each case lacks variability in the data needed to answer the question. The questions on favorite color, salary, and gas mileage each require a single piece of data, whether a fact or statistic. Therefore, variability is absent. Although the temperature question requires multiple pieces of data (the high temperature for each day), a single, distinct number is the answer. The height question requires two pieces of data, Bob's height and Ed's height, but no difference in variability exists between those two values. Therefore, this is not a statistical question. Statistical questions typically require calculations with data.

Consider the following statistical questions:

How many miles per gallon of gas does the 2016 Honda Civic get? To answer this question, data must be collected. This data should include miles driven and gallons used. Different cars, different drivers, and different driving conditions will produce different results. Therefore, variability exists in the data. To answer the question, the mean (average) value could be determined.

Are American men taller than German men? To answer this question, data must be collected. This data should include the heights of American men and the heights of German men. All American men are not the same height and all German men are not the same height. Some American men are taller than some German men and some German men are taller than some American men. Therefore, variability exists in the data. To answer the question, the median values for each group could be determined and compared.

The following are more examples of statistical questions: What proportion of 4th graders have a favorite color of blue? How much money do teachers make? Is it colder in Boston or Chicago?

An **experiment** is the method by which a hypothesis is tested using a controlled process called the scientific method. A cause and the effect of that cause are measured, and the hypothesis is accepted or rejected. Experiments are usually completed in a controlled environment where the results of a control population are compared to the results of a test population. The groups are selected using a randomization process in which each group has a representative mix of the population being tested. Finally, an **observational study** is similar to an experiment. However, this design is used when circumstances prevent or do not allow for a designated control group and experimental group (e.g., lack of funding or unrealistic expectations). Instead, existing control and test populations must be used, so this method has a lack of randomization.

Statistics involves making decisions and predictions about larger data sets based on smaller data sets. Basically, the information from one part or subset can help predict what happens in the entire data set or population at large. The entire process involves guessing, and the predictions and decisions may not be 100 percent correct all of the time; however, there is some truth to these predictions, and the decisions do have mathematical support. The smaller data set is called a **sample** and the larger data set (in which the decision is being made) is called a **population**. A **random sample** is used as the sample, which is an unbiased collection of data points that represents the population as well as it can. There are many methods of forming a random sample, and all adhere to the fact that every potential data point has a predetermined probability of being chosen.

Describing Distributions

Mean, Median, and Mode

The center of a set of data (statistical values) can be represented by its mean, median, or mode. These are sometimes referred to as measures of central tendency.

Mean

The first property that can be defined for this set of data is the **mean**. This is the same as the average. To find the mean, add up all the data points, then divide by the total number of data points. For example, suppose that in a class of 10 students, the scores on a test were 50, 60, 65, 65, 75, 80, 85, 85, 90, 100. Therefore, the average test score will be:

$$\frac{50 + 60 + 65 + 65 + 75 + 80 + 85 + 85 + 90 + 100}{10} = 75.5$$

The mean is a useful number if the distribution of data is normal (more on this later), which roughly means that the frequency of different outcomes has a single peak and is roughly equally distributed on both sides of that peak. However, it is less useful in some cases where the data might be split or where there are some outliers. **Outliers** are data points that are far from the rest of the data. For example, suppose there are 10 executives and 90 employees at a company. The executives make $1000 per hour, and the employees make $10 per hour.

Therefore, the average pay rate will be:

$$\frac{\$1000 \times 10 + \$10 \times 90}{100} = \$109 \text{ per hour}$$

In this case, this average is not very descriptive since it's not close to the actual pay of the executives *or* the employees.

Median

Another useful measurement is the **median**. In a data set, the median is the point in the middle. The middle refers to the point where half the data comes before it and half comes after, when the data is recorded in numerical order. For instance, these are the speeds of the fastball of a pitcher during the last inning that he pitched (in order from least to greatest):

90, 92, 93, 93, 95, 96, 97, 97, 97

There are nine total numbers, so the middle or *median* number is the 5th one, which is 95.

In cases where the number of data points is an even number, then the average of the two middle points is taken. In the previous example of test scores, the two middle points are 75 and 80. Since there is no single point, the average of these two scores needs to be found. The average is:

$$\frac{75 + 80}{2} = 77.5$$

The median is generally a good value to use if there are a few outliers in the data. It prevents those outliers from affecting the "middle" value as much as when using the mean.

Since an outlier is a data point that is far from most of the other data points in a data set, this means an outlier also is any point that is far from the median of the data set. The outliers can have a substantial effect on the mean of a data set, but they usually do not change the median or mode, or do not change them by a large quantity. For example, consider the data set (3, 5, 6, 6, 6, 8). This has a median of 6 and a mode of 6, with a mean of $\frac{34}{6} \approx 5.67$. Now, suppose a new data point of 1000 is added so that the data set is now (3, 5, 6, 6, 6, 8, 1000). The median and mode, which are both still 6, remain unchanged. However, the average is now $\frac{1034}{7}$, which is approximately 147.7. In this case, the median and mode will be better descriptions for most of the data points.

Outliers in a given data set are sometimes the result of an error by the experimenter, but oftentimes, they are perfectly valid data points that must be taken into consideration.

Mode
One additional measure to define for X is the **mode**. This is the data point that appears most frequently. If two or more data points all tie for the most frequent appearance, then each of them is considered a mode. In the case of the test scores, where the numbers were 50, 60, 65, 65, 75, 80, 85, 85, 90, 100, there are two modes: 65 and 85.

Quartiles and Percentiles
The **first quartile** of a set of data X refers to the largest value from the first ¼ of the data points. In practice, there are sometimes slightly different definitions that can be used, such as the median of the first half of the data points (excluding the median itself if there are an odd number of data points). The term also has a slightly different use: when it is said that a data point lies *in the first quartile*, it means it is less than or equal to the median of the first half of the data points. Conversely, if it lies *at* the first quartile, then it is equal to the first quartile.

When it is said that a data point lies in the **second quartile**, it means it is between the first quartile and the median.

The **third quartile** refers to data that lies between ½ and ¾ of the way through the data set. Again, there are various methods for defining this precisely, but the simplest way is to include all of the data that lie between the median and the median of the top half of the data.

Data that lies in the **fourth quartile** refers to all of the data above the third quartile.

Percentiles may be defined in a similar manner to quartiles. Generally, this is defined in the following manner:

If a data point lies **in the n-th percentile**, this means it lies in the range of the first n% of the data.

If a data point lies **at the n-th percentile**, then it means that $n\%$ of the data lies below this data point.

Standard Deviation

Given a data set X consisting of data points $(x_1, x_2, x_3, \ldots x_n)$, the **variance** of X is defined to be:

$$\frac{\sum_{i=1}^{n}(x_i - \bar{X})^2}{n}$$

This means that the variance of X is the average of the squares of the differences between each data point and the mean of X.

Given a data set X consisting of data points $(x_1, x_2, x_3, \ldots x_n)$, the **standard deviation** of X is defined to be:

$$s_x = \sqrt{\frac{\sum_{i=1}^{n}(x_i - \bar{X})^2}{n}}$$

In other words, the standard deviation is the square root of the variance.

Both the variance and the standard deviation are measures of how much the data tend to be spread out. When the standard deviation is low, the data points are mostly clustered around the mean. When the standard deviation is high, this generally indicates that the data are quite spread out, or else that there are a few substantial outliers.

As a simple example, compute the standard deviation for the data set (1, 3, 3, 5). First, compute the mean, which will be:

$$\frac{1 + 3 + 3 + 5}{4} = \frac{12}{4} = 3$$

Now, find the variance of X with the formula:

$$\sum_{i=1}^{4}(x_i - \bar{X})^2 = (1 - 3)^2 + (3 - 3)^2 + (3 - 3)^2 + (5 - 3)^2$$

$$-2^2 + 0^2 + 0^2 + 2^2 = 8$$

Therefore, the variance is $\frac{8}{4} = 2$. Taking the square root, the standard deviation will be $\sqrt{2}$.

Note that the standard deviation only depends upon the mean, not upon the median or mode(s). Generally, if there are multiple modes that are far apart from one another, the standard deviation will be high. A high standard deviation does not always mean there are multiple modes, however.

Describing a Set of Data

A set of data can be described in terms of its center, spread, shape and any unusual features. The center of a data set can be measured by its mean, median, or mode. The spread of a data set refers to how far the data points are from the center (mean or median). A data set with all its data points clustered around the center will have a small spread. A data set covering a wide range of values will have a large spread.

When a data set is displayed as a graph like the one below, the shape indicates if a sample is normally distributed, symmetrical, or has measures of skewness. When graphed, a data set with a normal distribution will resemble a bell curve.

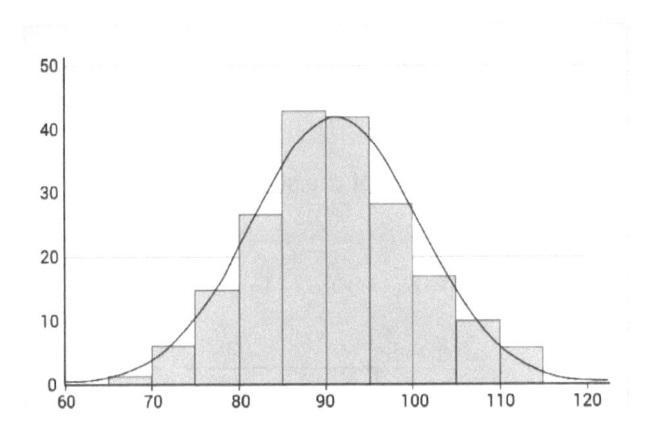

If the data set is symmetrical, each half of the graph when divided at the center is a mirror image of the other. If the graph has fewer data points to the right, the data is skewed right. If it has fewer data points to the left, the data is skewed left.

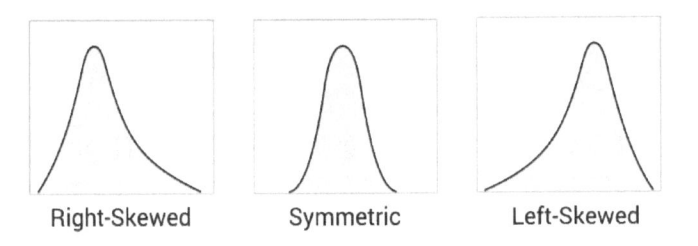

Right-Skewed   Symmetric   Left-Skewed

A description of a data set should include any unusual features such as gaps or outliers. A gap is a span within the range of the data set containing no data points. An outlier is a data point with a value either extremely large or extremely small when compared to the other values in the set.

The graphs above can be referred to as **unimodal** since they all have a single peak. In contrast, a bimodal graph has two peaks.

Practice Quiz

1. $4\frac{1}{3} + 3\frac{3}{4} =$
   a. $6\frac{5}{12}$
   b. $8\frac{1}{12}$
   c. $8\frac{2}{3}$
   d. $7\frac{7}{12}$

2. For which of the following are $x = 4$ and $x = -4$ solutions?
   a. $x^2 + 16 = 0$
   b. $x^2 + 4x - 4 = 0$
   c. $x^2 - 2x - 2 = 0$
   d. $x^2 - 16 = 0$

3. The total perimeter of a rectangle is 36 cm. If the length is 12 cm, what is the width?
   a. 3 cm
   b. 12 cm
   c. 6 cm
   d. 8 cm

4. A six-sided die is rolled. What is the probability that the roll is 1 or 2?
   a. $\frac{1}{6}$
   b. $\frac{1}{4}$
   c. $\frac{1}{3}$
   d. $\frac{1}{2}$

5. Dwayne has received the following scores on his math tests: 78, 92, 83, 97. What score must Dwayne get on his next math test to have an overall average of 90?
   a. 89
   b. 98
   c. 95
   d. 100

Answer Explanations

1. B: $4\frac{1}{3} + 3\frac{3}{4} = 4 + 3 + \frac{1}{3} + \frac{3}{4} = 7 + \frac{1}{3} + \frac{3}{4}$.

Adding the fractions gives:

$$\frac{1}{3} + \frac{3}{4} = \frac{4}{12} + \frac{9}{12} = \frac{13}{12} = 1 + \frac{1}{12}$$

Thus,

$$7 + \frac{1}{3} + \frac{3}{4} = 7 + 1 + \frac{1}{12} = 8\frac{1}{12}$$

2. D: Each value can be substituted into each equation. Choice A can be eliminated, since $4^2 + 16 = 32$. Choice B can be eliminated, since $4^2 + 4 \times 4 - 4 = 28$. C can be eliminated, since $4^2 - 2 \times 4 - 2 = 6$. But, plugging in either value into $x^2 - 16$, which gives:

$$(\pm 4)^2 - 16 = 16 - 16 = 0$$

3. C: The formula for the perimeter of a rectangle is $P = 2L + 2W$, where P is the perimeter, L is the length, and W is the width. The first step is to substitute all of the data into the formula:

$$36 = 2(12) + 2W$$

Simplify by multiplying 2×12:

$$36 = 24 + 2W$$

Simplifying this further by subtracting 24 on each side, which gives:

$$36 - 24 = 24 - 24 + 2W$$

$$12 = 2W$$

Divide by 2:

$$6 = W$$

The width is 6 cm. Remember to test this answer by substituting this value into the original formula:

$$36 = 2(12) + 2(6)$$

4. C: A die has an equal chance for each outcome. Since it has six sides, each outcome has a probability of $\frac{1}{6}$. The chance of a 1 or a 2 is therefore:

$$\frac{1}{6} + \frac{1}{6} = \frac{1}{3}$$

5. D: To find the average of a set of values, add the values together and then divide by the total number of values. In this case, include the unknown value of what Dwayne needs to score on his next test, in order to solve it.

$$\frac{78 + 92 + 83 + 97 + x}{5} = 90$$

Add the unknown value to the new average total, which is 5. Then multiply each side by 5 to simplify the equation, resulting in:

$$78 + 92 + 83 + 97 + x = 450$$

$$350 + x = 450$$

$$x = 100$$

Dwayne would need to get a perfect score of 100 in order to get an average of at least 90. Test this answer by substituting back into the original formula.

$$\frac{78 + 92 + 83 + 97 + 100}{5} = 90$$

Reading Comprehension

Literary Analysis

Style, Tone, and Mood

Style, tone, and mood are often thought to be the same thing. Though they're closely related, there are important differences to keep in mind. The easiest way to do this is to remember that style "creates and affects" tone and mood. More specifically, style is how the writer uses words to create the desired tone and mood for their writing.

Style

Style can include any number of technical writing choices. A few examples of style choices include:

- Sentence Construction: When presenting facts, does the writer use shorter sentences to create a quicker sense of the supporting evidence, or do they use longer sentences to elaborate and explain the information?

- Technical Language: Does the writer use jargon to demonstrate their expertise in the subject, or do they use ordinary language to help the reader understand things in simple terms?

- Formal Language: Does the writer refrain from using contractions such as *won't* or *can't* to create a more formal tone, or do they use a colloquial, conversational style to connect to the reader?

- Formatting: Does the writer use a series of shorter paragraphs to help the reader follow a line of argument, or do they use longer paragraphs to examine an issue in great detail and demonstrate their knowledge of the topic?

On the test, examine the writer's style and how their writing choices affect the way the text comes across.

Tone

Tone refers to the writer's attitude toward the subject matter. Tone is usually explained in terms of a work of fiction. For example, the tone conveys how the writer feels about their characters and the situations in which they're involved. Nonfiction writing is sometimes thought to have no tone at all; however, this is incorrect.

A lot of nonfiction writing has a neutral tone, which is an important tone for the writer to take. A neutral tone demonstrates that the writer is presenting a topic impartially and letting the information speak for itself. On the other hand, nonfiction writing can be just as effective and appropriate if the tone isn't neutral. For instance, take this example involving seat belts:

> Seat belts save more lives than any other automobile safety feature. Many studies show that airbags save lives as well; however, not all cars have airbags. For instance, some older cars don't. Furthermore, air bags aren't entirely reliable. For example, studies show that in 15% of accidents airbags don't deploy as designed, but, on the other hand, seat belt malfunctions are

extremely rare. The number of highway fatalities has plummeted since laws requiring seat belt usage were enacted.

In this passage, the writer mostly chooses to retain a neutral tone when presenting information. If the writer would instead include their own personal experience of losing a friend or family member in a car accident, the tone would change dramatically. The tone would no longer be neutral and would show that the writer has a personal stake in the content, allowing them to interpret the information in a different way. When analyzing tone, consider what the writer is trying to achieve in the text and how they *create* the tone using style.

Mood

Mood refers to the feelings and atmosphere that the writer's words create for the reader. Like tone, many nonfiction texts can have a neutral mood. To return to the previous example, if the writer would choose to include information about a person they know being killed in a car accident, the text would suddenly carry an emotional component that is absent in the previous example. Depending on how they present the information, the writer can create a sad, angry, or even hopeful mood. When analyzing the mood, consider what the writer wants to accomplish and whether the best choice was made to achieve that end.

Consistency

Whatever style, tone, and mood the writer uses, good writing should remain consistent throughout. If the writer chooses to include the tragic, personal experience above, it would affect the style, tone, and mood of the entire text. It would seem out of place for such an example to be used in the middle of a neutral, measured, and analytical text. To adjust the rest of the text, the writer needs to make additional choices to remain consistent. For example, the writer might decide to use the word *tragedy* in place of the more neutral *fatality*, or they could describe a series of car-related deaths as an *epidemic*. Adverbs and adjectives such as *devastating* or *horribly* could be included to maintain this consistent attitude toward the content. When analyzing writing, look for sudden shifts in style, tone, and mood, and consider whether the writer would be wiser to maintain the prevailing strategy.

Identify the Position and Purpose

When it comes to an author's writing, readers should always identify a position or stance. No matter how objective a text may seem, readers should assume the author has preconceived beliefs. One can reduce the likelihood of accepting an invalid argument by looking for multiple articles on the topic, including those with varying opinions. If several opinions point in the same direction and are backed by reputable peer-reviewed sources, it's more likely the author has a valid argument. Positions that run contrary to widely held beliefs and existing data should invite scrutiny. There are exceptions to the rule, so be a careful consumer of information.

Though themes, symbols, and motifs are buried deep within the text and can sometimes be difficult to infer, an author's purpose is usually obvious from the beginning. There are four purposes of writing: to inform, to persuade, to describe, and to entertain. Informative writing presents facts in an accessible way. Persuasive writing appeals to emotions and logic to inspire the reader to adopt a specific stance. Be wary of this type of writing, as it can mask a lack of objectivity with powerful emotion. Descriptive writing is designed to paint a picture in the reader's mind, while texts that entertain are often narratives designed to engage and delight the reader.

The various writing styles are usually blended, with one purpose dominating the rest. A persuasive text, for example, might begin with a humorous tale to make readers more receptive to the persuasive message, or a recipe in a cookbook designed to inform might be preceded by an entertaining anecdote that makes the recipes more appealing.

Identify Passage Characteristics

Writing can be classified under four passage types: narrative, expository, descriptive (sometimes called technical), and persuasive. Though these types are not mutually exclusive, one form tends to dominate the rest. By recognizing the *type* of passage you're reading, you gain insight into *how* you should read. When reading a narrative intended to entertain, sometimes you can read more quickly through the passage if the details are discernible. A technical document, on the other hand, might require a close read because skimming the passage might cause the reader to miss salient details.

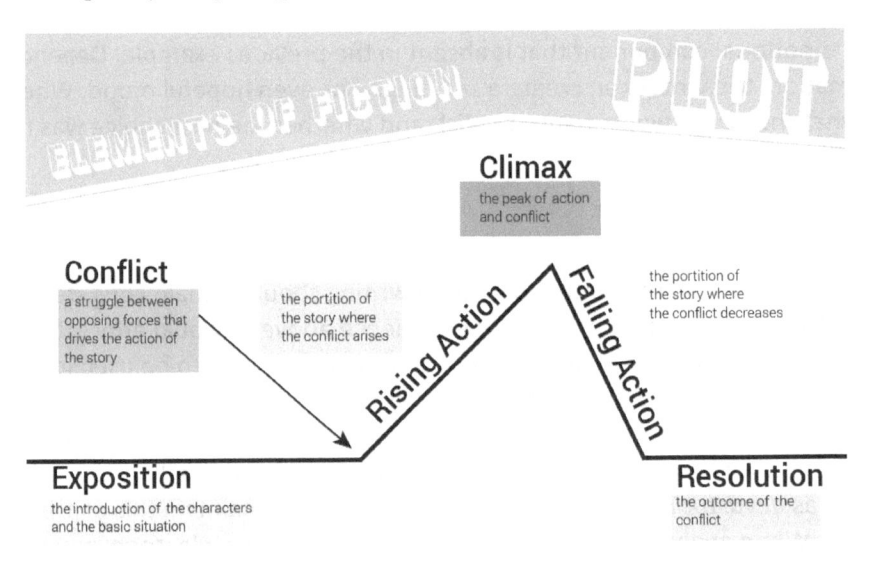

1. Narrative writing, at its core, is the art of storytelling. For a narrative to exist, certain elements must be present. First, it must have characters. While many characters are human, characters could be defined as anything that thinks, acts, and talks like a human. For example, many recent movies, such as *Lord of the Rings* and *The Chronicles of Narnia*, include animals, fantasy creatures, and even trees that behave like humans. Narratives also must have a plot or sequence of events. Typically, those events follow a standard plot diagram, but recent trends start *in medias res* or in the middle (nearer the climax). In this instance, foreshadowing and flashbacks often fill in plot details. Finally, along with characters and a plot, there must also be conflict. Conflict is usually divided into two types: internal and external. Internal conflict indicates the character is in turmoil. Think of an angel on one shoulder and the devil on the other, arguing it out. Internal conflicts are presented through the character's thoughts. External conflicts are visible. Types of external conflict include person versus person, person versus nature, person versus technology, person versus the supernatural, or person versus fate.

2. Expository writing is detached and to the point, while other types of writing — persuasive, narrative, and descriptive — are livelier. Since expository writing is designed to instruct or inform, it usually involves directions and steps written in second person ("you" voice) and lacks any persuasive or narrative elements. Sequence words such as *first*, *second*, and *third*, or *in the first place*, *secondly*, and

lastly are often given to add fluency and cohesion. Common examples of expository writing include instructor's lessons, cookbook recipes, and repair manuals.

3. Due to its empirical nature, technical writing is filled with steps, charts, graphs, data, and statistics. The goal of technical writing is to advance understanding in a field through the scientific method. Experts such as teachers, doctors, or mechanics use words unique to the profession in which they operate. These words, which often incorporate acronyms, are called *jargon*. Technical writing is a type of expository writing but is not meant to be understood by the general public. Instead, technical writers assume readers have received a formal education in a particular field of study and need no explanation as to what the jargon means. Imagine a doctor trying to understand a diagnostic reading for a car or a mechanic trying to interpret lab results. Only professionals with proper training will fully comprehend the text.

4. Persuasive writing is designed to change opinions and attitudes. The topic, stance, and arguments are found in the thesis, positioned near the end of the introduction. Later supporting paragraphs offer relevant quotations, paraphrases, and summaries from primary or secondary sources, which are then interpreted, analyzed, and evaluated. The goal of persuasive writers is not to stack quotes but to develop original ideas by using sources as a starting point. Good persuasive writing makes powerful arguments with valid sources and thoughtful analysis. Poor persuasive writing is riddled with bias and logical fallacies. Sometimes logical and illogical arguments are sandwiched together in the same text. Therefore, readers should display skepticism when reading persuasive arguments.

Interpret Influences of Historical Context

Studying historical literature is fascinating. It reveals a snapshot in time of people, places, and cultures; a collective set of beliefs and attitudes that no longer exist. Writing changes as attitudes and cultures evolve. Beliefs previously considered immoral or wrong may be considered acceptable today. Researching the historical period of an author gives the reader perspective. The dialogue in Jane Austen's *Pride and Prejudice*, for example, is indicative of social class during the Regency era. Similarly, the stereotypes and slurs in *The Adventures of Huckleberry Finn* were a result of common attitudes and beliefs in the late 1800s, attitudes now found to be reprehensible.

Recognizing Cultural Themes

Regardless of culture, place, or time, certain themes are universal to the human condition. Because humans experience joy, rage, jealousy, and pride, certain themes span centuries. For example, Shakespeare's *Macbeth,* as well as modern works like *The 50th Law* by rapper 50 Cent and Robert Greene or the Netflix series *House of Cards* all feature characters who commit atrocious acts because of ambition. Similarly, *The Adventures of Huckleberry Finn*, published in the 1880s, and *The Catcher in the Rye*, published in the 1950s, both have characters who lie, connive, and survive on their wits.

Moviegoers know whether they are seeing an action, romance or horror film, and are often disappointed if the movie doesn't fit into the conventions of a particular category. Similarly, categories or genres give readers a sense of what to expect from a text. Some of the most basic genres in literature include books, short stories, poetry, and drama. Many genres can be split into sub-genres. For example, the sub-genres of historical fiction, realistic fiction, and fantasy all fit under the fiction genre.

Each genre has a unique way of approaching a particular theme. Books and short stories use plot, characterization, and setting, while poems rely on figurative language, sound devices, and symbolism. Dramas reveal plot through dialogue and the actor's voice and body language.

Paragraph Comprehension

Topic Versus the Main Idea

It is very important to know the difference between the topic and the main idea of the text. Even though these two are similar because they both present the central point of a text, they have distinctive differences. A *topic* is the subject of the text; it can usually be described in a one- to two-word phrase and appears in the simplest form. On the other hand, the *main idea* is more detailed and provides the author's central point of the text. It can be expressed through a complete sentence and can be found in the beginning, middle, or end of a paragraph. In most nonfiction books, the first sentence of the passage usually (but not always) states the main idea. Take a look at the passage below to review the topic versus the main idea.

Cheetahs

Cheetahs are one of the fastest mammals on land, reaching up to 70 miles an hour over short distances. Even though cheetahs can run as fast as 70 miles an hour, they usually only have to run half that speed to catch up with their choice of prey. Cheetahs cannot maintain a fast pace over long periods of time because they will overheat their bodies. After a chase, cheetahs need to rest for approximately 30 minutes prior to eating or returning to any other activity.

In the example above, the topic of the passage is "Cheetahs" simply because that is the subject of the text. The main idea of the text is "Cheetahs are one of the fastest mammals on land but can only maintain this fast pace for short distances." While it covers the topic, it is more detailed and refers to the text in its entirety. The text continues to provide additional details called *supporting details,* which will be discussed in the next section.

Supporting Details

Supporting details help readers better develop and understand the main idea. Supporting details answer questions like *who, what, where, when, why,* and *how.* Different types of supporting details include examples, facts and statistics, anecdotes, and sensory details.

Persuasive and informative texts often use supporting details. In persuasive texts, authors attempt to make readers agree with their point of view, and supporting details are often used as "selling points." If authors make a statement, they should support the statement with evidence in order to adequately persuade readers. Informative texts use supporting details such as examples and facts to inform readers. Take another look at the previous "Cheetahs" passage to find examples of supporting details.

Cheetahs

Cheetahs are one of the fastest mammals on land, reaching up to 70 miles an hour over short distances. Even though cheetahs can run as fast as 70 miles an hour, they usually only have to run half that speed to catch up with their choice of prey. Cheetahs cannot maintain a fast pace over long periods of time because they will overheat their bodies. After a chase, cheetahs need to rest for approximately 30 minutes prior to eating or returning to any other activity.

In the example above, supporting details include:

- Cheetahs reach up to 70 miles per hour over short distances.
- They usually only have to run half that speed to catch up with their prey.
- Cheetahs will overheat their bodies if they exert a high speed over longer distances.
- Cheetahs need to rest for 30 minutes after a chase.

Look at the diagram below (applying the cheetah example) to help determine the hierarchy of topic, main idea, and supporting details.

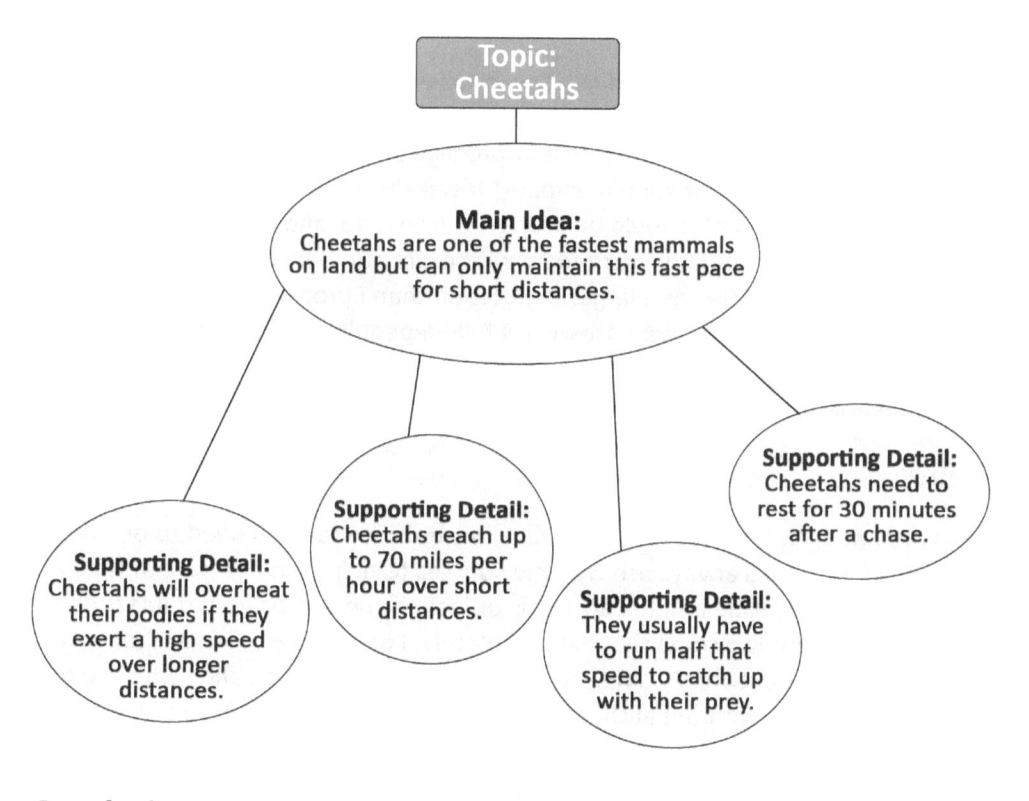

Drawing Conclusions

When drawing conclusions about texts or passages, readers should do two main things: 1) Use the information that they already know and 2) Use the information they have learned from the text or passage. Authors write with an intended purpose, and it is the reader's responsibility to understand and form logical conclusions of authors' ideas. It is important to remember that the reader's conclusions should be supported by information directly from the text. Readers cannot simply form conclusions based off of only information they already know.

There are several ways readers can draw conclusions from authors' ideas, such as note taking, text evidence, text credibility, writing a response to text, directly stated information versus implications, outlining, summarizing, and paraphrasing. Let's take a look at each important strategy to help readers draw logical conclusions.

Note Taking

When readers take notes throughout texts or passages, they are jotting down important facts or points that the author makes. Note taking is a useful record of information that helps readers understand the text or passage and respond to it. When taking notes, readers should keep lines brief and filled with pertinent information so that they are not rereading a large amount of text, but rather just key points, elements, or words. After readers have completed a text or passage, they can refer to their notes to help them form a conclusion about the author's ideas in the text or passage.

Text Evidence

Text evidence is the information readers find in a text or passage that supports the main idea or point(s) in a story. In turn, text evidence can help readers draw conclusions about the text or passage. The information should be taken directly from the text or passage and placed in quotation marks. Text evidence provides readers with information to support ideas about the text so that they do not rely simply on their own thoughts. Details should be precise, descriptive, and factual. Statistics are a great piece of text evidence because they provide readers with exact numbers and not just a generalization. For example, instead of saying "Asia has a larger population than Europe," authors could provide detailed information such as, "In Asia there are over 4 billion people, whereas in Europe there are a little over 750 million." More definitive information provides better evidence to readers to help support their conclusions about texts or passages.

Text Credibility

Credible sources are important when drawing conclusions because readers need to be able to trust what they are reading. Authors should always use credible sources to help gain the trust of their readers. A text is *credible* when it is believable and the author is objective and unbiased. If readers do not trust an author's words, they may simply dismiss the text completely. For example, if an author writes a persuasive essay, he or she is outwardly trying to sway readers' opinions to align with their own. Readers may agree or disagree with the author, which may, in turn, lead them to believe that the author is credible or not credible. Also, readers should keep in mind the source of the text. If readers review a journal about astronomy, would a more reliable source be a NASA employee or a medical doctor? Overall, text credibility is important when drawing conclusions, because readers want reliable sources that support the decisions they have made about the author's ideas.

Writing a Response to Text

Once readers have determined their opinions and validated the credibility of a text, they can then reflect on the text. Writing a response to a text is one way readers can reflect on the given text or passage. When readers write responses to a text, it is important for them to rely on the evidence within the text to support their opinions or thoughts. Supporting evidence such as facts, details, statistics, and quotes directly from the text are key pieces of information readers should reflect upon or use when writing a response to text.

Directly Stated Information Versus Implications

Engaged readers should constantly self-question while reviewing texts to help them form conclusions. Self-questioning is when readers review a paragraph, page, passage, or chapter and ask themselves, "Did I understand what I read?," "What was the main event in this section?," "Where is this taking

place?," and so on. Authors can provide clues or pieces of evidence throughout a text or passage to guide readers toward a conclusion. This is why active and engaged readers should read the text or passage in its entirety before forming a definitive conclusion. If readers do not gather all the pieces of evidence needed, then they may jump to an illogical conclusion.

At times, authors directly state conclusions while others simply imply them. Of course, it is easier if authors outwardly provide conclusions to readers because it does not leave any information open to interpretation. On the other hand, implications are things that authors do not directly state but can be assumed based off of information they provided. If authors only imply what may have happened, readers can form a menagerie of ideas for conclusions. For example, look at the following statement: "Once we heard the sirens, we hunkered down in the storm shelter." In this statement, the author does not directly state that there was a tornado, but clues such as "sirens" and "storm shelter" provide insight to the readers to help form that conclusion.

Outlining

An outline is a system used to organize writing. When reading texts, outlining is important because it helps readers organize important information in a logical pattern using roman numerals. Usually, outlines start with the main idea(s) and then branch out into subgroups or subsidiary thoughts of subjects. Not only do outlines provide a visual tool for readers to reflect on how events, characters, settings, or other key parts of the text or passage relate to one another, but they can also lead readers to a stronger conclusion.

The sample below demonstrates what a general outline looks like.

I. Main Topic 1
   a. Subtopic 1
   b. Subtopic 2
      1. Detail 1
      2. Detail 2
II. Main Topic 2
   a. Subtopic 1
   b. Subtopic 2
      1. Detail 1
      2. Detail 2

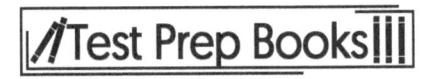

Summarizing

At the end of a text or passage, it is important to summarize what the readers read. Summarizing is a strategy in which readers determine what is important throughout the text or passage, shorten those ideas, and rewrite or retell it in their own words. A summary should identify the main idea of the text or passage. Important details or supportive evidence should also be accurately reported in the summary. If writers provide irrelevant details in the summary, it may cloud the greater meaning of the summary in the text. When summarizing, writers should not include their opinions, quotes, or what they thought the author should have said. A clear summary provides clarity of the text or passage to the readers. Let's review the checklist of items writers should include in their summary.

Summary Checklist
- Title of the story
- Someone: Who is or are the main character(s)?
- Wanted: What did the character(s) want?
- But: What was the problem?
- So: How did the character(s) solve the problem?
- Then: How did the story end? What was the resolution?

Paraphrasing

Another strategy readers can use to help them fully comprehend a text or passage is paraphrasing. Paraphrasing is when readers take the author's words and put them into their own words. When readers and writers paraphrase, they should avoid copying the text—that is plagiarism. It is also important to include as many details as possible when restating the facts. Not only will this help readers and writers recall information, but by putting the information into their own words, they demonstrate whether or not they fully comprehend the text or passage. Look at the example below showing an original text and how to paraphrase it.

Original Text: Fenway Park is home to the beloved Boston Red Sox. The stadium opened on April 20, 1912. The stadium currently seats over 37,000 fans, many of whom travel from all over the country to experience the iconic team and nostalgia of Fenway Park.

Paraphrased: On April 20, 1912, Fenway Park opened. Home to the Boston Red Sox, the stadium now seats over 37,000 fans. Many spectators travel to watch the Red Sox and experience the spirit of Fenway Park.

Paraphrasing, summarizing, and quoting can often cross paths with one another. Review the chart below showing the similarities and differences between the three strategies.

Paraphrasing	Summarizing	Quoting
Uses own words	Puts main ideas into own words	Uses words that are identical to text
References original source	References original source	Requires quotation marks
Uses own sentences	Shows important ideas of source	Uses author's own words and ideas

Inferences in a Text

Readers should be able to make *inferences*. Making an inference requires the reader to read between the lines and look for what is implied rather than what is explicitly stated. That is, using information that is known from the text, the reader is able to make a logical assumption about information that is not explicitly stated but is probably true. Read the following passage:

"Hey, do you wanna meet my new puppy?" Jonathan asked.

"Oh, I'm sorry but please don't—" Jacinta began to protest, but before she could finish, Jonathan had already opened the passenger side door of his car and a perfect white ball of fur came bouncing towards Jacinta.

"Isn't he the cutest?" beamed Jonathan.

"Yes—achoo!—he's pretty—aaaachooo!!—adora—aaa—aaaachoo!" Jacinta managed to say in between sneezes. "But if you don't mind, I—I—achoo!—need to go inside."

Which of the following can be inferred from Jacinta's reaction to the puppy?
a. she hates animals
b. she is allergic to dogs
c. she prefers cats to dogs
d. she is angry at Jonathan

An inference requires the reader to consider the information presented and then form their own idea about what is probably true. Based on the details in the passage, what is the best answer to the question? Important details to pay attention to include the tone of Jacinta's dialogue, which is overall polite and apologetic, as well as her reaction itself, which is a long string of sneezes. Answer choices (a) and (d) both express strong emotions ("hates" and "angry") that are not evident in Jacinta's speech or actions. Answer choice (c) mentions cats, but there is nothing in the passage to indicate Jacinta's feelings about cats. Answer choice (b), "she is allergic to dogs," is the most logical choice—based on the fact that she began sneezing as soon as a fluffy dog approached her, it makes sense to guess that Jacinta might be allergic to dogs. So even though Jacinta never directly states, "Sorry, I'm allergic to dogs!" using the clues in the passage, it is still reasonable to guess that this is true.

Making inferences is crucial for readers of literature because literary texts often avoid presenting complete and direct information to readers about characters' thoughts or feelings, or they present this information in an unclear way, leaving it up to the reader to interpret clues given in the text. In order to make inferences while reading, readers should ask themselves:

- What details are being presented in the text?
- Is there any important information that seems to be missing?
- Based on the information that the author *does* include, what else is probably true?
- Is this inference reasonable based on what is already known?

Apply Information

A natural extension of being able to make an inference from a given set of information is also being able to apply that information to a new context. This is especially useful in non-fiction or informative writing. Considering the facts and details presented in the text, readers should consider how the same information might be relevant in a different situation. The following is an example of applying an inferential conclusion to a different context:

> Often, individuals behave differently in large groups than they do as individuals. One example of this is the psychological phenomenon known as the bystander effect. According to the bystander effect, the more people who witness an accident or crime occur, the less likely each individual bystander is to respond or offer assistance to the victim. A classic example of this is the murder of Kitty Genovese in New York City in the 1960s. Although there were over thirty witnesses to her killing by a stabber, none of them intervened to help Kitty or contact the police.

Considering the phenomenon of the bystander effect, what would probably happen if somebody tripped on the stairs in a crowded subway station?

a. Everybody would stop to help the person who tripped
b. Bystanders would point and laugh at the person who tripped
c. Someone would call the police after walking away from the station
d. Few if any bystanders would offer assistance to the person who tripped

This question asks readers to apply the information they learned from the passage, which is an informative paragraph about the bystander effect. According to the passage, this is a concept in psychology that describes the way people in groups respond to an accident—the more people are present, the less likely any one person is to intervene. While the passage illustrates this effect with the example of a woman's murder, the question asks readers to apply it to a different context—in this case, someone falling down the stairs in front of many subway passengers. Although this specific situation is not discussed in the passage, readers should be able to apply the general concepts described in the paragraph. The definition of the bystander effect includes any instance of an accident or crime in front of a large group of people. The question asks about a situation that falls within the same definition, so the general concept should still hold true: in the midst of a large crowd, few individuals are likely to actually respond to an accident. In this case, answer choice (d) is the best response.

Author's Use of Language

Authors utilize a wide range of techniques to tell a story or communicate information. Readers should be familiar with the most common of these techniques. Techniques of writing are also commonly known as rhetorical devices.

Types of Appeals

In nonfiction writing, authors employ argumentative techniques to present their opinion to readers in the most convincing way. Persuasive writing usually includes at least one type of appeal: an appeal to logic (logos), emotion (pathos), or credibility and trustworthiness (ethos). When a writer appeals to logic, they are asking readers to agree with them based on research, evidence, and an established line of reasoning. An author's argument might also appeal to readers' emotions, perhaps by including personal stories and anecdotes (a short narrative of a specific event). A final type of appeal, appeal to authority, asks the reader to agree with the author's argument on the basis of their expertise or credentials. Consider three different approaches to arguing the same opinion:

Logic (Logos)

This is an example of an appeal to logic:

> Our school should abolish its current ban on campus cell phone use. The ban was adopted last year as an attempt to reduce class disruptions and help students focus more on their lessons. However, since the rule was enacted, there has been no change in the number of disciplinary problems in class. Therefore, the rule is ineffective and should be done away with.

The author uses evidence to disprove the logic of the school's rule (the rule was supposed to reduce discipline problems, but the number of problems has not been reduced; therefore, the rule is not working) and call for its repeal.

Emotion (Pathos)

An author's argument might also appeal to readers' emotions, perhaps by including personal stories and anecdotes. The next example presents an appeal to emotion. By sharing the personal anecdote of one student and speaking about emotional topics like family relationships, the author invokes the reader's empathy in asking them to reconsider the school rule.

> Our school should abolish its current ban on campus cell phone use. If students aren't able to use their phones during the school day, many of them feel isolated from their loved ones. For example, last semester, one student's grandmother had a heart attack in the morning. However, because he couldn't use his cell phone, the student didn't know about his grandmother's accident until the end of the day—when she had already passed away, and it was too late to say goodbye. By preventing students from contacting their friends and family, our school is placing undue stress and anxiety on students.

Credibility (Ethos)

Finally, an appeal to authority includes a statement from a relevant expert. In this case, the author uses a doctor in the field of education to support the argument. All three examples begin from the same opinion—the school's phone ban needs to change—but rely on different argumentative styles to persuade the reader.

> Our school should abolish its current ban on campus cell phone use. According to Dr. Bartholomew Everett, a leading educational expert, "Research studies show that cell phone usage has no real impact on student attentiveness. Rather, phones provide a valuable technological resource for learning. Schools need to learn how to integrate this new technology into their curriculum." Rather than banning phones altogether, our school should follow the advice of experts and allow students to use phones as part of their learning.

Rhetorical Questions

Another commonly used argumentative technique is asking rhetorical questions, questions that do not actually require an answer but that push the reader to consider the topic further.

> I wholly disagree with the proposal to ban restaurants from serving foods with high sugar and sodium contents. Do we really want to live in a world where the government can control what we eat? I prefer to make my own food choices.

Here, the author's rhetorical question prompts readers to put themselves in a hypothetical situation and imagine how they would feel about it.

Figurative Language

Literary texts also employ rhetorical devices. Figurative language like simile and metaphor is a type of rhetorical device commonly found in literature. In addition to rhetorical devices that play on the *meanings* of words, there are also rhetorical devices that use the *sounds* of words. These devices are most often found in poetry but may also be found in other types of literature and in non-fiction writing like speech texts.

Alliteration and *assonance* are both varieties of sound repetition. Other types of sound repetition include: anaphora, repetition that occurs at the beginning of the sentences; epiphora, repetition occurring at the end of phrases; antimetabole, repetition of words in reverse order; and antiphrasis, a form of denial of an assertion in a text.

Alliteration refers to the repetition of the first sound of each word. Recall Robert Burns' opening line:

> My love is like a red, red rose

This line includes two instances of alliteration: "love" and "like" (repeated *L* sound), as well as "red" and "rose" (repeated *R* sound). Next, assonance refers to the repetition of vowel sounds, and can occur anywhere within a word (not just the opening sound).

Here is the opening of a poem by John Keats:

> When I have fears that I may cease to be

> Before my pen has glean'd my teeming brain

Assonance can be found in the words "fears," "cease," "be," "glean'd," and "teeming," all of which stress the long *E* sound. Both alliteration and assonance create a harmony that unifies the writer's language.

Another sound device is *onomatopoeia*, or words whose spelling mimics the sound they describe. Words such as "crash," "bang," and "sizzle" are all examples of onomatopoeia. Use of onomatopoetic language adds auditory imagery to the text.

Readers are probably most familiar with the technique of *pun*. A pun is a play on words, taking advantage of two words that have the same or similar pronunciation. Puns can be found throughout Shakespeare's plays, for instance:

> Now is the winter of our discontent
> Made glorious summer by this son of York

These lines from *Richard III* contain a play on words. Richard III refers to his brother, the newly crowned King Edward IV, as the "son of York," referencing their family heritage from the house of York. However, while drawing a comparison between the political climate and the weather (times of political trouble were the "winter," but now the new king brings "glorious summer"), Richard's use of the word "son" also implies another word with the same pronunciation, "sun"—so Edward IV is also like the sun, bringing light, warmth, and hope to England. Puns are a clever way for writers to suggest two meanings at once.

Counterarguments

If an author presents a differing opinion or a counterargument in order to refute it, the reader should consider how and why the information is being presented. It is meant to strengthen the original argument and shouldn't be confused with the author's intended conclusion, but it should also be considered in the reader's final evaluation.

Authors can also use bias if they ignore the opposing viewpoint or present their side in an unbalanced way. A strong argument considers the opposition and finds a way to refute it. Critical readers should look for an unfair or one-sided presentation of the argument and be skeptical, as a bias may be present. Even if this bias is unintentional, if it exists in the writing, the reader should be wary of the validity of the argument. Readers should also look for the use of stereotypes, which refer to specific groups. Stereotypes are often negative connotations about a person or place, and should always be avoided. When a critical reader finds stereotypes in a piece of writing, they should be critical of the argument, and consider the validity of anything the author presents. Stereotypes reveal a flaw in the writer's thinking and may suggest a lack of knowledge or understanding about the subject.

Meaning of Words in Context

There will be many occasions in one's reading career in which an unknown word or a word with multiple meanings will pop up. There are ways of determining what these words or phrases mean that do not require the use of the dictionary, which is especially helpful during a test where one may not be available. Even outside of the exam, knowing how to derive an understanding of a word via context clues will be a critical skill in the real world. The context is the circumstances in which a story or a passage is happening, and can usually be found in the series of words directly before or directly after the word or phrase in question. The clues are the words that hint towards the meaning of the unknown word or phrase.

There may be questions that ask about the meaning of a particular word or phrase within a passage. There are a couple ways to approach these kinds of questions:

1. Define the word or phrase in a way that is easy to comprehend (using context clues).
2. Try out each answer choice in place of the word.

To demonstrate, here's an example from *Alice in Wonderland*:

> Alice was beginning to get very tired of sitting by her sister on the bank, and of having nothing to do: once or twice she <u>peeped</u> into the book her sister was reading, but it had no pictures or conversations in it, "and what is the use of a book," thought Alice, "without pictures or conversations?"

Q: As it is used in the selection, the word <u>peeped</u> means:

Using the first technique, before looking at the answers, define the word "peeped" using context clues and then find the matching answer. Then, analyze the entire passage in order to determine the meaning, not just the surrounding words.

To begin, imagine a blank where the word should be and put a synonym or definition there: "once or twice she _____ into the book her sister was reading." The context clue here is the book. It may be tempting to put "read" where the blank is, but notice the preposition word, "into." One does not read *into* a book, one simply reads a book, and since reading a book requires that it is seen with a pair of eyes, then "look" would make the most sense to put into the blank: "once or twice she <u>looked </u>into the book her sister was reading."

Once an easy-to-understand word or synonym has been supplanted, readers should check to make sure it makes sense with the rest of the passage. What happened after she looked into the book? She thought to herself how a book without pictures or conversations is useless. This situation in its entirety makes sense.

Now check the answer choices for a match:
- a. To make a high-pitched cry
- b. To smack
- c. To look curiously
- d. To pout

Since the word was already defined, Choice *C* is the best option.

Using the second technique, replace the figurative blank with each of the answer choices and determine which one is the most appropriate. Remember to look further into the passage to clarify that they work, because they could still make sense out of context.
- a. Once or twice she <u>made a high pitched cry</u> into the book her sister was reading
- b. Once or twice she <u>smacked</u> into the book her sister was reading
- c. Once or twice she <u>looked curiously</u> into the book her sister was reading
- d. Once or twice she <u>pouted</u> into the book her sister was reading

For Choice *A*, it does not make much sense in any context for a person to yell into a book, unless maybe something terrible has happened in the story. Given that afterward Alice thinks to herself how useless a book without pictures is, this option does not make sense within context.

For Choice *B*, smacking a book someone is reading may make sense if the rest of the passage indicates a reason for doing so. If Alice was angry or her sister had shoved it in her face, then maybe smacking the book would make sense within context. However, since whatever she does with the book causes her to think, "what is the use of a book without pictures or conversations?" then answer Choice *B* is not an appropriate answer. Answer Choice *C* fits well within context, given her subsequent thoughts on the

matter. Answer Choice *D* does not make sense in context or grammatically, as people do not "pout into" things.

This is a simple example to illustrate the techniques outlined above. There may, however, be a question in which all of the definitions are correct and also make sense out of context, in which the appropriate context clues will really need to be honed in on in order to determine the correct answer. For example, here is another passage from *Alice in Wonderland*:

> . . . but when the Rabbit actually took a watch out of its waistcoat pocket, and looked at it, and then hurried on, Alice <u>started</u> to her feet, for it flashed across her mind that she had never before seen a rabbit with either a waistcoat-pocket or a watch to take out of it, and burning with curiosity, she ran across the field after it, and was just in time to see it pop down a large rabbit-hole under the hedge.

Q: As it is used in the passage, the word started means
   a. To turn on
   b. To begin
   c. To move quickly
   d. To be surprised

All of these words qualify as a definition of "start," but using context clues, the correct answer can be identified using one of the two techniques above. It's easy to see that one does not turn on, begin, or be surprised to one's feet. The selection also states that she "ran across the field after it," indicating that she was in a hurry. Therefore, to move quickly would make the most sense in this context.

The same strategies can be applied to vocabulary that may be completely unfamiliar. In this case, focus on the words before or after the unknown word in order to determine its definition. Take this sentence, for example:

> Sam was such a <u>miser</u> that he forced Andrew to pay him twelve cents for the candy, even though he had a large inheritance and he knew his friend was poor.

Unlike with assertion questions, for vocabulary questions, it may be necessary to apply some critical thinking skills that may not be explicitly stated within the passage. Think about the implications of the passage, or what the text is trying to say. With this example, it is important to realize that it is considered unusually stingy for a person to demand so little money from someone instead of just letting their friend have the candy, especially if this person is already wealthy. Hence, a <u>miser</u> is a greedy or stingy individual.

Questions about complex vocabulary may not be explicitly asked, but this is a useful skill to know. If there is an unfamiliar word while reading a passage and its definition goes unknown, it is possible to miss out on a critical message that could inhibit the ability to appropriately answer the questions. Practicing this technique in daily life will sharpen this ability to derive meanings from context clues with ease.

Practice Quiz

Directions: Assume each passage below to be true. Then, pick the answer choice that can be inferred only from the passage itself. Some of the other answer choices might make sense, but only one of them can be derived solely from the passage.

1. The news exclusively covers important current events. Reality television stars are never covered on the news, except when they become pregnant.
   a. Reality television stars never qualify as an important current event.
   b. All current events involve reality television stars.
   c. Some pregnancies are important current events.
   d. All pregnancies are important current events.

2. In 2015, 28 countries, including Estonia, Portugal, Slovenia, and Latvia, scored significantly higher than the United States on standardized high school math tests. In the 1960s, the United States consistently ranked first in the world. Today, the United States spends more than $800 billion dollars on education, which exceeds the next highest country by more than $600 billion dollars. The United States also leads the world in spending per school-aged child by an enormous margin.
   a. Outspending other countries on education has benefits beyond standardized math tests.
   b. The United States' education system is corrupt and broken.
   c. The standardized math tests are not representative of American academic prowess.
   d. Spending more money does not guarantee success on standardized math tests.

3. Dwight is the manager of a mid-sized regional paper company. The company's sales have declined for seven consecutive quarters. All of the paper company's regional and national competitors have experienced a similar loss in revenue. Dwight instituted a mass layoff and successfully kept his company out of bankruptcy.
   a. Mass layoffs were the only way to keep the company out of bankruptcy.
   b. The paper industry is experiencing a fundamental change in demand.
   c. Mid-sized regional paper companies will no longer exist in ten years.
   d. National paper companies poached Dwight's customers, causing the decline in sales.

4. The Founding Fathers created the Electoral College to balance state-based and population-based representation. The Electoral College allows only electors to vote for the President, and the number of electors per state is equal to the state's total number of representatives and senators. Each political party nominates its electors based on their own methods. In the past, some electors have cast ballots for candidates that did not receive the most popular votes.
   a. The number of states' representatives is based on population, while the number of senators is equal for every state.
   b. The Founding Fathers considered the majority to be a mob that cannot be trusted.
   c. The Electoral College is not a direct election.
   d. The winning candidate must receive the majority of the electors' votes.

5. Bill is capable of reading two pages per minute, typing one hundred words per minute, and speaking twenty words per minute. All lawyers can read two pages per minute, and some philosophers can read two pages per minute. Only secretaries can type one hundred words per minute. Many chief executive officers can speak twenty words per minute, and few doctors can speak more than twenty words per minute.

  a. Bill is a lawyer.
  b. Bill is a philosopher.
  c. Bill is a secretary.
  d. Bill is a chief executive officer.

Answer Explanations

1. C: The correct answer must necessarily be true. The second sentence of the argument says that reality television stars are never covered on the news, except when they become pregnant; therefore, pregnant reality television stars are covered on the news. The first sentence of the argument says that the news exclusively covers important current events. As a result, pregnant reality television stars must be important current events. Choice *C* is the correct answer, because *some* pregnancies are important current events. Choice *D* is too broad since the exception for pregnant reality television stars only allows for the possibility of the news covering the event. In addition, the argument does not mention pregnancies of non-reality television stars. Choices *A* and *B* are contradicted by the argument's second sentence.

2. D: Outspending other countries on education could have other benefits, but there is no reference to this in the passage, so Choice *A* is incorrect. Choice *B* is incorrect because the author does not mention corruption. Choice *C* is incorrect because there is nothing in the passage stating that the tests are not genuinely representative. Choice *D* is accurate because spending more money has not brought success. The United States already spends the most money, and the country is not excelling on these tests. Choice *D* is the correct answer.

3. B: The passage is clear that it is not only Dwight's mid-sized regional paper company that is struggling, but all of its regional and national competitors are as well. The paper industry is clearly undergoing a massive downturn, and Choice *B* provides an explanation—a fundamental change in demand. Thus, Choice *B* is the correct answer. Choice *A* is incorrect, because it is unclear whether mass layoffs were the *only* way to keep the company out of bankruptcy. Choice *C* speculates without any justification. Choice *D* is unsupported by the passage, and in addition, it is unlikely since the national paper companies are experiencing similar struggles.

4. C: The correct answer will be something that can be directly inferred from the closed world of the passage. The passage states that only electors vote for the president, and political parties nominate electors according to their own methods. As such, it can be properly inferred that the Electoral College is not a direct election. Thus, Choice *C* is the correct answer. The other answers are factually correct, but they cannot be inferred from the language contained in the passage.

5. C: Bill meets the standards for all of the professions, so he *could* be any of the five professions. However, we need the profession that *must* be true. Only secretaries can type one hundred words per minute, and Bill can type one hundred words per minute; therefore, Bill must be a secretary. If Bill were not a secretary, then he would not be able to type one hundred words per minute. Thus, Choice *C* is the correct answer. All of the other answer choices are only possibilities.

Mechanical Comprehension

The *Mechanical Comprehension (MC)* section tests a candidate's knowledge of mechanics and physical principles. These include concepts of force, energy, and work, and how they're used to predict the functioning of tools and machines. This knowledge is important for a successful career in the military. A good score on the MC test shows that a candidate has a solid background for learning how to use tools and machines properly. This is extremely important for the efficient, safe completion of most tasks a future soldier, sailor, or airman must undertake during their service.

The test problems in the MC section of the exam focus on understanding physical principles, but they are *qualitative* in nature rather than *quantitative*. This means the problems involve predicting the *behavior* of a system (such as the direction it moves) rather than calculating a specific measurement (such as its velocity). The figure below shows a sample problem similar to those on the MC test:

Mechanical Comprehension Sample Test Problem

Question 1.

Extending the reach of this crane will shift its

- ○ **A.** total weight
- ○ **B.** allowable speed
- ○ **C.** center of gravity
- ○ **D.** center of buoyancy

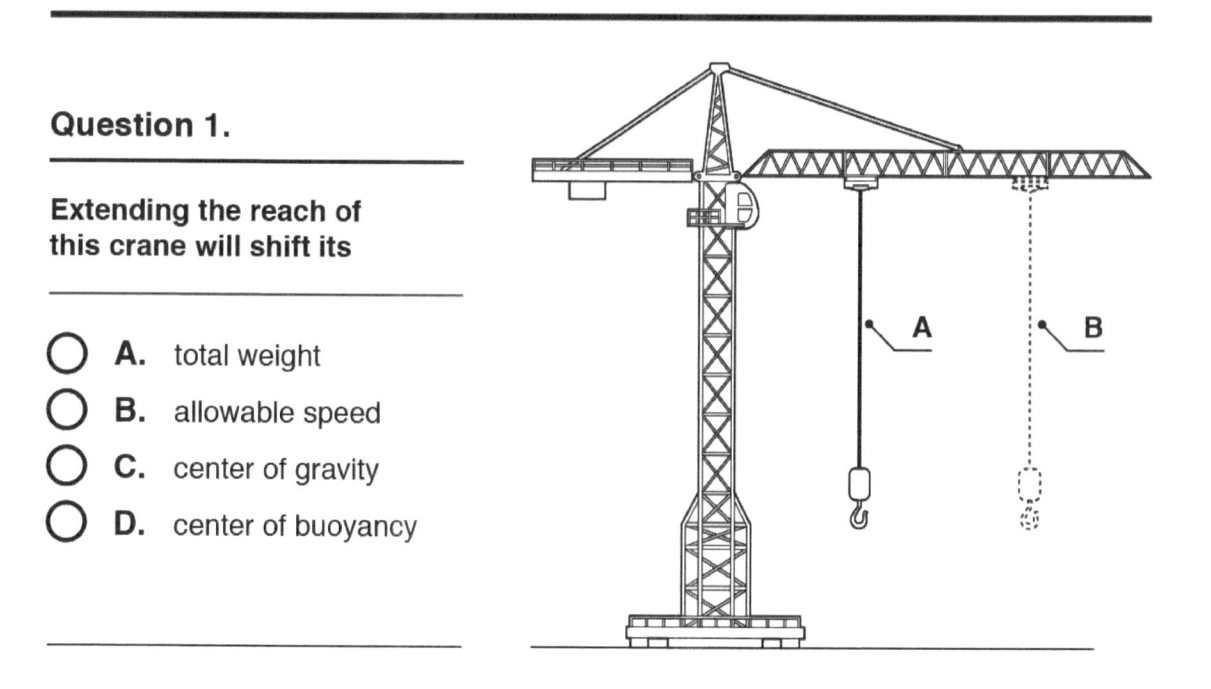

The sample problem pictures a system of a crane lifting a weight, and below the picture is a question. On the exam, it's *very important* to read these questions *carefully.* This question involves completing the following sentence: *Extending the reach of this crane will shift its _____.* After the sentence, four possible answers are provided.

The correct answer is *C, center of gravity.* In this sample problem, it's easy to guess the correct answer simply by eliminating the rest. Answer *A* is incorrect because moving the load out along the crane's boom won't change its weight, just like moving a bodybuilder's arm that's holding a dumbbell won't change the combined weight of the bodybuilder and the dumbbell. Answer *B* is incorrect because the crane isn't moving. That leaves Answers *C* and *D*, but *D* is incorrect because buoyancy is only involved in

systems with a liquid (the buoyancy of air is negligible). Therefore, through the process of elimination, *C* is the correct answer.

Review of Physics and Mechanical Principles

The proper use of tools and machinery depends on an understanding of basic physics, which includes the study of motion and the interactions of *mass*, *force*, and *energy*. These terms are used every day, but their exact meanings are difficult to define. In fact, they're usually defined in terms of each other.

The matter in the universe (atoms and molecules) is characterized in terms of its *mass*, which is measured in kilograms in the *International System of Units (SI)*. The amount of mass that occupies a given volume of space is termed *density*.

Mass occupies space, but it's also a component that inversely relates to acceleration when a force is applied to it. This *force* is the application of *energy* to an object with the intent of changing its position (mainly its acceleration).

To understand *acceleration*, it's necessary to relate it to displacement and velocity. The *displacement* of an object is simply the distance it travels. The *velocity* of an object is the distance it travels in a unit of time, such as miles per hour or meters per second:

$$Velocity = \frac{Distance\ Traveled}{Time\ Required}$$

There's often confusion between the words "speed" and "velocity." Velocity includes speed *and* direction. For example, a car traveling east and another traveling west can have the same speed of 30 miles per hour (mph), but their velocities are different. If movement eastward is considered positive, then movement westward is negative. Thus, the eastbound car has a velocity of 30 mph while the westbound car has a velocity of -30 mph.

The fact that velocity has a *magnitude* (speed) and a direction makes it a vector quantity. A *vector* is an arrow pointing in the direction of motion, with its length proportional to its magnitude.

Vectors can be added geometrically as shown below. In this example, a boat is traveling east at 4 *knots* (nautical miles per hour) and there's a current of 3 knots (thus a slow boat and a very fast current). If the boat travels in the same direction as the current, it gets a "lift" from the current and its speed is 7 knots. If the boat heads *into* the current, it has a forward speed of only 1 knot (4 knots – 3 knots = 1 knot) and makes very little headway. As shown in the figure below, the current is flowing north across the boat's path. Thus, for every 4 miles of progress the boat makes eastward, it drifts 3 miles to the north.

Working with Velocity Vectors

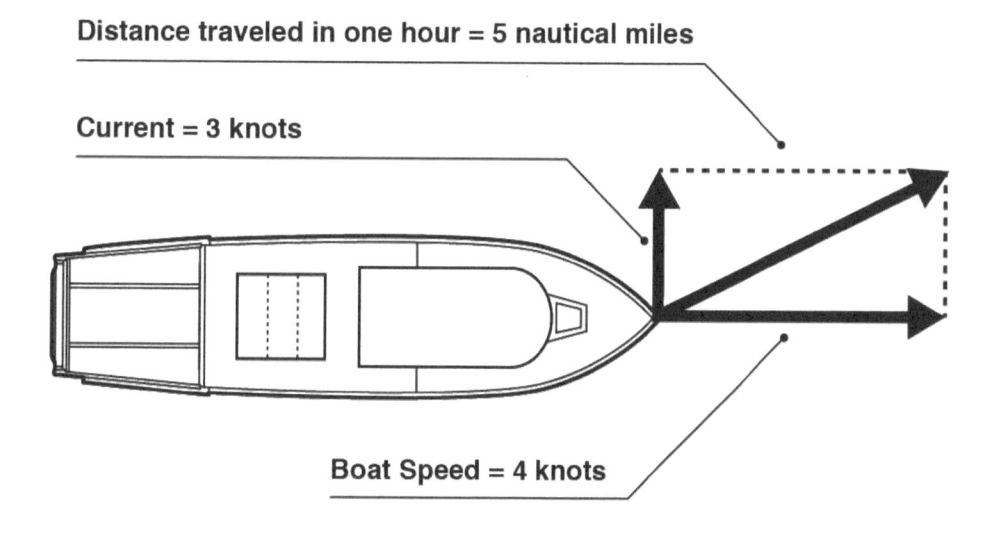

Distance traveled in one hour = 5 nautical miles

Current = 3 knots

Boat Speed = 4 knots

The total distance traveled is calculated using the *Pythagorean Theorem* for a right triangle, which should be memorized as follows:

$$a^2 + b^2 = c^2 \text{ or } c = \sqrt{a^2 + b^2}$$

The problem above was set up using a Pythagorean triple which is made up of positive integers which fit the rule of $a^2 + b^2 = c^2$. In this case, the integers are 3, 4, and 5.

Another example where velocity and speed are different is with a car traveling around a bend in the road. The speed is constant along the road, but the direction (and therefore the velocity) changes continuously.

The *acceleration* of an object is the change in its velocity in a given period of time:

$$Acceleration = \frac{Change\ in\ Velocity}{Time\ Required}$$

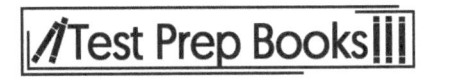

For example, a car starts at rest and then reaches a velocity of 70 mph in 8 seconds. What's the car's acceleration in feet per second squared? First, the velocity must be converted from miles per hour to feet per second:

$$70\frac{\text{miles}}{\text{hour}} \times \frac{5{,}280 \text{ feet}}{\text{mile}} \times \frac{\text{hour}}{3600 \text{ seconds}} = 102.67 \text{ feet/second}$$

Starting from rest, the acceleration is:

$$Acceleration = \frac{102.67\frac{\text{feet}}{\text{second}} - 0\frac{\text{feet}}{\text{second}}}{8 \text{ seconds}} = 12.8 \text{ feet/second}^2$$

Newton's Laws

Isaac Newton's three laws of motion describe how the acceleration of an object is related to its mass and the forces acting on it. The three laws are:

1. Unless acted on by a force, a body at rest tends to remain at rest; a body in motion tends to remain in motion with a constant velocity and direction.

2. A force that acts on a body accelerates it in the direction of the force. The larger the force, the greater the acceleration; the larger the mass, the greater its inertia (resistance to movement and acceleration).

3. Every force acting on a body is resisted by an equal and opposite force.

To understand Newton's laws, it's necessary to understand forces. These forces can push or pull on a mass, and they have a magnitude and a direction. Forces are represented by a vector, which is the arrow lined up along the direction of the force with its tip at the point of application. The magnitude of the force is represented by the length of the vector.

The figure below shows a mass acted on or "pushed" by two equal forces (shown here by vectors of the same length). Both vectors "push" along the same line through the center of the mass, but in opposite directions. What happens?

A Mass Acted on by Equal and Opposite Forces

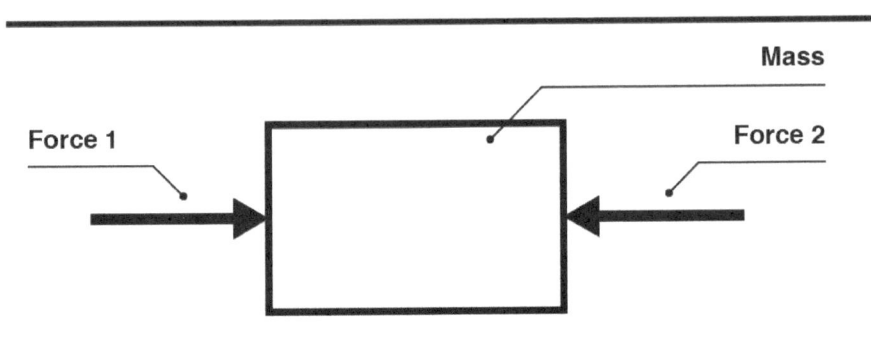

According to Newton's third law, every force on a body is resisted by an equal and opposite force. In the figure above, Force 1 acts on the left side of the mass. The mass pushes back. Force 2 acts on the right

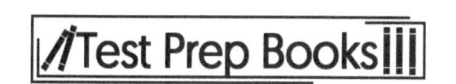

side, and the mass pushes back against this force too. The net force on the mass is zero, so according to Newton's first law, there's no change in the *momentum* (the mass times its velocity) of the mass. Therefore, if the mass is at rest before the forces are applied, it remains at rest. If the mass is in motion with a constant velocity, its momentum doesn't change. So, what happens when the net force on the mass isn't zero, as shown in the figure below?

A Mass Acted on by Unbalanced Forces

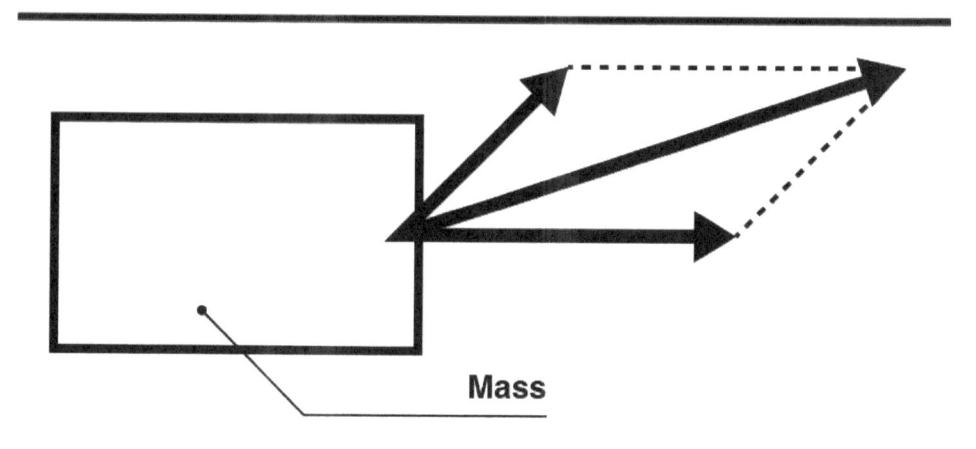

Mass

Notice that the forces are vector quantities and are added geometrically the same way that velocity vectors are manipulated.

Here in the figure above, the mass is pulled by two forces acting to the right, so the mass accelerates in the direction of the net force. This is described by Newton's second law:

$$Force = Mass \times Acceleration$$

The force (measured in *newtons*) is equal to the product of the mass (measured in kilograms) and its acceleration (measured in meters per second squared or meters per second, per second). A better way to look at the equation is dividing through by the mass:

$$Acceleration = Force/Mass$$

This form of the equation makes it easier to see that the acceleration of an object varies directly with the net force applied and inversely with the mass. Thus, as the mass increases, the acceleration is reduced for a given force. To better understand, think of how a baseball accelerates when hit by a bat. Now imagine hitting a cannonball with the same bat and the same force. The cannonball is more massive than the baseball, so it won't accelerate very much when hit by the bat.

In addition to forces acting on a body by touching it, gravity acts as a force at a distance and causes all bodies in the universe to attract each other. The *force of gravity (F_g)* is proportional to the masses of the two objects (*m* and *M*) and inversely proportional to the square of the distance (r^2) between them (and *G* is the proportionality constant). This is shown in the following equation:

$$F_g = G \frac{mM}{r^2}$$

The force of gravity is what causes an object to fall to Earth when dropped. Understanding gravity helps explain the difference between mass and weight. Mass is a property of an object that remains the same while it's intact, no matter where it's located. A 10-kilogram cannonball has the same mass on Earth as it does on the moon. On Earth, it *weighs* 98.1 newtons because of the attractive force of gravity, so it accelerates at 9.81 m/s². However, on the moon, the same cannonball has a weight of only about 16 newtons. This is because the gravitational attraction on the moon is approximately one-sixth that on Earth. Although Earth still attracts the body on the moon, it's so far away that its force is negligible.

For Americans, there's often confusion when talking about mass because the United States still uses "pounds" as a measurement of weight. In the traditional system used in the United States, the unit of mass is called a *slug*. It's derived by dividing the weight in pounds by the acceleration of gravity (32 feet/s²); however, it's rarely used today. To avoid future problems, test takers should continue using SI units and *remember to express mass in kilograms and weight in Newtons*.

Another way to understand Newton's second law is to think of it as an object's change in momentum, which is defined as the product of the object's mass and its velocity:

$$Momentum = Mass \times Velocity$$

Which of the following has the greater momentum: a pitched baseball, a softball, or a bullet fired from a rifle?

A bullet with a mass of 5 grams (0.005 kilograms) is fired from a rifle with a muzzle velocity of 2200 mph. Its momentum is calculated as:

$$2200 \frac{miles}{hour} \times \frac{5,280 \ feet}{mile} \times \frac{m}{3.28 \ feet} \times \frac{hour}{3600 \ seconds} \times 0.005 kg = 4.92 \frac{kg. \, m}{seconds}$$

A softball has a mass between 177 grams and 198 grams and is thrown by a college pitcher at 50 miles per hour. Taking an average mass of 188 grams (0.188 kilograms), a softball's momentum is calculated as:

$$50 \frac{miles}{hour} \times \frac{5280 \ feet}{mile} \times \frac{m}{3.28 \ feet} \times \frac{hour}{3600 \ seconds} \times 0.188 kg = 4.19 \frac{kg. \, m}{seconds}$$

That's only slightly less than the momentum of the bullet. Although the speed of the softball is considerably less, its mass is much greater than the bullet's.

A professional baseball pitcher can throw a 145-gram baseball at 100 miles per hour. A similar calculation (try doing it!) shows that the pitched hardball has a momentum of about 6.48 kg.m/seconds. That's more momentum than a speeding bullet!

So why is the bullet more harmful than the hard ball? It's because the force that it applies acts on a much smaller area.

Changing the expression of Newton's second law of motion yields a new expression.

$$Force(F) = ma = m \times \frac{\Delta v}{\Delta t}$$

If both sides of the expression are multiplied by the change in time, the law produces the impulse equation.

$$F\Delta t = m\Delta v$$

This equation shows that the amount of force during a length of time creates an impulse. This means that if a force acts on an object during a given amount of time, it will have a determined impulse. However, if the same change in velocity happens over a longer amount of time, the required force is much smaller, due to the conservation of momentum.

$$p = mv$$

In the case of the rifle, the force created by the pressure of the charge's explosion in its shell pushes the bullet, accelerating it until it leaves the barrel of the gun with its *muzzle velocity* (the speed the bullet has when it leaves the muzzle). After leaving the gun, the bullet doesn't accelerate because the gas pressure is exhausted. The bullet travels with a constant velocity in the direction it's fired (ignoring the force exerted against the bullet by friction and drag).

Similarly, the pitcher applies a force to the ball by using their muscles when throwing. Once the ball leaves the pitcher's fingers, it doesn't accelerate and the ball travels toward the batter at a constant speed (again ignoring friction and drag). The speed is constant, but the velocity can change if the ball travels along a curve.

Projectile Motion

According to Newton's first law, if no additional forces act on the bullet or ball, it travels in a straight line. This is also true if the bullet is fired in outer space. However, here on Earth, the force of gravity continues to act so the motion of the bullet or ball is affected.

What happens when a bullet is fired from the top of a hill using a rifle held perfectly horizontal? Ignoring air resistance, its horizontal velocity remains constant at its muzzle velocity. Its vertical velocity (which is zero when it leaves the gun barrel) increases because of gravity's acceleration. Each passing second, the bullet traces out the same distance horizontally while increasing distance vertically (shown in the figure below). In the end, the projectile traces out a *parabolic curve*.

Projectile Path for a Bullet Fired Horizontally from a Hill (Ignoring Air Resistance)

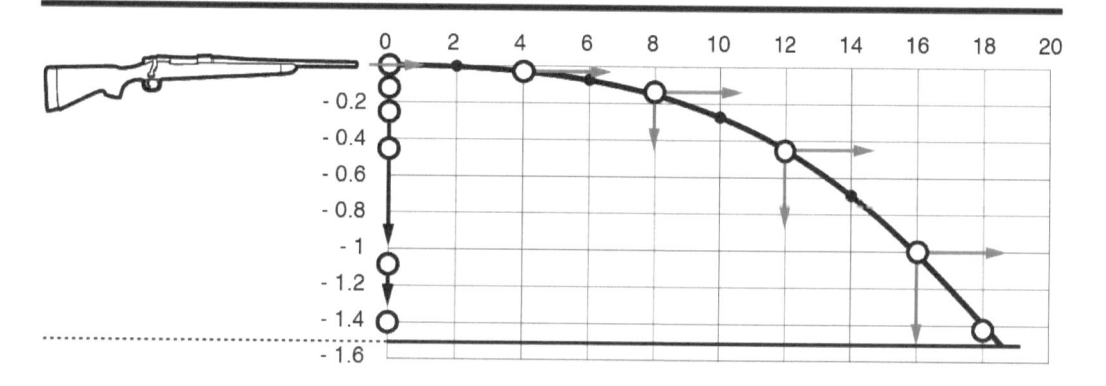

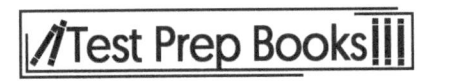

This vertical, downward acceleration is why a pitcher must put an arc on the ball when throwing across home plate. Otherwise the ball will fall at the batter's feet.

It's also interesting to note that if an artillery crew simultaneously drops one cannonball and fires another one horizontally, the two cannonballs will hit the ground at the same time since both balls are accelerating at the same rate and experience the same changes in vertical velocity.

What if air resistance is taken into account? This is best answered by looking at the horizontal and vertical motions separately.

The horizontal velocity is no longer constant because the initial velocity of the projectile is continually reduced by the resistance of the air. This is a complex problem in fluid mechanics, but it's sufficient to note that that the projectile doesn't fly as far before landing as predicted from the simple theory.

The vertical velocity is also reduced by air resistance. However, unlike the horizontal motion where the propelling force is zero after the cannonball is fired, the downward force of gravity acts continuously. The downward velocity increases every second due to the acceleration of gravity. As the velocity increases, the resisting force (called *drag*) increases with the square of the velocity. If the projectile is fired or dropped from a sufficient height, it reaches a terminal velocity such that the upward drag force equals the downward force of gravity. When that occurs, the projectile falls at a constant rate.

This is the same principle that's used for a parachute. Its drag (caused by its shape that scoops up air) is sufficient enough to slow down the fall of the parachutist to a safe velocity, thus avoiding a fatal crash on the ground.

So, what's the bottom line? If the vertical height isn't too great, a real projectile will fall short of the theoretical point of impact. However, if the height of the fall is significant and the drag of the object results in a small terminal fall velocity, then the projectile can go further than the theoretical point of impact.

What if the projectile is launched from a moving platform? In this case, the platform's velocity is added to the projectile's velocity. That's why an object dropped from the mast of a moving ship lands at the base of the mast rather than behind it. However, to an observer on the shore, the object traces out a parabolic arc.

Angular Momentum

In the previous examples, all forces acted through the center of the mass, but what happens if the forces aren't applied through the same line of action, like in the figure below?

A Mass Acted on by Forces Out of Line with Each Other

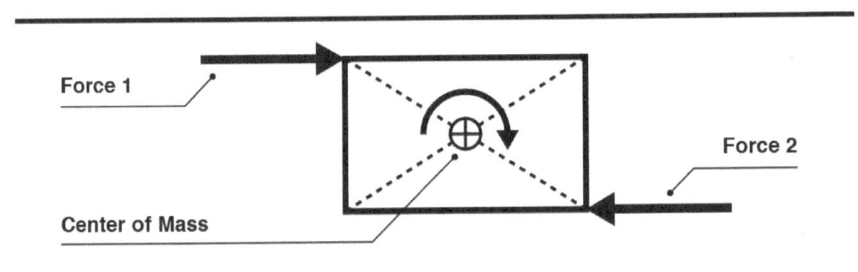

When this happens, the two forces create *torque* and the mass rotates around its center of gravity. In the figure above, the center of gravity is the center of the rectangle ("Center of Mass"), which is determined by the two, intersecting main diagonals. The center of an irregularly shaped object is found by hanging it from two different edges, and the center of gravity is at the intersection of the two "plumb lines."

Newton's second law still applies when the forces form a moment pair, but it must be expressed in terms of angular acceleration and the moment of inertia. The *moment of inertia* is a measure of the body's resistance to rotation, similar to the mass's resistance to linear acceleration. The more compact the body, the less the moment of inertia and the faster it rotates, much like how an ice skater spinning with outstretched arms will speed up as the arms are brought in close to the body.

The concept of torque is important in understanding the use of wrenches and is likely to be on the test. The concept of torque and moment/lever arm will be taken up again below, when the physics of simple machines is presented.

Energy and Work

The previous examples of moving boats, cars, bullets, and baseballs are examples of simple systems that are thought of as particles with forces acting through their center of gravity. They all have one property in common: *energy*. The energy of the system results from the forces acting on it and is considered its ability to do work.

Work or the energy required to do work (which are the same) is calculated as the product of force and distance traveled along the line of action of the force. It's measured in *foot-pounds* in the traditional system (which is still used in workshops and factories) and in *newton meters (N·m)* in the International System of Units (SI), which is the preferred system of measurement today.

Potential and Kinetic Energy

Energy can neither be created nor destroyed, but it can be converted from one form to another. There are many forms of energy, but it's useful to start with mechanical energy and potential energy.

The *potential energy* of an object is equal to the work that's required to lift it from its original elevation to its current elevation. This is calculated as the weight of the object or its downward force (mass times the acceleration of gravity) multiplied by the distance (*y*) it is lifted above the reference elevation or "datum." This is written:

$$PE = mgy$$

The mechanical or *kinetic energy* of a system is related to its mass and velocity and involves the energy of motion:

$$KE = \frac{1}{2}mv^2$$

The *total energy* is the sum of the kinetic energy and the potential energy, both of which are measured in foot-pounds or newton meters.

If a weight with a mass of 10 kilograms is raised up a ladder to a height of 10 meters, it has a potential energy of:

$$10 \text{ m} \times 10 \text{ kg} \times 9.81 \frac{\text{m}}{\text{s}^2} = 981 \text{ N} \cdot \text{m}$$

This is approximately 1000 newton meters if the acceleration of gravity (9.81 m/s^2) is rounded to 10 m/s^2, which is accurate enough for most earth-bound calculations. It has zero kinetic energy because it's at rest, with zero velocity.

If the weight is dropped from its perch, it accelerates downward so that its velocity and kinetic energy increase as its potential energy is "used up" or, more precisely, converted to kinetic energy.

When the weight reaches the bottom of the ladder, just before it hits the ground, it has a kinetic energy of 981 N·m (ignoring small losses due to air resistance).

When the 10-kilogram weight hits the ground, its potential energy (which was measured *from* the ground) and its velocity are both zero, so its kinetic energy is also zero. What's happened to the energy? It's dissipated into heat, noise, and kicking up some dust. It's important to remember that energy can neither be created nor destroyed, so it can only change from one form to another.

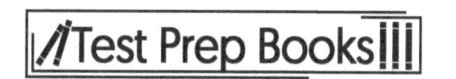

The conversion between potential and kinetic energy works the same way for a pendulum. If it's raised and held at its highest position, it has maximum potential energy but zero kinetic energy.

Potential and Kinetic Energy for a Swinging Pendulum

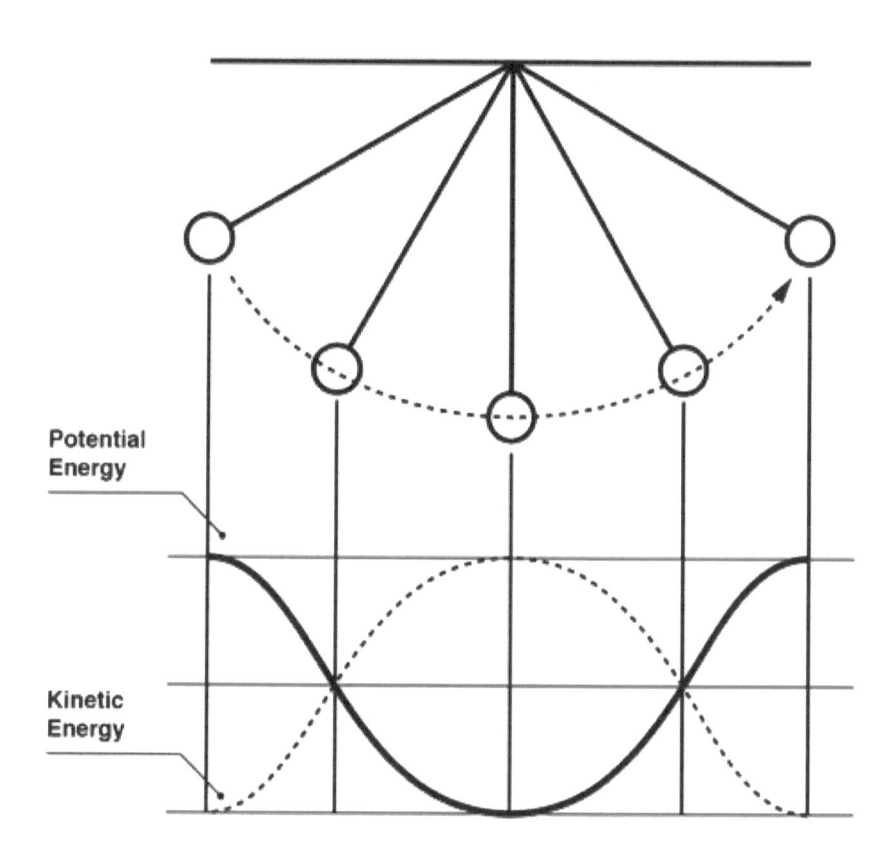

When the pendulum is released from its highest position (see left side of the figure above), it swings down so that its kinetic energy increases as its potential energy decreases. At the bottom of its swing, the pendulum is moving at its maximum velocity with its maximum kinetic energy. As the pendulum swings past the bottom of its path, its velocity slows down as its potential energy increases.

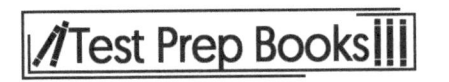

Work

The released potential energy of a system can be used to do *work*.

For instance, most of the energy lost by letting a weight fall freely can be recovered by hooking it up to a pulley to do work by pulling another weight back up (as shown in the figure below).

Using the Energy of a Falling Weight to Raise Another Weight

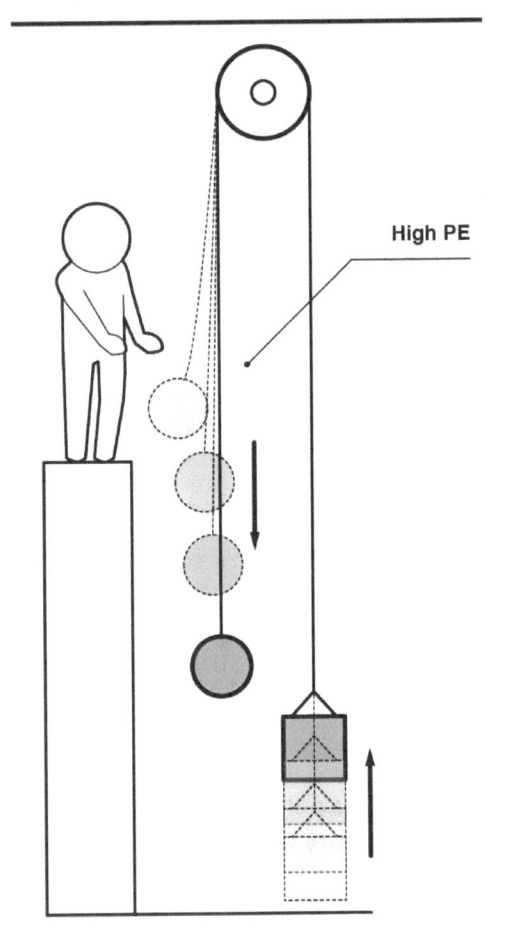

In other words, the potential energy expended to lower the weight is used to do the work of lifting another object. Of course, in a real system, there are losses due to friction. The action of pulleys will be discussed later in this study guide.

Since *energy* is defined as *the capacity to do work*, energy and work are measured in the same units:

$$Energy = Work = Force \times Distance$$

Force is measured in *newtons (N)*. Distance is measured in meters. The units of work are *newton meters (N·m)*. The same is true for kinetic energy and potential energy.

Another way to store energy is to compress a spring. Energy is stored in the spring by stretching or compressing it. The work required to shorten or lengthen the spring is given by the equation:

$$F = k \times d$$

Here, "d" is the length in meters and "k" is the resistance of the spring constant (measured in N × m), which is a constant as long as the spring isn't stretched past its elastic limit. The resistance of the spring is constant, but the force needed to compress the spring increases with each millimeter it's pushed.

The potential energy stored in the spring is equal to the work done to compress it, which is the total force times the change in length.

The potential energy in the spring is stored by locking it into place, and the work energy used to compress it is recovered when the spring is unlocked. It's the same when dropping a weight from a height—the energy doesn't have to be wasted. In the case of the spring, the energy is used to propel an object.

Potential and Kinetic Energy of a Spring

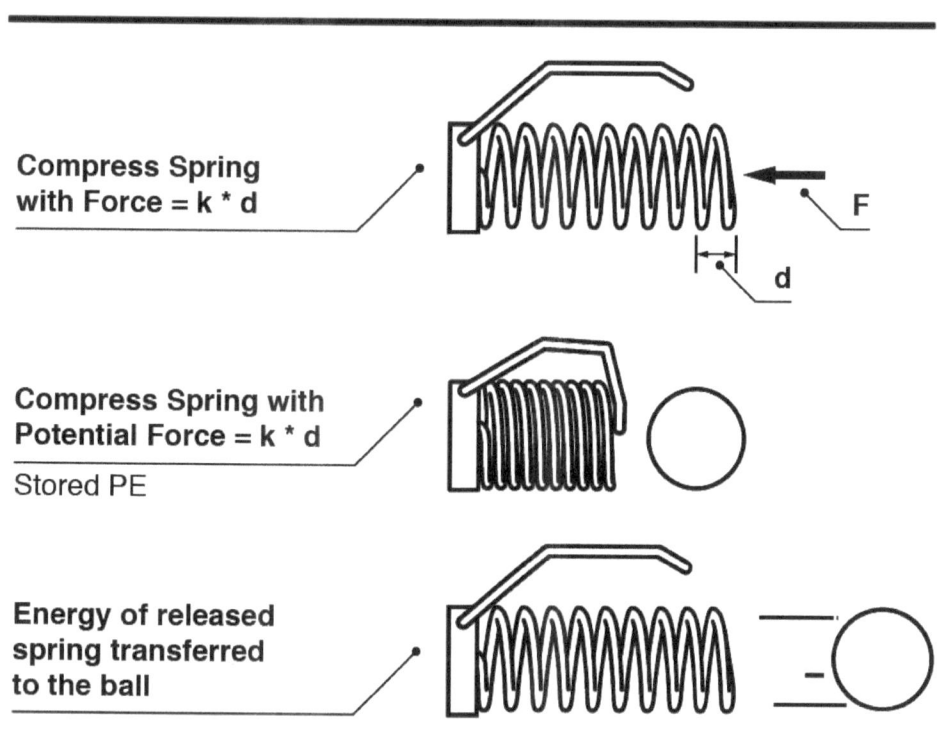

Pushing a block horizontally along a rough surface requires work. In this example, the work needs to overcome the force of friction, which opposes the direction of the motion and equals the weight of the block times a *friction factor (f)*. The friction factor is greater for rough surfaces than smooth surfaces,

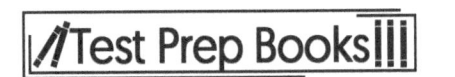

and it's usually greater *before* the motion starts than after it has begun to slide. These terms are illustrated in the figure below.

Pushing a Block Horizontally Against the Force of Friction

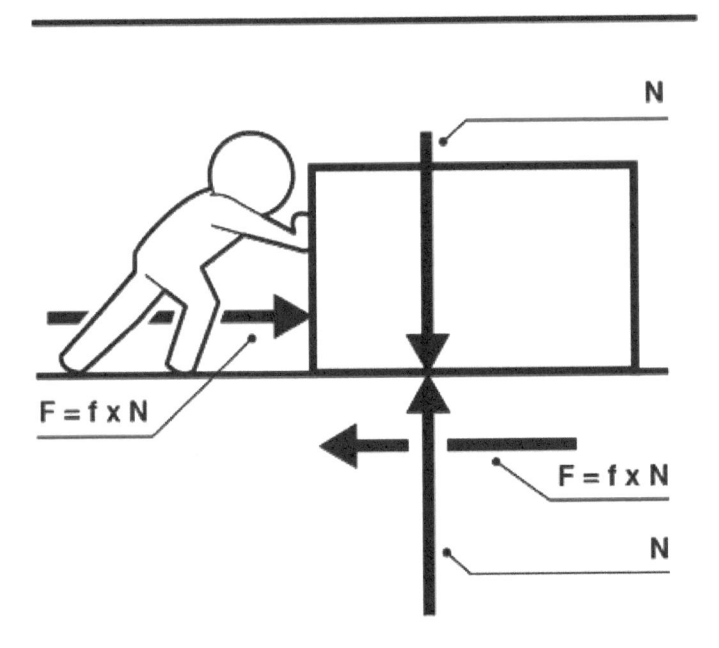

When pushing a block, there's no increase in potential energy since the block's elevation doesn't change. Expending the energy to overcome friction is "wasted" in the generation of heat. Yet, to move a block from point A to point B, an energy cost must be paid. However, friction isn't always a hindrance. In fact, it's the force that makes the motion of a wheel possible.

Heat energy can also be created by burning organic fuels, such as wood, coal, natural gas, and petroleum. All of these are derived from plant matter that's created using solar energy and photosynthesis. The chemical energy or *"heat"* liberated by the combustion of these fuels is used to warm buildings during the winter or even melt metal in a foundry. The heat is also used to generate steam, which can drive engines or turn turbines to generate electric energy.

In fact, work and heat are interchangeable. This fact was first recognized by gun founders when they were boring out cast, brass cannon blanks. The cannon blanks were submerged in a water bath to reduce friction, yet as the boring continued, the water bath boiled away!

Later, the amount of work needed to raise the temperature of water was measured by an English physicist (and brewer) named James Prescott Joule. The way that Joule measured the mechanical equivalent of heat is illustrated in the figure below. This setup is similar to the one in the figure above with the pulley, except instead of lifting another weight, the falling weight's potential energy is converted to the mechanical energy of the rotating vertical shaft. This turns the paddles, which churns the water to increase its temperature. Through a long series of repeated measurements, Joule showed

that 4186 N·m of work was necessary to raise the temperature of one kilogram of water by one degree Celsius, no matter how the work was delivered.

Device Measuring the Mechanical Energy Needed to Increase the Temperature of Water

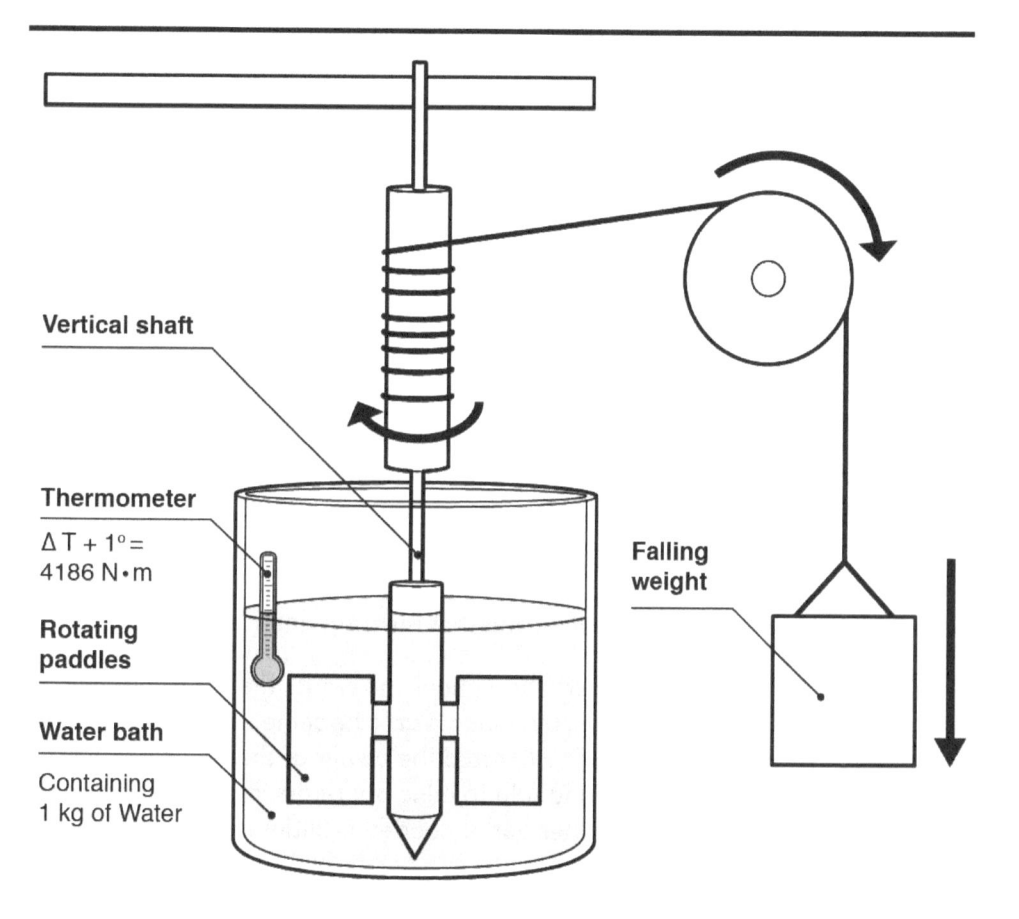

Vertical shaft

Thermometer

$\Delta T + 1° =$ 4186 N·m

Falling weight

Rotating paddles

Water bath

Containing 1 kg of Water

In recognition of this experiment, the newton meter is also called a *"joule."* Linking the names for work and heat to the names of two great physicists is truly appropriate because heat and work being interchangeable is of the greatest practical importance. These two men were part of a very small, select group of scientists for whom units of measurement have been named: Marie Curie for radioactivity, Blaise Pascal for pressure, James Watt for power, Andre Ampere for electric current, and only a few others.

Just as mechanical work is converted into heat energy, heat energy is converted into mechanical energy in the reverse process. An example of this is a closely fitting piston supporting a weight and mounted in a cylinder where steam enters from the bottom.

In this example, water is heated into steam in a boiler, and then the steam is drawn off and piped into a cylinder. Steam pressure builds up in the piston, exerting a force in all directions. This is counteracted by the tensile strength of the cylinder; otherwise, it would burst. The pressure also acts on the exposed face of the piston, pushing it upwards against the load (displacing it) and thus doing work.

Work developed from the pressure acting over the area exerts a force on the piston as described in the following equation:

$$Work = Pressure \times Piston\ Area \times Displacement$$

Here, the work is measured in newton meters, the pressure in newtons per square meter or *pascals (Pa)*, and the piston displacement is measured in meters.

Since the volume enclosed between the cylinder and piston increases with the displacement, the work can also be expressed as:

$$Work = Pressure \times \Delta Volume$$

If steam with a pressure slightly greater than this value is piped into the cylinder, it slowly lifts the load. If steam at a much higher pressure is suddenly admitted to the cylinder, it throws the load into the air. This is the principle used to steam-catapult airplanes off the deck of an aircraft carrier.

Power

Power is defined as the rate at which work is done, or the time it takes to do a given amount of work. In the International System of Units (SI), work is measured in *newton meters (N·m)* or *joules (J)*. Power is measured in joules/second or *watts (W)*.

For example, to raise a 1-kilogram mass one meter off the ground, it takes approximately 10 newton meters of work (approximating the gravitational acceleration of 9.81 m/s^2 as 10 m/s^2). To do the work in 10 seconds, it requires 1 watt of power. Doing it in 1 second requires 10 watts of power. Essentially, *doing it faster means dividing by a smaller number*, and that means greater power.

Although SI units are preferred for technical work throughout the world, the old traditional (or English) unit of measuring power is still used. Introduced by *James Watt* (the same man for whom the SI unit of power "watt" is named), the unit of *horsepower (HP)* rated the power of the steam engines that he and his partner (Matthew Boulton) manufactured and sold to mine operators in 18th century England. The mine operators used these engines to pump water out of flooded facilities in the beginning of the Industrial Revolution.

To provide a measurement that the miners would be familiar with, Watt and Boulton referenced the power of their engines with the "power of a horse."

Watt's measurements showed that, on average, a well-harnessed horse could lift a 330-pound weight 100 feet up a well in one minute (330 pounds is the weight of a 40-gallon barrel filled to the brim). Remembering that power is expressed in terms of energy or work per unit time, horsepower came to be measured as:

$$1\ HP = \frac{100\ feet \times 330\ pounds}{1\ minute} \times \frac{1\ minute}{60\ seconds} = 550\ foot\ pounds/second$$

A horse that pulled the weight up faster, or pulled up more weight in the same time, was a more *powerful* horse than Watt's "average horse."

Hundreds of millions of engines of all types have been built since Watt and Boulton started manufacturing their products, and the unit of *horsepower* has been used throughout the world to this

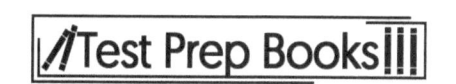

day. Of course, modern technicians and engineers still need to convert horsepower to watts to work with SI units. An approximate conversion is 1 HP = 746 W.

Fluids

In addition to the behavior of solid particles acted on by forces, it is important to understand the behavior of fluids. Fluids include both liquids and gasses. The best way to understand fluid behavior is to contrast it with the behavior of solids, as shown in the figure below.

First, consider a block of ice, which is solid water. If it is set down inside a large box it will exert a force on the bottom of the box due to its weight as shown on the left, in Part A of the figure. The solid block exerts a pressure on the bottom of the box equal to its total weight divided by the area of its base:

$$Pressure = Weight\ of\ block/Area\ of\ base$$

That pressure acts only in the area directly under the block of ice.

If the same mass of ice is melted, it behaves much differently. It still has the same weight as before because its mass hasn't changed. However, the volume has decreased because liquid water molecules are more tightly packed together than ice molecules, which is why ice floats (it is less dense).

The Behavior of Solids and Liquids Compared

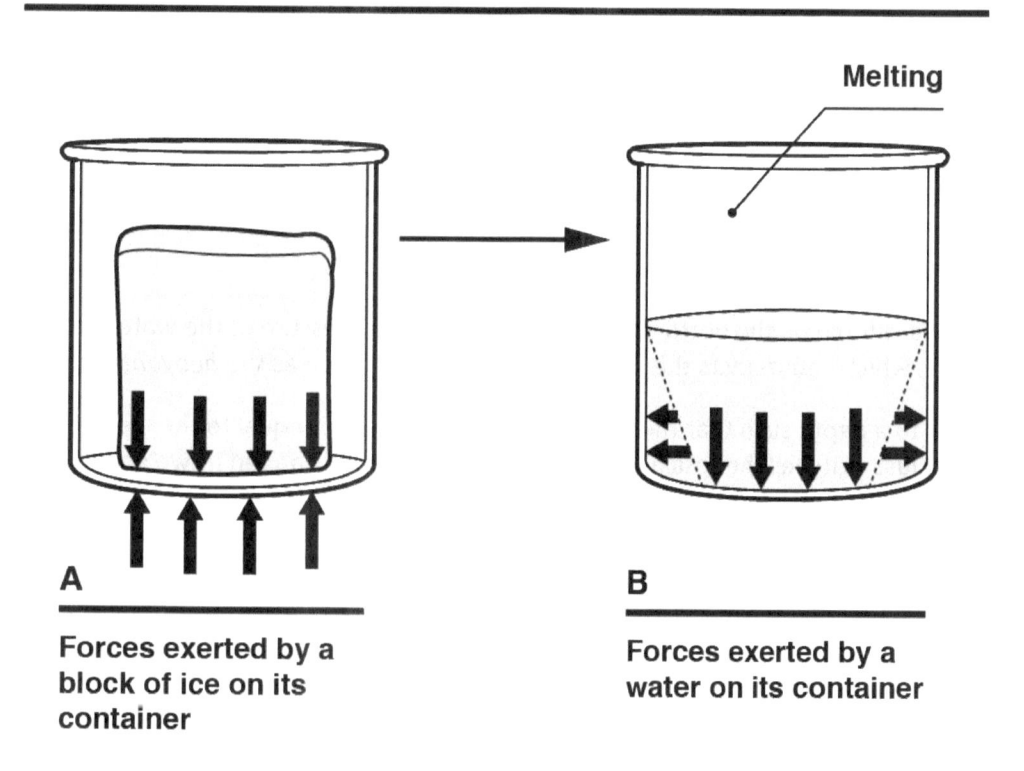

A

Forces exerted by a block of ice on its container

B

Forces exerted by a water on its container

The melted ice (now water) conforms to the shape of the container. This means that the fluid exerts pressure not only on the base, but on the sides of the box at the water line and below. Actually, pressure in a liquid is exerted in all directions, but all the forces in the interior of the fluid cancel each

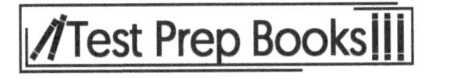

other out, so that a net force is only exerted on the walls. Note also that the pressure on the walls increases with the depth of the water.

The fact that the liquid exerts pressure in all directions is part of the reason some solids float in liquids. Consider the forces acting on a block of wood floating in water, as shown in the figure below.

Floatation of a Block of Wood

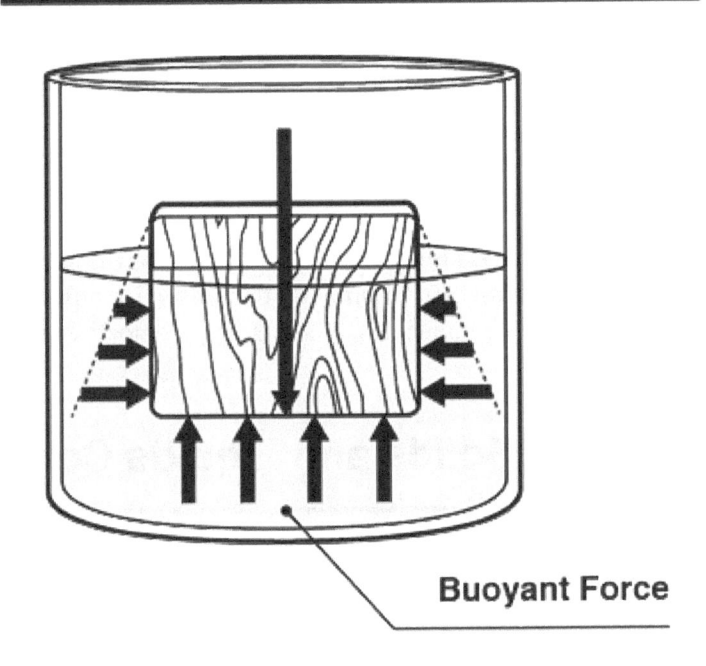

Buoyant Force

The block of wood is submerged in the water and pressure acts on its bottom and sides as shown. The weight of the block tends to force it down into the water. The force of the pressure on the left side of the block just cancels the force of the pressure on the right side.

There is a net upward force on the bottom of the block due to the pressure of the water acting on that surface. This force, which counteracts the weight of the block, is known as the *buoyant force*.

The block will sink to a depth such that the buoyant force of the water (equal to the weight of the volume displaced) just matches the total weight of the block. This will happen if two conditions are met:

1. The body of water is deep enough to float the block
2. The density of the block is less than the density of the water

If the body of water is not deep enough, the water pressure on the bottom side of the block won't be enough to develop a buoyant force equal to the block's weight. The block will be "beached" just like a boat caught at low tide.

If the density of the block is greater than the density of the fluid, the buoyant force acting on the bottom of the boat will not be sufficient to counteract the total weight of the block. That's why a solid steel block will sink in water.

If steel is denser than water, how can a steel ship float? The steel ship floats because it's hollow. The volume of water displaced by its steel shell (hull) is heavier than the entire weight of the ship and its contents (which includes a lot of empty space). In fact, there's so much empty space within a steel ship's hull that it can bob out of the water and be unstable at sea if some of the void spaces (called ballast tanks) aren't filled with water. This provides more weight and balance (or "trim") to the vessel.

The discussion of buoyant forces on solids holds for liquids as well. A less dense liquid can float on a denser liquid if they're *immiscible* (do not mix). For instance, oil can float on water because oil isn't as dense as the water. Fresh water can float on salt water for the same reason.

Pascal's law states that a change in pressure, applied to an enclosed fluid, is transmitted undiminished to every portion of the fluid and to the walls of its containing vessel. This principle is used in the design of hydraulic jacks, as shown in the figure below.

A force (F_1) is exerted on a small "driving" piston, which creates pressure on the hydraulic fluid. This pressure is transmitted through the fluid to a large cylinder. While the pressure is the same everywhere in the oil, the pressure action on the area of the larger cylinder creates a much higher upward force (F_2).

Illustration of a Hydraulic Jack Exemplifying Pascal's Law

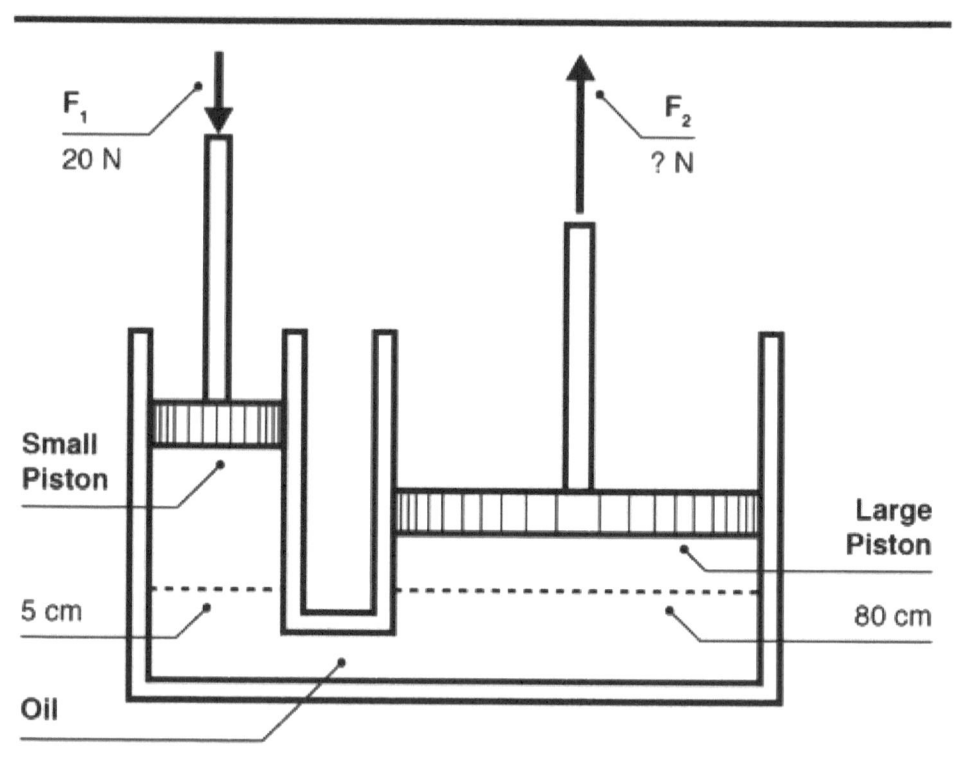

Looking again at the figure above, suppose the diameter of the small cylinder is 5 centimeters and the diameter of the large cylinder is 80 centimeters. If a force of 20 newtons (N) is exerted on the small driving piston, what's the value of the upward force F_2? In other words, what weight can the large piston support?

The pressure within the system is created from the force F_1 acting over the area of the piston:

$$P = \frac{F_1}{A} = \frac{20\ \text{N}}{\pi\,(0.05\ \text{m})^2/4} = 10{,}185\ \text{Pa}$$

The same pressure acts on the larger piston, creating the upward force, F_2:

$$F_2 = P \times A = 10{,}185\ \text{Pa} \times \pi \times (0.8\ \text{m})^2/4 = 5120\ \text{N}$$

Because a liquid has no internal shear strength, it can be transported in a pipe or channel between two locations. A fluid's "rate of flow" is the volume of fluid that passes a given location in a given amount of time and is expressed in $m^3/second$. The *flow rate* (Q) is determined by measuring the *area of flow* (A) in m^2, and the *flow velocity* (v) in *m/s*:

$$Q = v \times A$$

This equation is called the *Continuity Equation*. It's one of the most important equations in engineering and should be memorized.

It's important to understand that, for a given flow rate, a smaller pipe requires a higher velocity.

The energy of a flow system is evaluated in terms of potential and kinetic energy, the same way the energy of a falling weight is evaluated. The total energy of a fluid flow system is divided into potential energy of elevation, and pressure and the kinetic energy of velocity. *Bernoulli's Equation* states that, for a constant flow rate, the total energy of the system (divided into components of elevation, pressure, and velocity) remains constant. This is written as:

$$Z + \frac{P}{\rho g} + \frac{v^2}{2g} = Constant$$

Each of the terms in this equation has dimensions of meters. The first term is the *elevation energy*, where Z is the elevation in meters. The second term is the *pressure energy*, where P is the pressure, ρ is the density, and g is the acceleration of gravity. The dimensions of the second term are also in meters. The third term is the *velocity energy*, also expressed in meters.

For a fixed elevation, the equation shows that, as the pressure increases, the velocity decreases. In the other case, as the velocity increases, the pressure decreases.

The use of the Bernoulli Equation is illustrated in the figure below. The total energy is the same at Sections 1 and 2. The area of flow at Section 1 is greater than the area at Section 2. Since the flow rate is the same at each section, the velocity at Point 2 is higher than at Point 1:

$$Q = V_1 \times A_1 = V_2 \times A_2, \qquad V_2 = V_1 \times \frac{A_1}{A_2}$$

Finally, since the total energy is the same at the two sections, the pressure at Point 2 is less than at Point 1. The tubes drawn at Points 1 and 2 would actually have the water levels shown in the figure; the

pressure at each point would support a column of water of a height equal to the pressure divided by the unit weight of the water ($h = P/\rho g$).

An Example of Using the Bernoulli Equation

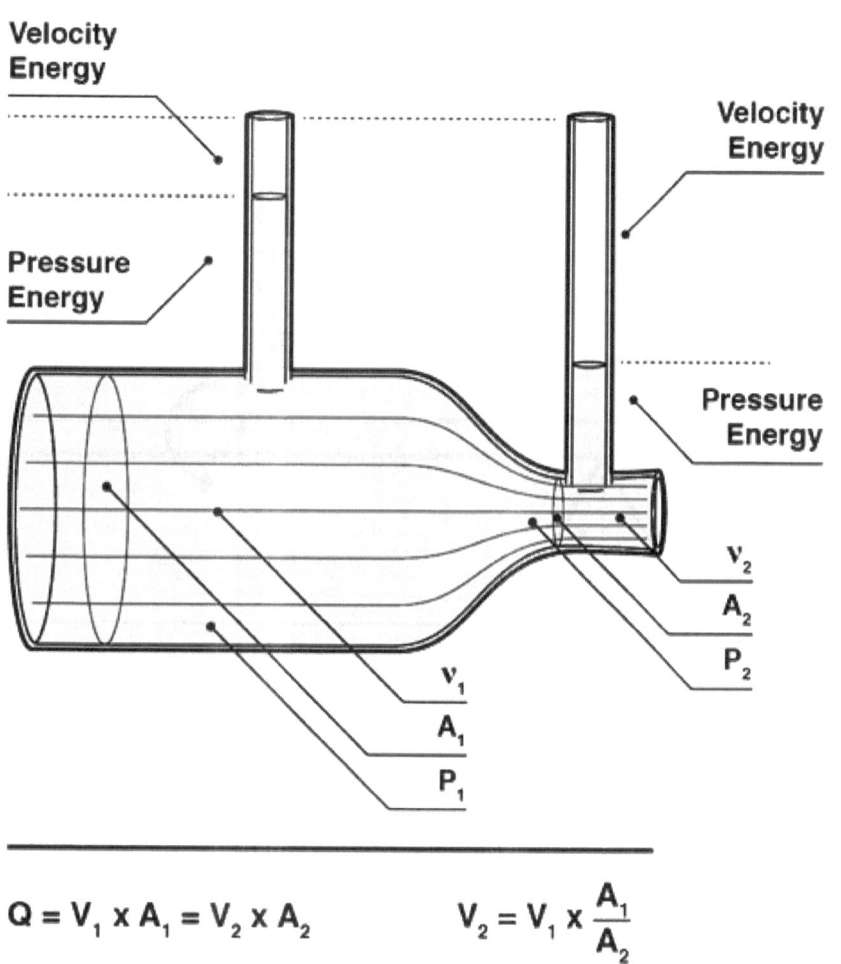

$$Q = V_1 \times A_1 = V_2 \times A_2 \qquad V_2 = V_1 \times \frac{A_1}{A_2}$$

Machines

Now that the basic physics of work and energy have been discussed, the common machines used to do the work can be discussed in more detail.

A *machine* is a device that: transforms energy from one form to another, multiplies the force applied to do work, changes the direction of the resultant force, or increases the speed at which the work is done.

The details of how energy is converted into work by a system are extremely complicated but, no matter how complicated the "linkage" between the components, every system is composed of certain elemental or simple machines. These are discussed briefly in the following sections.

Levers

The simplest machine is a *lever*, which consists of two pieces or components: a *bar* (or beam) and a *fulcrum* (the pivot-point around which motion takes place). As shown below, the *effort* acts at a distance (L_1) from the fulcrum and the *load* acts at a distance (L_2) from the fulcrum.

Components of a Lever

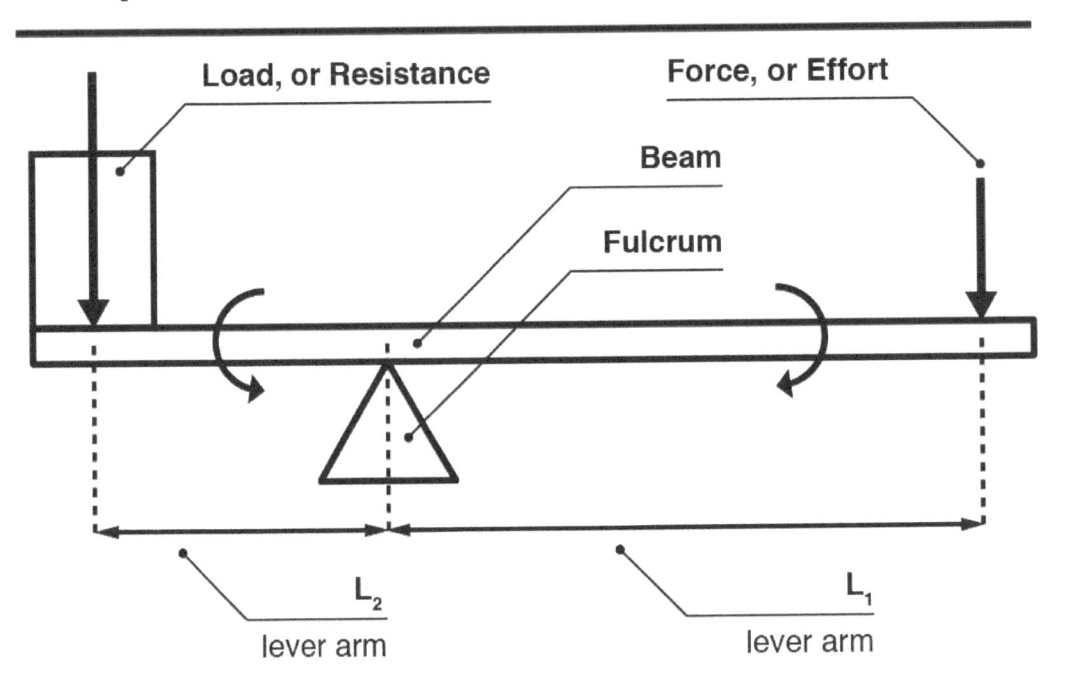

These lengths L_1 and L_2 are called *lever arms*. When the lever is balanced, the load (R) times its lever arm (L_2) equals the effort (F) times its lever arm (L_1). The force needed to lift the load is:

$$F = R \times \frac{L_2}{L_1}$$

This equation shows that as the lever arm L_1 is increased, the force required to lift the resisting load (R) is reduced. This is why Archimedes, one of the leading ancient Greek scientists, said, "Give me a lever long enough, and a place to stand, and I can move the Earth."

The ratio of the moment arms is the so-called "mechanical advantage" of the simple lever; the effort is multiplied by the mechanical advantage. For example, a 100-kilogram mass (a weight of approximately 1000 N) is lifted with a lever like the one in the figure below, with a total length of 3 meters, and the fulcrum situated 50 centimeters from the left end. What's the force needed to balance the load?

$$F = 1000 \text{ N} \times \frac{0.5 \text{ meters}}{2.5 \text{ meters}} = 200 \text{ N}$$

Depending on the location of the load and effort with respect to the fulcrum, three "classes" of lever are recognized. In each case, the forces can be analyzed as described above.

The Three Classes of Levers

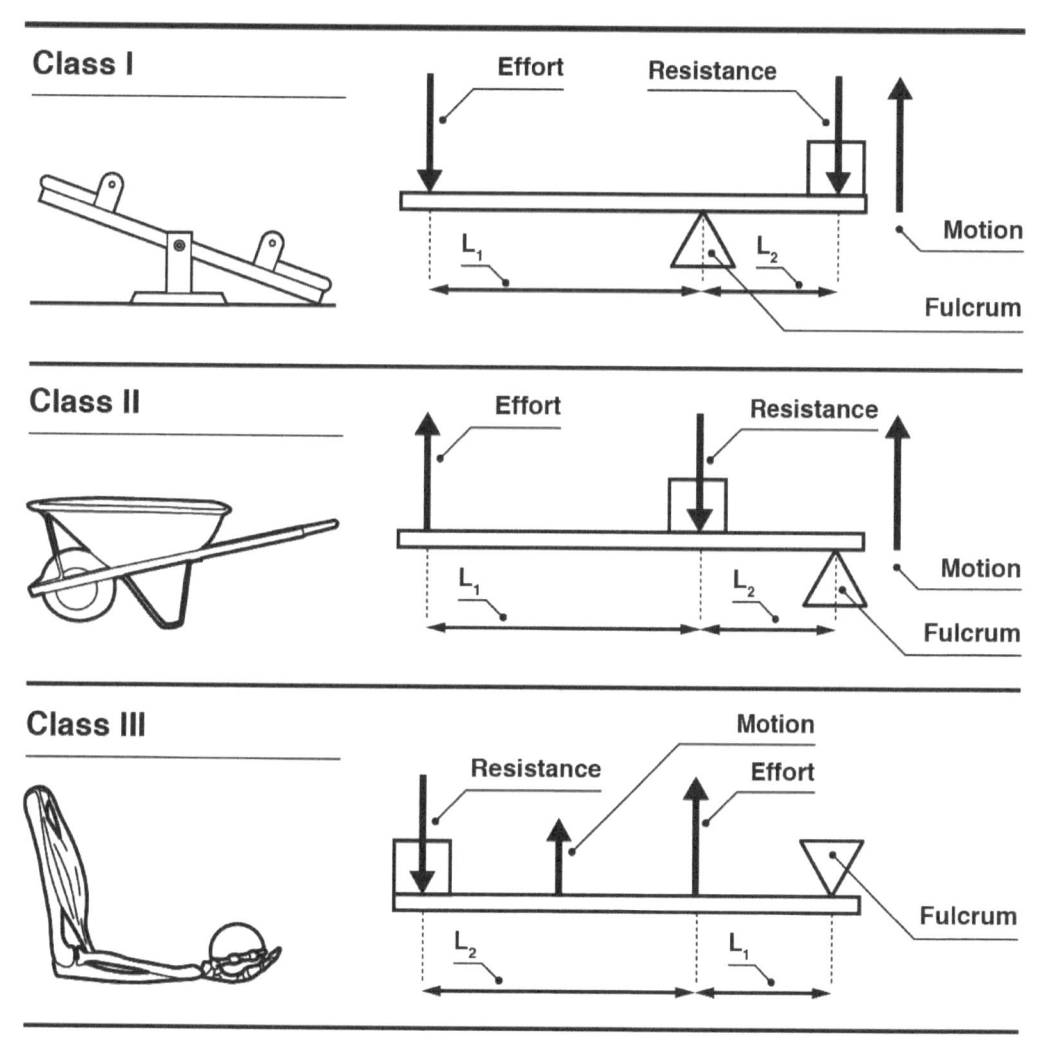

As seen in the figure, a *Class I* lever has the fulcrum positioned between the effort and the load. Examples of Class I levers include see-saws, balance scales, crow bars, and scissors. As explained above, the force needed to balance the load is $F = R \times (L_2/L_1)$, which means that the mechanical advantage is L_2/L_1. The crane boom shown back in the first figure in this section was a Class I lever, where the tower acted as the fulcrum and the counterweight on the left end of the boom provided the effort.

For a *Class II* lever, the load is placed between the fulcrum and the effort. A wheel barrow is a good example of a Class II lever. The mechanical advantage of a Class II lever is $(L_1 + L_2)/L_2$.

For a *Class III* lever, the effort is applied at a point between the fulcrum and the load, which increases the speed at which the load is moved. A human arm is a Class III lever, with the elbow acting as the fulcrum. The mechanical advantage of a Class III lever is $(L_1 + L_2)/L_1$.

Wheels and Axles

The wheel and axle is a special kind of lever. The *axle*, to which the load is applied, is set perpendicular to the *wheel* through its center. Effort is then applied along the rim of the wheel, either with a cable running around the perimeter or with a *crank* set parallel to the axle.

The mechanical advantage of the wheel and axle is provided by the moment arm of the perimeter cable or crank. Using the center of the axle (with a radius of *r*) as the fulcrum, the resistance of the load (*L*) is just balanced by the effort (*F*) times the wheel radius:

$$F \times R = L \times r \quad \text{or} \quad F = L \times \frac{r}{R}$$

This equation shows that increasing the wheel's radius for a given shaft reduces the required effort to carry the load. Of course, the axle must be made of a strong material or it'll be twisted apart by the applied torque. This is why steel axles are used.

Gears, Belts, and Cams

The functioning of a wheel and axle can be modified with the use of gears and belts. *Gears* are used to change the direction or speed of a wheel's motion.

The direction of a wheel's motion can be changed by using *beveled gears*, with the shafts set at right angles to each other, as shown in part *A* in the figure below.

The speed of a wheel can be changed by meshing together *spur gears* with different diameters. A small gear (A) is shown driving a larger gear (B) in the middle section *(B)* in the figure below. The gears rotate in opposite directions; if the driver, Gear A, moves clockwise, then Gear B is driven counter-clockwise. Gear B rotates at half the speed of the driver, Gear A. In general, the change in speed is given by the ratio of the number of teeth in each gear:

$$\frac{Rev_{Gear\ B}}{Rev_{Gear\ A}} = \frac{\text{Number of Teeth in A}}{\text{Number of Teeth in B}}$$

Rather than meshing the gears, *belts* are used to connect them as shown in part *(C)*.

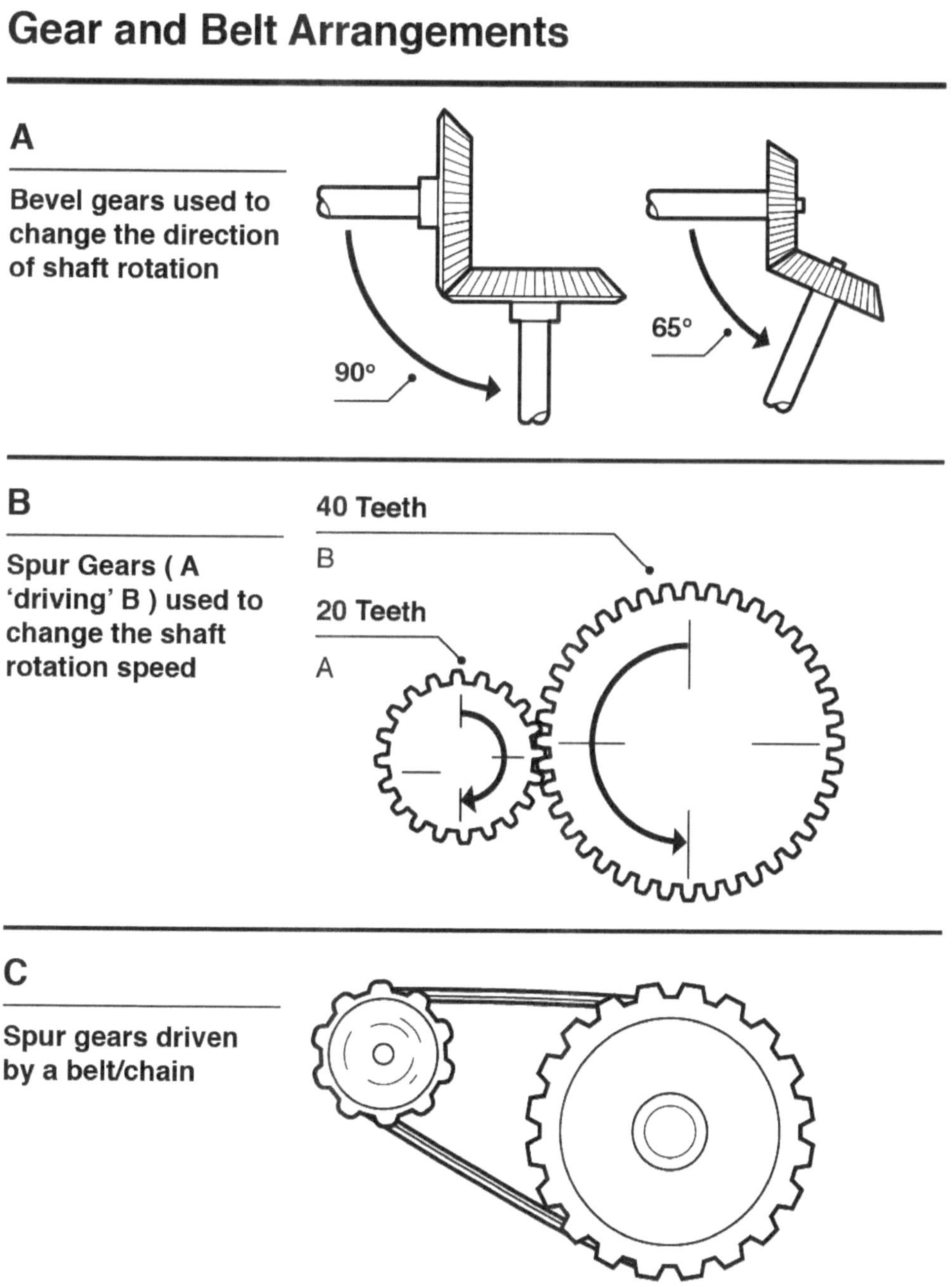

Gear and Belt Arrangements

A

Bevel gears used to change the direction of shaft rotation

90°

65°

B

Spur Gears (A 'driving' B) used to change the shaft rotation speed

40 Teeth

B

20 Teeth

A

C

Spur gears driven by a belt/chain

Gears can change the speed and direction of the axle rotation, but the rotary motion is maintained. To convert the rotary motion of a gear train into linear motion, it's necessary to use a *cam* (a type of off-centered wheel shown in the figure below, where rotary shaft motion lifts the valve in a vertical direction.

Conversion of Rotary to Vertical Linear Motion with a Cam

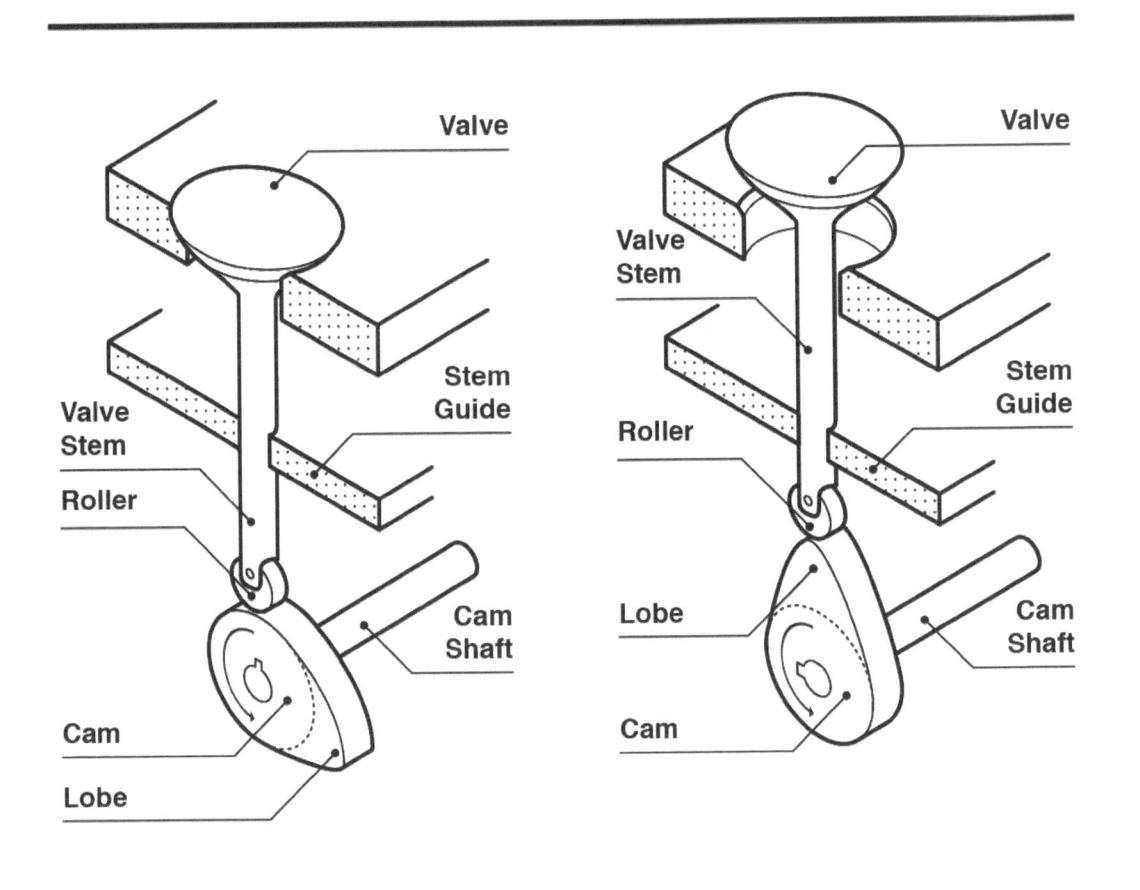

Pulleys

A *pulley* looks like a wheel and axle, but provides a mechanical advantage in a different way. A *fixed pulley* was shown previously as a way to capture the potential energy of a falling weight to do useful work by lifting another weight. As shown in part *A* in the figure below, the fixed pulley is used to change the direction of the downward force exerted by the falling weight, but it doesn't provide any mechanical advantage.

The lever arm of the falling weight (A) is the distance between the rim of the fixed pulley and the center of the axle. This is also the length of the lever arm acting on the rising weight (B), so the ratio of the two arms is 1:0, meaning there's no mechanical advantage. In the case of a wheel and axle, the mechanical advantage is the ratio of the wheel radius to the axle radius.

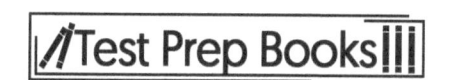

A *moving pulley*, which is really a Class II lever, provides a mechanical advantage of 2:1 as shown below on the right side of the figure *(B)*.

Fixed-Block Versus Moving-Block Pulleys

A

Single Fixed Block with No
Mechanical Advantage

B

Single Moving Block with 2:1
Mechanical Advantage

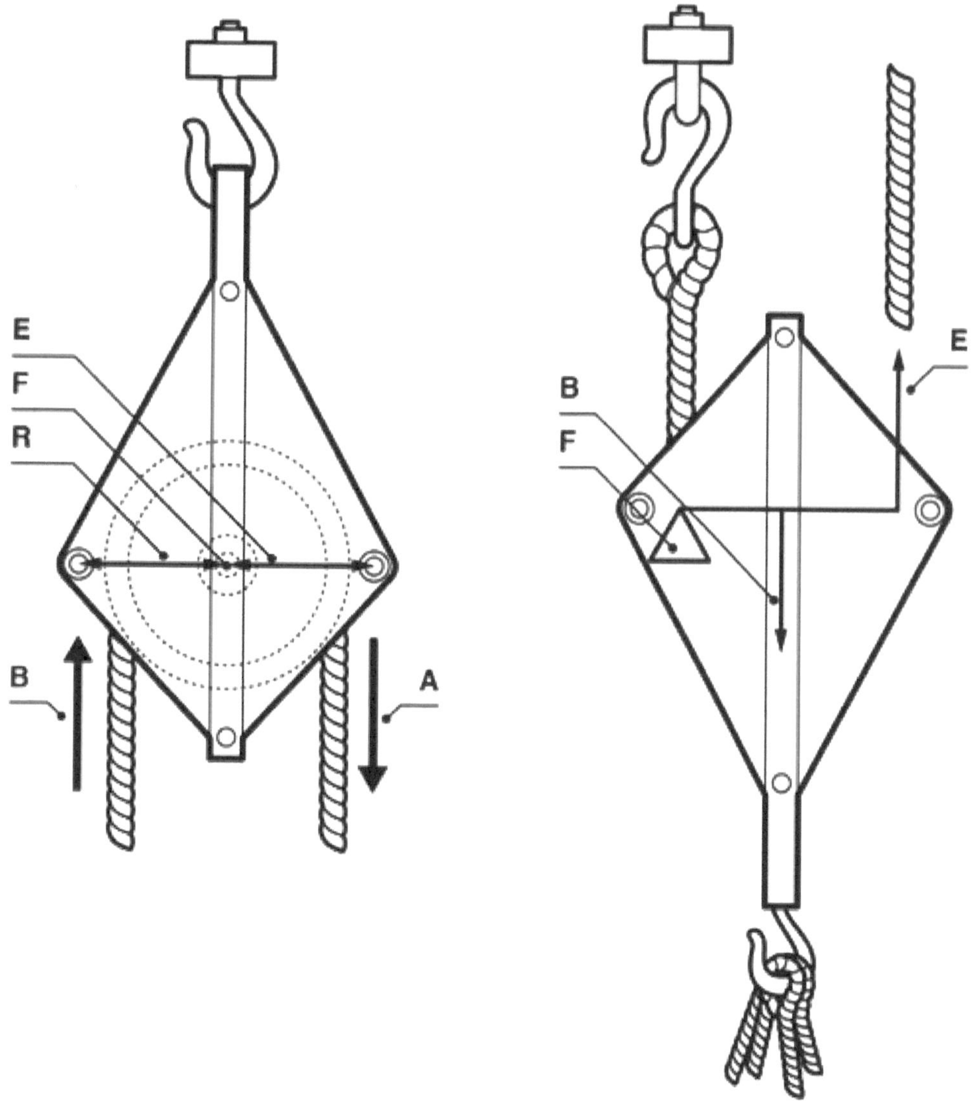

As demonstrated by the rigs in the figure below, using a wider moving block with multiple sheaves can achieve a greater mechanical advantage.

Single-Acting and Double-Acting Block and Tackles

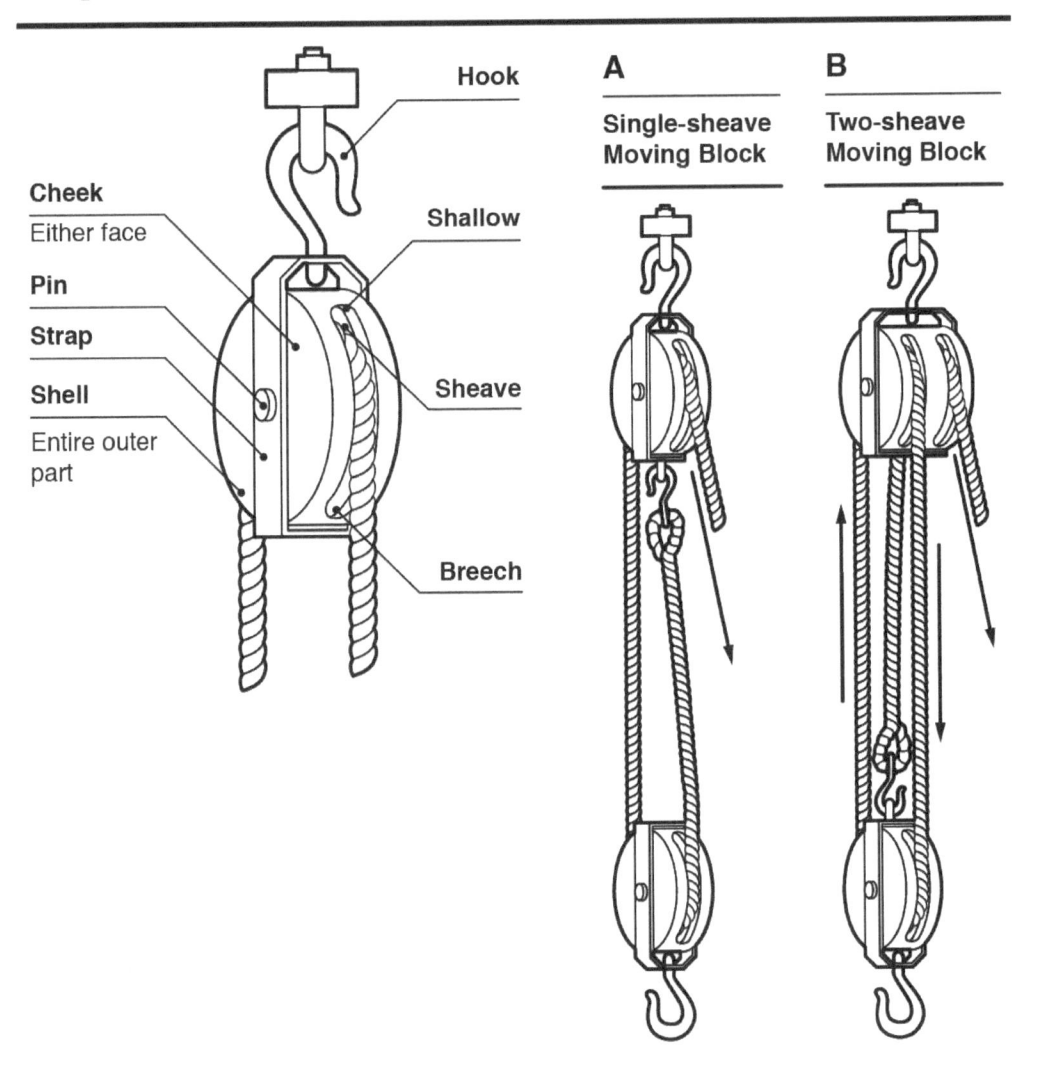

The mechanical advantage of the multiple-sheave block and tackle is approximated by counting the number of ropes going to and from the moving block. For example, there are two ropes connecting the moving block to the fixed block in part *A* of the figure above, so the mechanical advantage is 2:1. There are three ropes connecting the moving and fixed blocks in part *B*, so the mechanical advantage is 3:1. The advantage of using a multiple-sheave block is the increased hauling power obtained, but there's a cost; the weight of the moving block must be overcome, and a multiple-sheave block is significantly heavier than a single-sheave block.

Ramps
The *ramp* (or inclined plane) has been used since ancient times to move massive, extremely heavy objects up to higher positions, such as in the pyramids of the Middle East and Central America.

For example, to lift a barrel straight up to a height (*H*) requires a force equal to its weight (*W*). However, the force needed to lift the barrel is reduced by rolling it up a ramp, as shown below. So, if the ramp is *D* meters long and *H* meters high, the force (*F*) required to roll the weight (*W*) up the ramp is:

$$F = \frac{H}{D} \times W$$

Definition Sketch for a Ramp or Inclined Plane

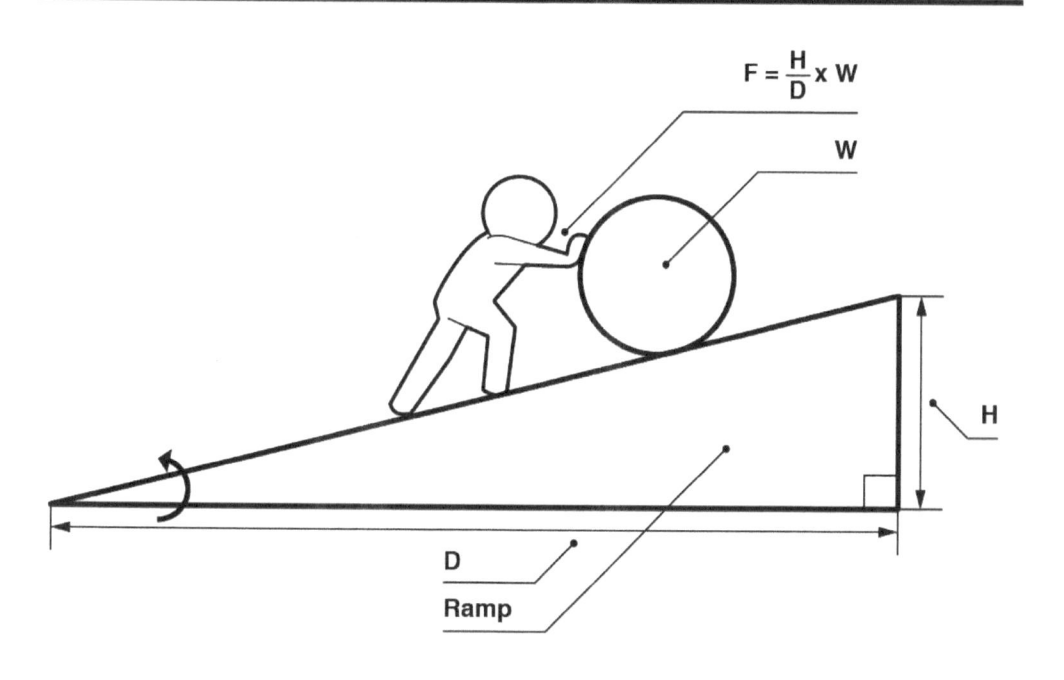

For a fixed height and weight, the longer the ramp, the less force must be applied. Remember, though, that the useful work done (in *N-m*) is the same in either case and is equal to $W \times H$.

Wedges

If an incline or ramp is imagined as a right triangle like in the figure above, then a *wedge* would be formed by placing two inclines (ramps) back to back (or an isosceles triangle). A wedge is one of the six simple machines and is used to cut or split material. It does this by being driven for its full length into the material being cut. This material is then forced apart by a distance equal to the base of the wedge. Axes, chisels, and knives work on the same principle.

Screws

Screws are used in many applications, including vises and jacks. They are also used to fasten wood and other materials together. A screw is thought of as an inclined plane wrapped around a central cylinder. To visualize this, one can think of a barbershop pole, or cutting the shape of an incline (right triangle) out of a sheet of paper and wrapping it around a pencil (as in part *A* in the figure below). Threads are

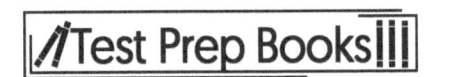

made from steel by turning round stock on a lathe and slowly advancing a cutting tool (a wedge) along it, as shown in part *B*.

Definition Sketch for a Screw and Its Use in a Car Jack

A

Theoretical formation of a thread

P
Pitch

B

Cutting a steel screw thread on a lathe

Workpiece rotation

L
Lenght of threaded portion

P
Pitch

D
Diameter

Turning tool

Feed direction

C

Use of a screw in a car jack

P
Pitch

R

$2 \pi R$

The application of a simple screw in a car jack is shown in part *C* in the figure above. The mechanical advantage of the jack is derived from the pitch of the screw winding. Turning the handle of the jack one revolution raises the screw by a height equal to the *screw pitch (p)*. If the handle has a length *R*, the

distance the handle travels is equal to the circumference of the circle it traces out. The theoretical mechanical advantage of the jack's screw is:

$$MA = \frac{F}{L} = \frac{p}{2\pi R} \quad \text{so} \quad F = L \times \frac{p}{2\pi R}$$

For example, the theoretical force (F) required to lift a car with a mass (L) of 5000 kilograms, using a jack with a handle 30 centimeters long and a screw pitch of 0.5 cm, is given as:

$$F \cong 50{,}000 \text{ N} \times \frac{0.5 \text{ cm}}{6.284 * 30 \text{ cm}} \cong 130 \text{ N}$$

The theoretical value of mechanical advantage doesn't account for friction, so the actual force needed to turn the handle is higher than calculated.

Practice Quiz

1. Which of the following is NOT a vector quantity?
   a. Force
   b. Work
   c. Acceleration
   d. Gravity

2. A small gear has 12 teeth and a large one has 24 teeth. They are connected. If the large gear makes 8 rotations, how many does the small gear make?
   a. 4
   b. 8
   c. 12
   d. 16

3. In a shipping warehouse, there is a block and tackle pulley system with two ropes connecting the moving block to the fixed block. How much force is required to lift each 20-kg pallet of garden mulch?
   a. 5 kg
   b. 10 kg
   c. 20 kg
   d. 40 kg

4. How is the mechanical advantage of a system, such as a simple machine, calculated?
   a. Output force times input force
   b. Input force minus output force
   c. Output force divided by input force
   d. Input force divided by output force

5. Cameron is trying to build an aquarium on top of his coffee table. The tank is 3 feet deep, 2 feet wide, and 4 feet long. It will be filled to the top with water. If each cubic foot of water is about 62.5 pounds, approximately how much pressure is exerted at the bottom of the tank in pounds per square inch?
   a. 1.3 psi
   b. 3.0 psi
   c. 187.5 psi
   d. 1500 psi

Answer Explanations

1. B: Vector quantities have a magnitude and direction, while scalar quantities just involve a magnitude. Work is a scalar quantity. Force and acceleration are vector quantities, as is gravity, because gravity is a type of force.

2. D: The gears are in a 2:1 ratio because there are twice as many teeth in the large gear relative to the small one. Therefore, the smaller gear makes two revolutions for each complete revolution of the large gear.

$$8 \times 2 = 16$$

3. B: The mechanical advantage of the multiple-sheave block and tackle is approximated by counting the number of ropes going to and from the moving block. Because there are two ropes connecting the moving block to the fixed block in warehouse's pulley system, the mechanical advantage is 2:1. This means that the force needed to lift the item is one-half the weight of the item. Thus, 10 kg of force is needed to lift a 20-kg bag of garden mulch.

4. C: The mechanical advantage of a system or machine can be calculated by comparing the force input by the user to the machine with the force output from the use of the machine such that:

$$Mechanical\ Advantage\ = \frac{output\ force}{input\ force}$$

$$MA\ = \frac{F_{out}}{F_{in}}$$

Therefore, Choice *C* is correct.

5. A: The pressure experienced at the bottom of the tank can be calculated by first calculating the weight of the water in the tank. This is found by multiplying the volume in cubic feet by the weight of water per cubic foot. The volume of a rectangular prism is *V = l x w x d.* Thus, the tank has a volume of:

$$4 \times 2 \times 3 = 24 \text{ cubic feet}$$

This value is then multiplied by 62.5 pounds per cubic foot:

$$24 \times 62.5 = 1,500 \text{ pounds of water}$$

Next, we need to find the surface area of the bottom of the tank in square inches. The area of the rectangular base is $A = l \times w$. We need to convert our 4-foot length to inches ($4 \times 12 = 48$ inches) and 2-foot length to inches ($2 \times 12 = 24$ inches). Then,

$$Area = 48 \times 24 = 1,152 \text{ square inches}$$

Lastly, we divide the weight of the water by the total surface area of the bottom of the tank:

$$\frac{1,500 \text{ pounds}}{1,152 \text{ square inches}} = 1.302 \text{ psi}$$

Practice Test

Math Skills

1. If a is even, and b is odd, indicate which expression below must result in an even number.
   a. ab
   b. $3a + 5b$
   c. $-5a + 7b^2$
   d. $b^{a+b} + a^2$

2. In an office, there are 50 workers. A total of 60% of the workers are women, and the chances of a woman wearing a skirt is 50%. If no men wear skirts, how many workers are wearing skirts?
   a. 12
   b. 15
   c. 16
   d. 20

3. Four people split a bill. The first person pays for $\frac{1}{5}$, the second person pays for $\frac{1}{4}$, and the third person pays for $\frac{1}{3}$. What fraction of the bill does the fourth person pay?
   a. $\frac{13}{60}$
   b. $\frac{47}{60}$
   c. $\frac{1}{4}$
   d. $\frac{4}{15}$

4. Shawna buys $2\frac{1}{2}$ gallons of paint. If she uses $\frac{1}{3}$ of it on the first day, how much does she have left?
   a. $1\frac{5}{6}$ gallons
   b. $1\frac{1}{2}$ gallons
   c. $1\frac{2}{3}$ gallons
   d. 2 gallons

5. Jessica buys 10 cans of paint. Red paint costs $1 per can and blue paint costs $2 per can. In total, she spends $16. How many red cans did she buy?
   a. 2
   b. 3
   c. 4
   d. 5

6. What is the value of $x^2 - 2xy + 2y^2$ when $x = 2, y = 3$?
   a. 8
   b. 10
   c. 12
   d. 14

7. $(2x - 4y)^2 =$
   a. $4x^2 - 16xy + 16y^2$
   b. $4x^2 - 8xy + 16y^2$
   c. $4x^2 - 16xy - 16y^2$
   d. $2x^2 - 8xy + 8y^2$

8. If $x > 3$, then $\frac{x^2 - 6x + 9}{x^2 - x - 6} =$
   a. $\frac{x+2}{x-3}$

   b. $\frac{x-2}{x-3}$

   c. $\frac{x-3}{x+3}$

   d. $\frac{x-3}{x+2}$

9. The square and circle have the same center. The circle has a radius of r. What is the area of the shaded region?

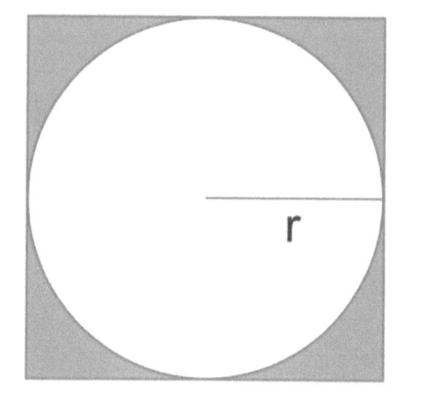

   a. $r^2 - \pi r^2$
   b. $4r^2 - 2\pi r$
   c. $(4 - \pi)r^2$
   d. $(\pi - 1)r^2$

10. Which of the following is the result of simplifying the expression: $\frac{4a^{-1}b^3}{a^4b^{-2}} \times \frac{3a}{b}$?

   a. $12a^3b^5$

   b. $12\frac{b^4}{a^4}$

   c. $\frac{12}{a^4}$

   d. $7\frac{b^4}{a}$

11. If $\log_{10} x = 2$, then x is

   a. 4

   b. 20

   c. 100

   d. 1000

12. Find the determinant of the matrix $\begin{bmatrix} -4 & 2 \\ 3 & -1 \end{bmatrix}$.

   a. -10

   b. -2

   c. 0

   d. 2

13. If $n = 2^2$, and $m = n^2$, then m^n equals?

   a. 2^{12}

   b. 2^{10}

   c. 2^{18}

   d. 2^{16}

14. What is the y-intercept of $y = x^{5/3} + (x - 3)(x + 1)$?

   a. 3.5

   b. 7.6

   c. -3

   d. -15.1

15. Which of the following augmented matrices represents the system of equations below?
$$2x - 3y + z = -5$$
$$4x - y - 2z = -7$$
$$-x + 2z = -1$$

a. $\begin{bmatrix} 2 & -3 & 1 & -5 \\ 4 & -1 & -2 & -7 \\ -1 & 0 & 2 & -1 \end{bmatrix}$

b. $\begin{bmatrix} 2 & 4 & -1 \\ -3 & -1 & 0 \\ 1 & -2 & 2 \\ -5 & -7 & -1 \end{bmatrix}$

c. $\begin{bmatrix} 2 & 4 & -1 & -5 \\ -3 & -1 & 0 & -7 \\ 2 & -2 & 2 & -1 \end{bmatrix}$

d. $\begin{bmatrix} 2 & -3 & 1 \\ 4 & -1 & -2 \\ -1 & 0 & 2 \end{bmatrix}$

16. What are the zeros of the function: $f(x) = x^3 + 4x^2 + 4x$?
a. -2
b. 0, -2
c. 2
d. 0, 2

17. What is the solution to the radical equation $\sqrt[3]{2x + 11} + 9 = 12$?
a. −8
b. 8
c. 0
d. 12

18. What is the solution to the following system of equations?
$$x^2 - 2x + y = 8$$
$$x - y = -2$$
a. $(-2, 3)$
b. There is no solution.
c. $(-2, 0) (1, 3)$
d. $(-2, 0) (3, 5)$

19. If $f(x) = \left(\frac{1}{2}\right)^x$ and $a < b$, then which of the following must be true?
a. $f(a) < f(b)$
b. $f(a) > f(b)$
c. $f(a) + f(b) = 0$
d $3f(a) = f(b)$

20. The sequence $\{a_n\}$ is defined by the relation $a_{n+1} = 3a_n - 1, a_1 = 1$. Find a_3.
    a. 2
    b. 4
    c. 5
    d. 15

21. A figure skater is facing north when she begins to spin to her right. She spins 2250 degrees. Which direction is she facing when she finishes her spin?
    a. North
    b. Southwest
    c. East
    d. West

22. Choose the answer below that could possibly be side lengths for a right triangle.
    a. 7, 8, 12
    b. 3, 4, 5
    c. 4, 8, 13
    d. 5, 7, 11

23. A rectangle has a length that is 5 feet longer than three times its width. If the perimeter is 90 feet, what is the length in feet?
    a. 10
    b. 20
    c. 25
    d. 35

24. Nina has a jar where she puts her loose change at the end of each day. There are 13 quarters, 25 dimes, 18 nickels, and 30 pennies in the jar. If she chooses a coin at random, what is the probability that the coin will not be a penny or a dime?
    a. 0.36
    b. 0.64
    c. 0.56
    d. 0.34

25. Six people apply to work for Janice's company, but she only needs four workers. How many different groups of four employees can Janice choose?
    a. 6
    b. 10
    c. 15
    d. 36

26. Two cards are drawn from a shuffled deck of 52 cards. What's the probability that both cards are kings if the first card isn't replaced after it's drawn?

    a. $\frac{1}{169}$

    b. $\frac{1}{221}$

    c. $\frac{1}{13}$

    d. $\frac{4}{13}$

27. How many different ways can we order the letters a, b, and c?

    a. 3
    b. 6
    c. 9
    d. 12

28. Five students take a test. The scores of the first four students are 80, 85, 75, and 60. If the median score is 80, which of the following could NOT be the score of the fifth student?

    a. 100
    b. 85
    c. 80
    d. 60

29. Ten students take a test. Five students get a 50. Four students get a 70. If the average score is 55, what was the last student's score?

    a. 20
    b. 40
    c. 50
    d. 60

30. For a group of 20 men, the median weight is 180 pounds and the range is 30 pounds. If each man gains 10 pounds, which of the following would be true?

    a. The median weight will increase, and the range will remain the same.
    b. The median weight and range will both remain the same.
    c. The median weight will stay the same, and the range will increase.
    d. The median weight and range will both increase.

Reading Comprehension

Directions: Assume each passage below to be true. Then, pick the answer choice that can be inferred only from the passage itself. Some of the other answer choices might make sense, but only one of them can be derived solely from the passage.

1. Kate has to buy a camera for her trip. She is hiking the Appalachian trail, and it is a requirement that she must pack the lightest equipment possible. Kate has to choose between the compact system camera (CSC) and the digital single-lens reflex camera (DSLR).

    a. Kate has two issues: buying a camera and figuring out which equipment to pack.

    b. Kate will either buy the CSC or the DSLR, depending on which one is smaller.

    c. Kate won't buy the DSLR because it's way too expensive.

    d. Kate is worried that both cameras might be too large to fit in her pack.

2. He adopted a kitten before he went to work. He worried about her all morning. However, he wasn't too concerned when he got a call later that evening about his son being suspended.

    a. He was so worried about the two events he couldn't focus on his work.

    b. His son getting suspended didn't bother him because he was used to it.

    c. The two major events did not occur at the same time.

    d. He worried more about his son being suspended than the kitten.

3. Hard water occurs when rainwater mixes with minerals from rock and soil. Hard water has a high mineral count, including calcium and magnesium. The mineral deposits from hard water can stain hard surfaces in bathrooms and kitchens as well as clog pipes. Hard water can stain dishes, ruin clothes, and reduce the life of any appliances it touches, such as hot water heaters, washing machines, and humidifiers.

    a. Hard water has the ability to reduce the life of a dishwasher.

    b. Hard water is the worst thing to wash your clothes with.

    c. The mineral count in hard water isn't as hard as they say.

    d. Things other than hard water can clog pipes, such as hair and oil.

4. Coaches of kids' sports teams are increasingly concerned about the behavior of parents at games. Parents are screaming and cursing at coaches, officials, players, and other parents. Physical fights have even broken out at games. Parents need to be reminded that coaches are volunteers, giving up their time and energy to help kids develop in their chosen sport. The goal of kids' sports teams is to learn and develop skills, but it's also to have fun. When parents are out of control at games and practices, it takes the fun out of the sport.

    a. Physical fights break out at every single game.

    b. Parents are adding stress to the kids during the game.

    c. Forming a union would help coaches out in their position.

    d. Coaches help kids on sports teams develop their skills and have fun.

5. Tornadoes are dangerous funnel clouds that occur during a large thunderstorm. When warm, humid air near the ground meets cold, dry air from above, a column of the warm air can be drawn up into the clouds. Winds at different altitudes blowing at different speeds make the column of air rotate. As the spinning column of air picks up speed, a funnel cloud is formed. This funnel cloud moves rapidly and haphazardly. Rain and hail inside the cloud cause it to touch down, creating a tornado.

    a. Tornadoes are formed from a mixture of cold and warm air.

    b. Tornadoes are the most dangerous of extreme weather patterns.

    c. Scientists still aren't exactly sure why tornadoes form.

    d. Scientists continue to study tornadoes to improve radar detection and warning times.

6. Digestion begins in the mouth where teeth grind up food and saliva breaks it down, making it easier for the body to absorb. Next, the food moves to the esophagus, and it is pushed into the stomach. The stomach is where food is stored and broken down further by acids and digestive enzymes, preparing it for passage into the intestines.

    a. Food waste is passed into the large intestine.

    b. Nutrients pass into the blood stream while in the small intestine.

    c. As soon as you chew your food, it travels to the esophagus.

    d. Food travels to the esophagus after it is pushed into the stomach.

7. Jordan is the leader of the group, which means she must decide whether the topic for their presentation will be over climate change or food deserts. Jordan is also very diplomatic. Miguel is an expert in climate change while Kennedy has experience growing up in a food desert.

    a. Jordan will pick climate change because it's a more important topic than food deserts.

    b. Jordan has to decide whether personal experience is more important than research experience.

    c. Jordan will flip a coin in order to decide which topic to present over.

    d. Jordan probably likes Kennedy more since she is more relatable, so she will pick her topic.

8. Vacationers looking for a perfect experience should opt out of Disney parks and try a trip on Disney Cruise Lines. While a park offers rides, characters, and show experiences, it also includes long lines, often very hot weather, and enormous crowds.

    a. Although Disney Cruise Lines is fun for the family, it has long lines, very hot weather, and enormous crowds.

    b. Families with small children should not go to Disney parks because there are too many people and the weather is too hot.

    c. At Disney parks, hot weather, lines, and crowds will be less extreme than at Disney Cruise Lines.

    d. At Disney Cruise Lines, hot weather, lines, and crowds will be less extreme than at Disney parks.

9. As summer approaches, drowning incidents will increase. Drowning happens very quickly and silently. Most people assume that drowning is easy to spot, but a person who is drowning doesn't make noise or wave their arms.

    a. Drowning happens silently and decreases as summer approaches.

    b. Drowning happens silently and increases as summer approaches.

    c. Each summer, more children drown than adults.

    d. Many people in summertime wave their arms and make a lot of noise.

10. Last year was the warmest ever recorded in the last 134 years. During that time period, the ten warmest years have all occurred since 2000.
    a. The hottest years in earth's history probably occurred during the dinosaurs' time.
    b. The next 134 years will be hotter than the past ten years.
    c. Out of the last 50 years, the ten warmest years have occurred since 2000.
    d. Burning fossil fuels is what caused the ten warmest years since 2000.

11. A famous children's author recently published a historical fiction novel under a pseudonym; however, it did not sell as many copies as her children's books. In her earlier years, she had majored in history and earned a graduate degree in Antebellum American History, which is the time frame of her new novel.
    a. The author's children's books are more popular than her historical fiction novel.
    b. It's ironic that the author majored in history yet did not sell many copies of her novel.
    c. The author did not sell many copies of her historical fiction novel because it was boring.
    d. Most children's authors cannot cross literary genres without being criticized.

12. Hannah started smoking when she was nineteen years old. The day after Hannah finished a book called *Smoke Free*, she quit. She has been cigarette-free for over a decade.
    a. Hannah will probably have respiratory problems as she gets older.
    b. *Smoke Free* is the reason Hannah quit smoking.
    c. Hannah is at least twenty-nine years old.
    d. Everyone who smokes should read the book *Smoke Free*.

13. Heat loss is proportional to surface area exposed. An elephant loses a great deal more heat than an anteater because it has a much greater surface area than an anteater.
    a. Surface area causes heat loss.
    b. Too much heat loss can be dangerous.
    c. Elephants lose more heat than anteaters.
    d. Anteaters lose more heat than elephants.

14. The landlord sent an interested tenant the following information about his three apartments for rent: A, B, and C. Apartment B was bigger than Apartment A. Apartment A was in front of Apartment C but above Apartment B. Apartment C was smaller than Apartment A.
    a. Apartment B was above Apartment C.
    b. Apartment B was smaller than Apartment C.
    c. Apartment A was the biggest apartment.
    d. Apartment B was in front of Apartment C.

15. People who argue that William Shakespeare is not responsible for the plays attributed to his name are known as anti-Stratfordians (from the name of Shakespeare's birthplace, Stratford-upon-Avon).
    a. Dr. Porter believes that William Shakespeare is responsible for writing his own plays. He is known as an anti-Stratfordian.
    b. Dr. Filigree believes that William Shakespeare is not responsible for writing his own plays. He is known as an anti-Stratfordian.
    c. Dr. Casings believes that Shakespeare was born somewhere other than Stratford. He is known as an anti-Stratfordian.
    d. Dr. Hendrix believes that Shakespeare died somewhere other than Stratford. He is known as an anti-Stratfordian.

16. Nina is allergic to dairy (which includes cheese and milk), and she doesn't eat any meat except for fish. At his barbecues, Oliver always invites Nina and consistently prepares her a meal that is suitable to her diet. Nina is planning on going to a barbecue later that Oliver is throwing.
  - a. Oliver prepares Nina a black bean burger with French fries.
  - b. Oliver prepares Nina grilled chicken with asparagus.
  - c. Oliver prepares Nina a cheese pizza with a side salad.
  - d. Oliver prepares Nina a veggie hot dog with a milkshake.

17. Samuel teaches at a high school in one of the biggest cities in the United States. His students come from diverse family backgrounds. Samuel observes that the best students in his class are from homes where parental supervision is minimal.
a. Samuel should write an academic paper based on his findings.
b. The parents of the bottom five students are probably the most involved.
c. In Samuel's observation, his best students have maximum interference from parents.
d. In Samuel's observation, his best students have minimal interference from parents.

18. Cynthia keeps to a strict vegetarian diet, which is part of her religion. She absolutely cannot have any meat or fish dishes. This is more than a preference; her body has never developed the enzymes to process meat or fish, so she becomes violently ill if she accidentally eats any of the offending foods.
  - a. Cynthia doesn't eat meat due to necessity.
  - b. Cynthia doesn't eat meat due to preference.
  - c. Cynthia doesn't eat meat due to preference as well as necessity.
  - d. Cynthia can develop a tolerance to meat by eating small pieces at a time.

19. Samantha wants to be a professional chef, so she started working at a nearby restaurant called *Chesapeake Cuisine*. Samantha also wants to go to college one day to study nutrition. Samantha's mom surprised her later that year by offering to send her to school, but she will only send her if Samantha goes to law school. Samantha cannot afford college on her own.
  - a. Samantha's mom is too controlling.
  - b. Samantha will never enjoy law school.
  - c. Samantha has to choose between going to law school and getting a degree in nutrition.
  - d. Samantha has to choose between working at *Chesapeake Cuisine* and going to law school.

20. Barbara had to have an exercise bike for $150 at a store, but soon found out there was a cheaper one online for $75. Barbara always went for cheaper machinery, but hardly ever returned items if she could help it.
  - a. Barbara bought both bikes but only used one of them.
  - b. Barbara returned the $150 bike and bought the $75 bike.
  - c. Barbara kept the $150 bike because she hated returning things.
  - d. Barbara did not want a bike that bad so she did not keep any of the bikes.

Mechanical Comprehension

1. A car is traveling at a constant velocity of 25 m/s. How long does it take the car to travel 45 kilometers in a straight line?
  - a. 1 hour
  - b. 3600 seconds
  - c. 1800 seconds
  - d. 900 seconds

2. A ship is traveling due east at a speed of 1 m/s against a current flowing due west at a speed of 0.5 m/s. How far has the ship travelled from its point of departure after two hours?
    a. 1.8 kilometers west of its point of departure
    b. 3.6 kilometers west of its point of departure
    c. 1.8 kilometers east of its point of departure
    d. 3.6 kilometers east of its point of departure

3. A car is driving along a straight stretch of highway at a constant speed of 60 km/hour when the driver slams the gas pedal to the floor, reaching a speed of 132 km/hour in 10 seconds. What's the average acceleration of the car after the engine is floored?
    a. 1 m/s^2
    b. 2 m/s^2
    c. 3 m/s^2
    d. 4 m/s^2

4. A spaceship with a mass of 100,000 kilograms is far away from any planet. To accelerate the craft at the rate of 0.5 m/sec^2, what is the rocket thrust?
    a. 98.1 N
    b. 25,000 N
    c. 50,000 N
    d. 75,000 N

5. The gravitational acceleration on Earth averages 9.81 m/s^2. An astronaut weighs 1962 N on Earth. The diameter of Earth is six times the diameter of its moon. What's the mass of the astronaut on Earth's moon?
    a. 100 kilograms
    b. 200 kilograms
    c. 300 kilograms
    d. 400 kilograms

6. A football is kicked so that it leaves the punter's toe at a horizontal angle of 45 degrees. Ignoring any spin or tumbling, at what point is the upward vertical velocity of the football at a maximum?

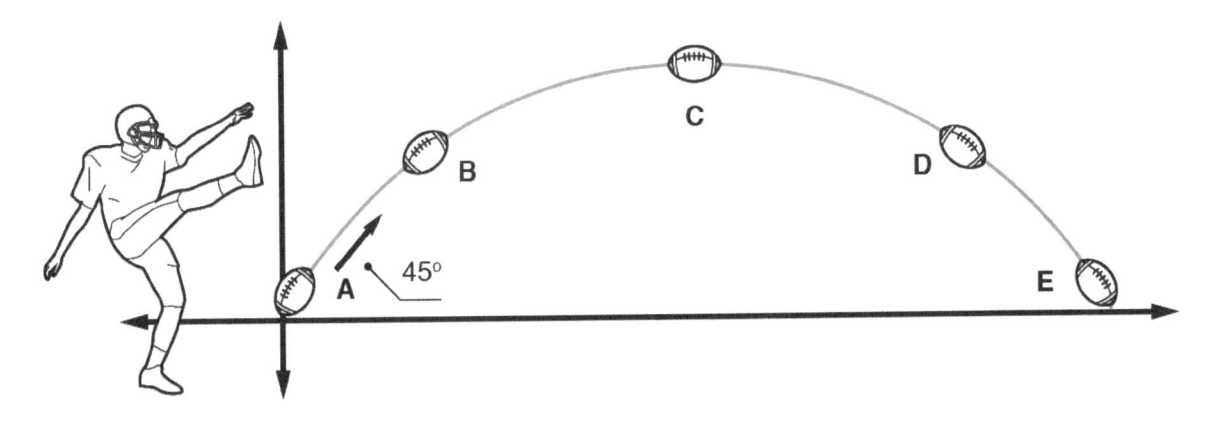

    a. At Point A
    b. At Point C
    c. At Points B and D
    d. At Points A and E

7. The skater is shown spinning in Figure (a), then bringing in her arms in Figure (b). Which sequence accurately describes what happens to her angular velocity?

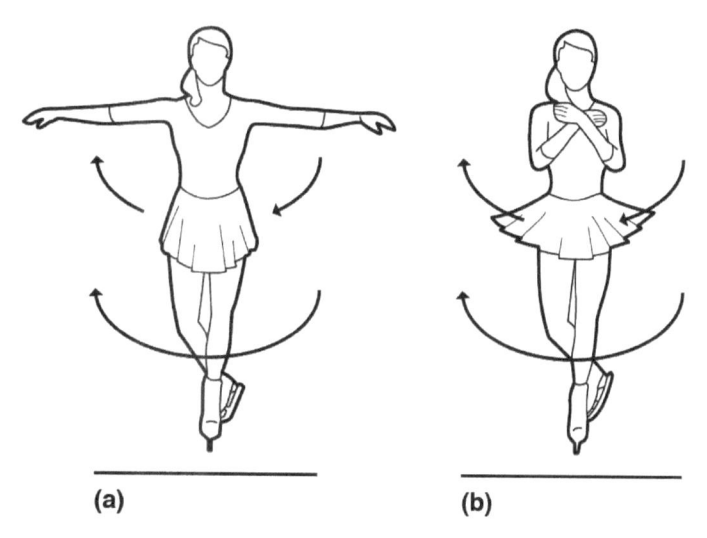

(a)   (b)

    a. Her angular velocity decreases from (a) to (b)
    b. Her angular velocity doesn't change from (a) to (b)
    c. Her angular velocity increases from (a) to (b)
    d. It's not possible to determine what happens to her angular velocity if her weight is unknown.

8. A cannonball is dropped from a height of 10 meters off of the ground. What is its approximate velocity just before it hits the ground?
    a. 9.81 m/s
    b. 14 m/s
    c. 32 m/s
    d. It can't be determined without knowing the cannonball's mass

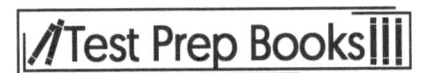

9. The pendulum is held at point A, and then released to swing to the right. At what point does the pendulum have the greatest kinetic energy?

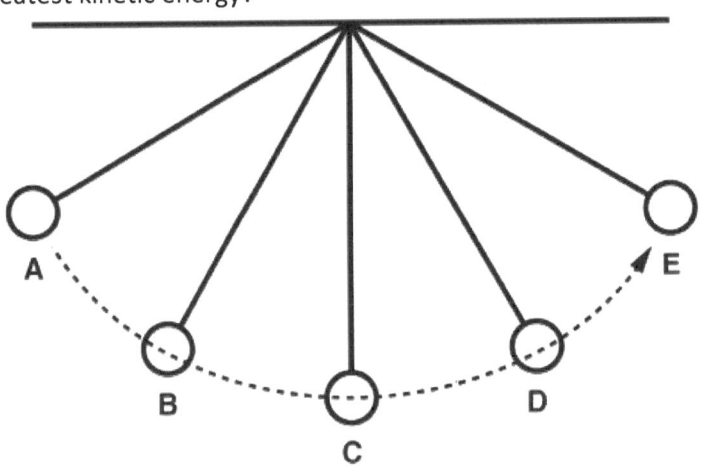

    a. At Point B
    b. At Point C
    c. At Point D
    d. At Point E

10. Which statement is true of the total energy of the pendulum?

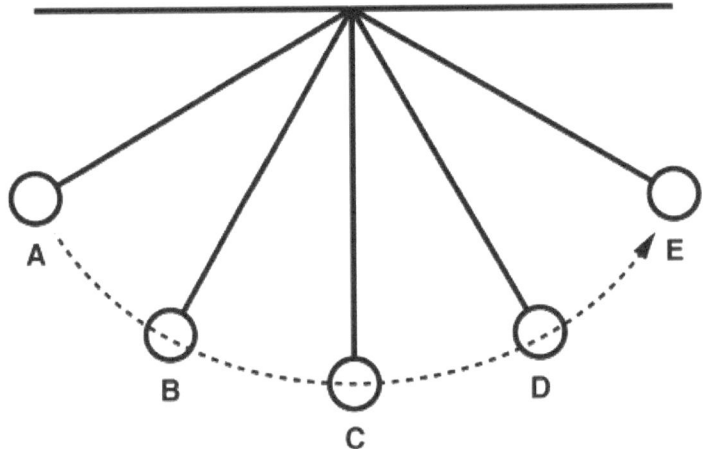

    a. Its total energy is at a maximum and equal at Points A and E.
    b. Its total energy is at a maximum at Point C.
    c. Its total energy is the same at Points A, B, C, D, and E.
    d. The total energy can't be determined without knowing the pendulum's mass.

11. How do you calculate the useful work performed in lifting a 10-kilogram weight from the ground to the top of a 2-meter ladder?
    a. $10 \text{ kg} \times 2 \text{ m} \times 32 \text{ m/s}^2$
    b. $10 \text{ kg} \times 2 \text{ m}^2 \times 9.81 \text{ m/s}$
    c. $10 \text{ kg} \times 2 \text{ m} \times 9.81 \text{ m/s}^2$
    d. It can't be determined without knowing the ground elevation

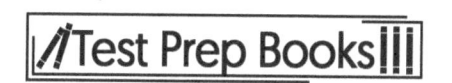

12. A steel spring is loaded with a 10-newton weight and is stretched by 0.5 centimeters. What is the deflection if it's loaded with two 10-newton weights?

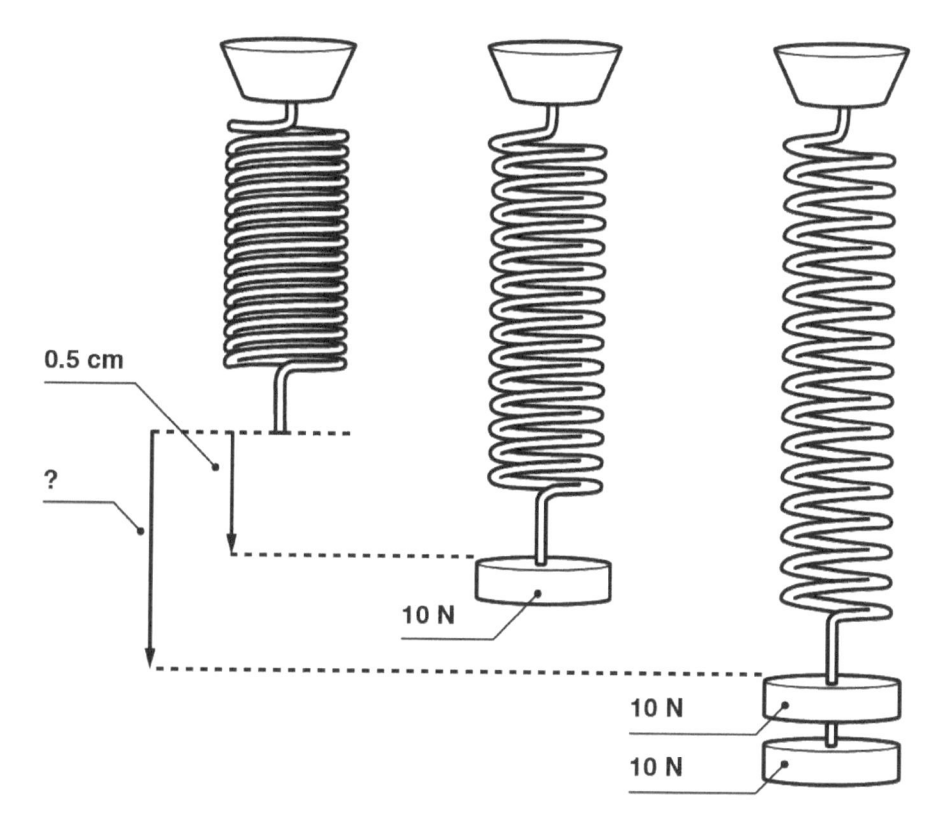

a. 0.5 centimeter
b. 1 centimeter
c. 2 centimeters
d. It can't be determined without knowing the Modulus of Elasticity of the steel.

13. A 1000-kilogram concrete block is resting on a wooden surface. Between these two materials, the coefficient of sliding friction is 0.4 and the coefficient of static friction is 0.5. How much more force is needed to get the block moving than to keep it moving?

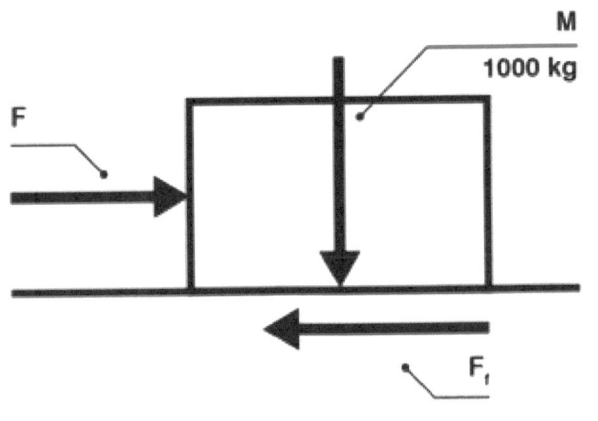

    a. 981 N
    b. 1962 N
    c. 3924 N
    d. 9810 N

14. The master cylinder (F1) of a hydraulic jack has a cross-sectional area of 0.1 m², and a force of 50 N is applied. What must the area of the drive cylinder (F2) be to support a weight of 800 N?

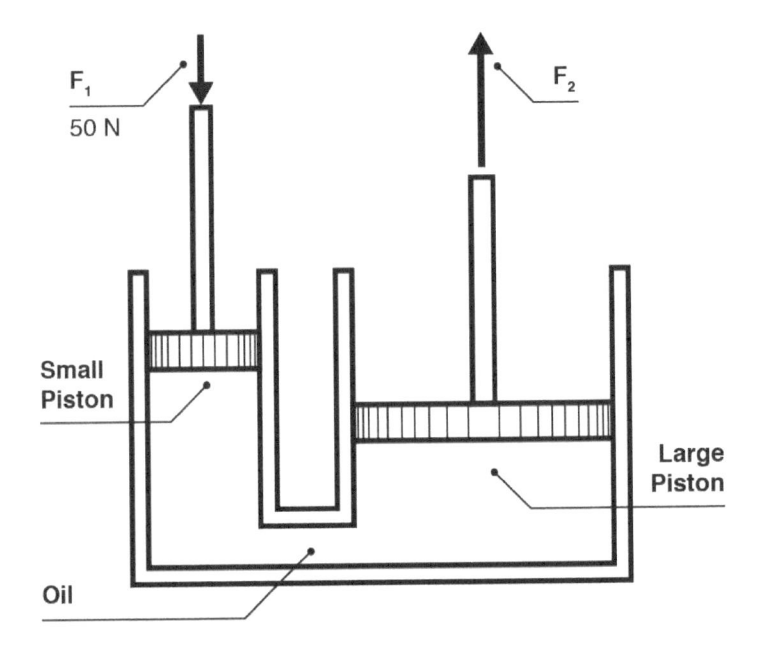

    a. 0.4 m²
    b. 0.8 m²
    c. 1.6 m²
    d. 3.2 m²

15. A gas with a volume V_1 is held down by a piston with a force of F newtons. The piston has an area of A. After heating the gas, it expands against the weight to a volume V_2. What was the work done?

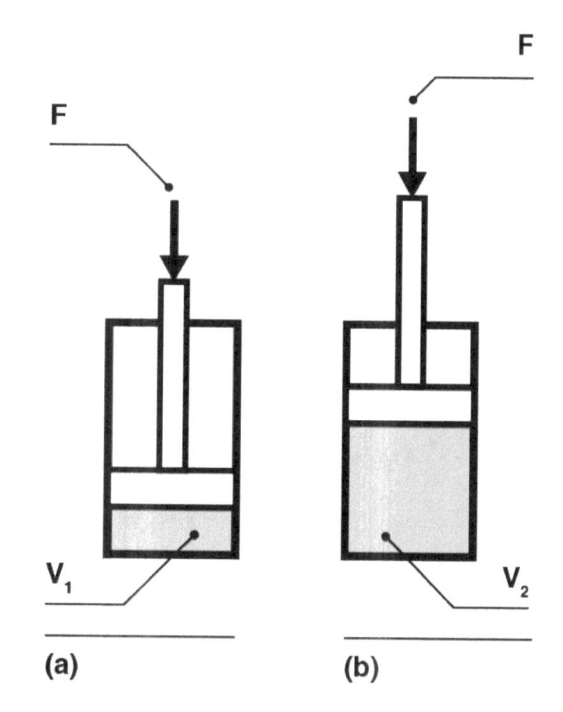

(a)   (b)

a. $\dfrac{F}{A}$

b. $\dfrac{F}{A} \times V_1$

c. $\dfrac{F}{A} \times V_2$

d. $\dfrac{F}{A} \times (V_2 - V_1)$

16. A 1000-kilogram weight is raised 30 meters in 10 minutes. What is the approximate power expended in the period?

    a. 1000 Kg × m/s²

    b. 500 N·m

    c. 500 J/s

    d. 100 watts

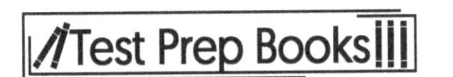

17. A 2-meter high, concrete block is submerged in a body of water 12 meters deep (as shown). Assuming that the water has a unit weight of 1000 N/m³, what is the pressure acting on the upper surface of the block?

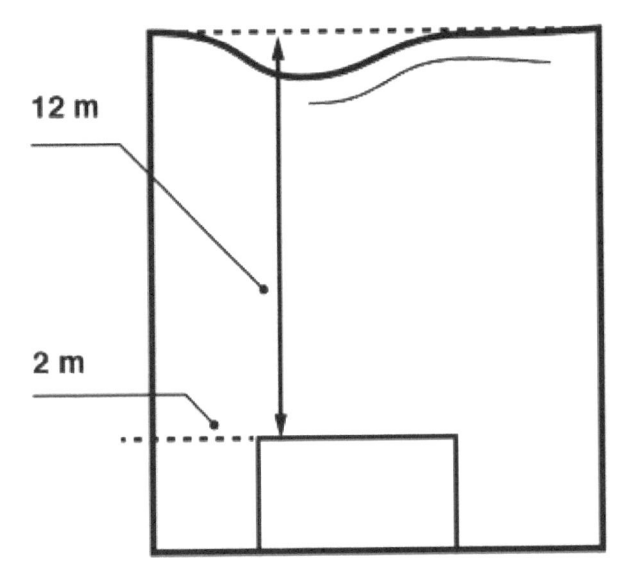

    a. 10,000 Pa
    b. 12,000 Pa
    c. 14,000 Pa
    d. It can't be calculated without knowing the top area of the block.

18. Closed Basins A and B each contain a 10,000-ton block of ice. The ice block in Basin A is floating in sea water. The ice block in Basin B is aground on a rock ledge (as shown). When all the ice melts, what happens to the water level in Basin A and Basin B?

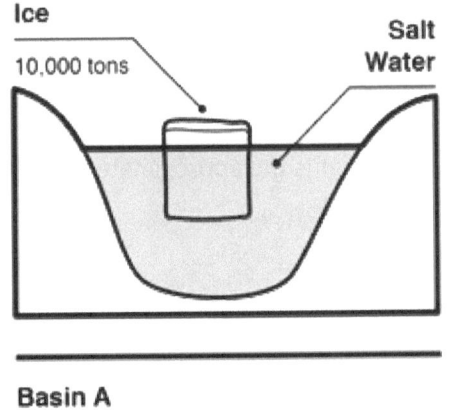 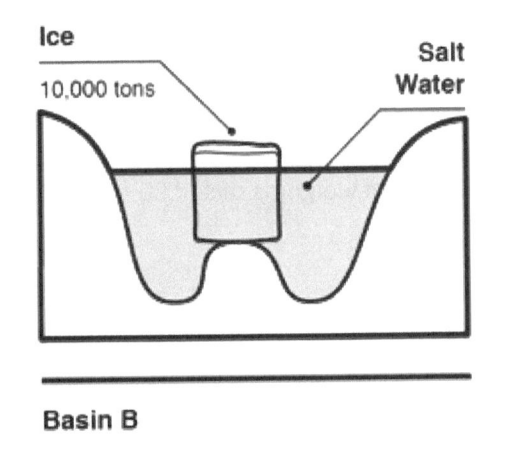

    a. Water level rises in A but not in B
    b. Water level rises in B but not in A
    c. Water level rises in neither A nor B
    d. Water level rises in both A and B

19. An official 10-lane Olympic pool is 50 meters wide by 25 meters long. How long does it take to fill the pool to the recommended depth of 3 meters using a pump with a 750 liter per second capacity?
    a. 2500 seconds
    b. 5000 seconds
    c. 10,000 seconds
    d. 100,000 seconds

20. Water is flowing in a rectangular canal 10 meters wide by 2 meters deep at a velocity of 3 m/s. The canal is half full. What is the flow rate?
    a. 30 m³/s
    b. 60 m³/s
    c. 90 m³/s
    d. 120 m³/s

21. A 150-kilogram mass is placed on the left side of the lever as shown. What force must be exerted on the right side (in the location shown by the arrow) to balance the weight of this mass?

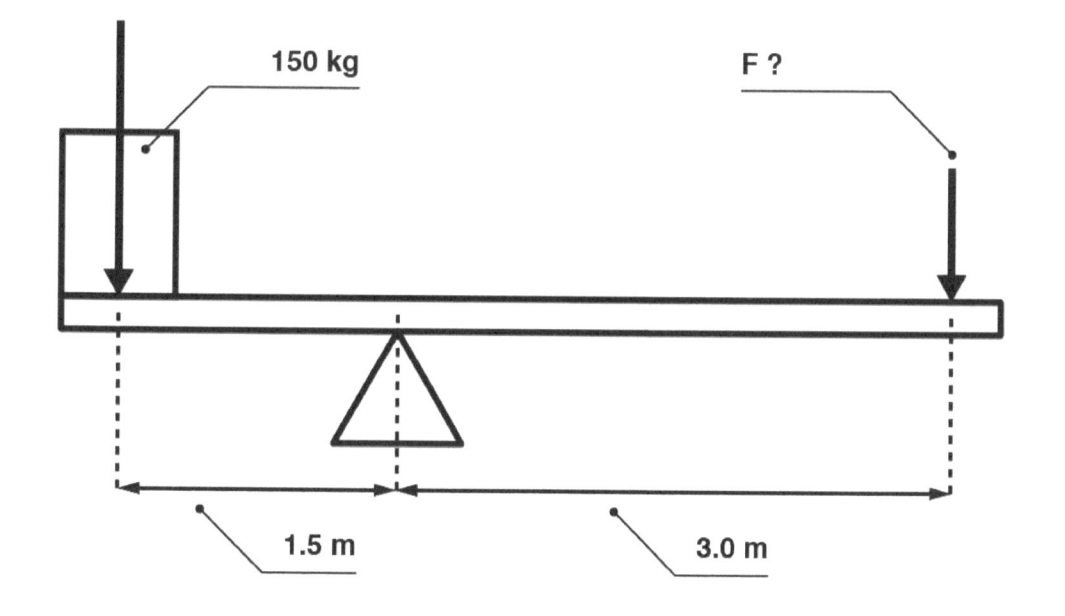

    a. 675 kg·m
    b. 735.75 N
    c. 1471.5 N
    d. 2207.25 N·m

22. For the wheel and axle assembly shown, the shaft radius is 20 millimeters and the wheel radius is 300 millimeters. What's the required effort to lift a 600 N load?

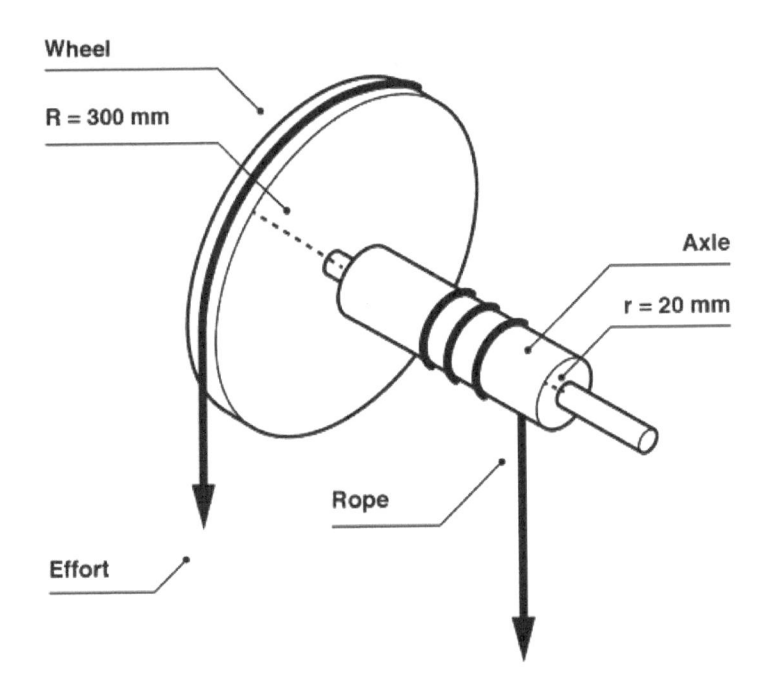

a. 10 N
b. 20 N
c. 30 N
d. 40 N

23. The driver gear (Gear A) turns clockwise at a rate of 60 RPM. In what direction does Gear B turn and at what rotational speed?

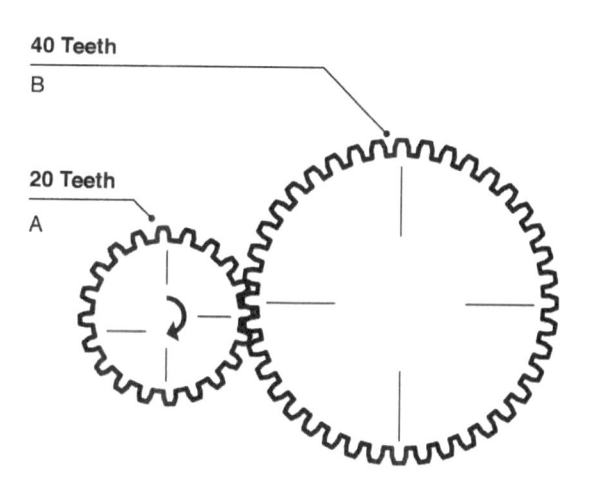

a. Clockwise at 120 RPM
b. Counterclockwise at 120 RPM
c. Clockwise at 30 RPM
d. Counterclockwise at 30 RPM

24. The three steel wheels shown are connected by rubber belts. The two wheels at the top have the same diameter, while the wheel below is twice their diameter. If the driver wheel at the upper left is turning clockwise at 60 RPM, at what speed and in which direction is the large bottom wheel turning?

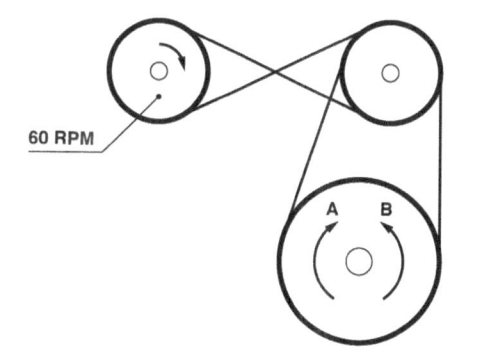

a. 30 RPM, clockwise (A)
b. 30 RPM, counterclockwise (B)
c. 120 RPM, clockwise (A)
d. 120 RPM, counterclockwise (B)

25. In case (a), both blocks are fixed. In case (b), the load is hung from a moveable block. Ignoring friction, what is the required force to move the blocks in both cases?

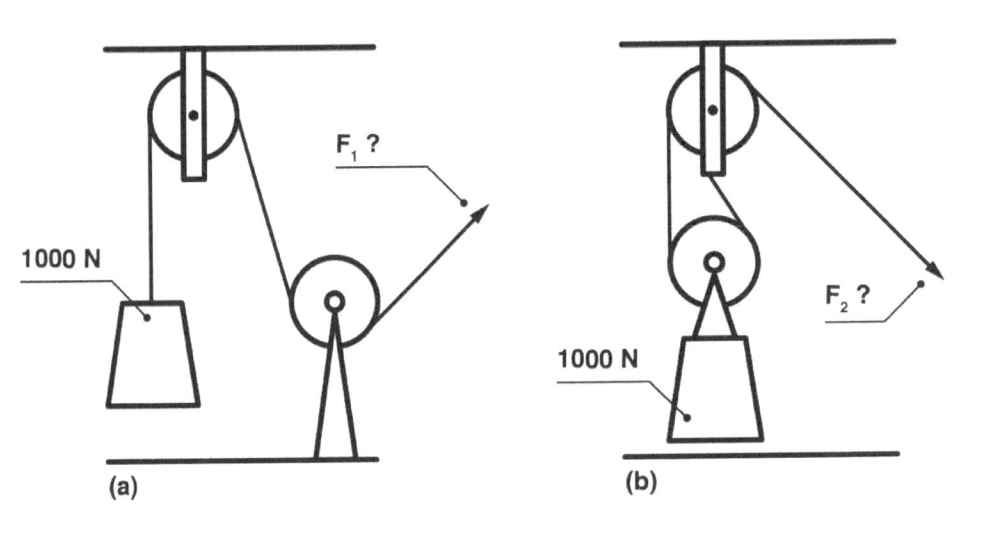

(a) (b)

a. $F_1 = 500$ N; $F_2 = 500$ N
b. $F_1 = 500$ N; $F_2 = 1000$ N
c. $F_1 = 1000$ N; $F_2 = 500$ N
d. $F_1 = 1000$ N; $F_2 = 1000$ N

26. Considering a gas in a closed system, at a constant volume, what will happen to the temperature if the pressure is increased?
    a. The temperature will stay the same.
    b. The temperature will decrease.
    c. The temperature will increase.
    d. It cannot be determined with the information given.

27. What is the current when a 3.0 V battery is wired across a lightbulb that has a resistance of 6.0 ohms?
    a. 0.5 A
    b. 18.0 A
    c. 0.5 J
    d. 18.0 J

28. According to Newton's Three Laws of Motion, which of the following is true?
    a. Two objects cannot exert a force on each other without touching.
    b. An object at rest has no inertia.
    c. The weight of an object is equal to the mass of an object multiplied by gravity.
    d. All of the above

29. What is the total mechanical energy of a system?
   a. The total potential energy
   b. The total kinetic energy
   c. Kinetic energy plus potential energy
   d. The total energy created by a system

30. What is the molarity of a solution made by dissolving 4.0 grams of NaCl into enough water to make 120 mL of solution? The atomic mass of Na is 23.0 g/mol and Cl is 35.5 g/mol.
   a. 0.34 M
   b. 0.57 M
   c. 0.034 M
   d. 0.057 M

Answer Explanations

Math Skills

1. A: If an even number is multiplied by an odd number, the result is always even. In Choice *B*, multiplying 3 by an even number still results in an even number. However, multiplying 5 by an odd number does not mean the result is even, so *B* is not included in the possible answers. Squaring an odd number results in an odd number, which means Choice *C* is not a possible answer. In Choice *D*, the value of an odd number raised to any power is still an odd number, and adding an even number to it remains odd, so Choice *D* is not a possible answer.

2. B: If 60% of 50 workers are women, then there are 30 women working in the office. If half of them are wearing skirts, then that means 15 women wear skirts. Since none of the men wear skirts, this means there are 15 people wearing skirts.

3. A: To find the fraction of the bill that the first three people pay, the fractions need to be added, which means finding the common denominator. The common denominator will be 60.

$$\frac{1}{5}+\frac{1}{4}+\frac{1}{3}=\frac{12}{60}+\frac{15}{60}+\frac{20}{60}=\frac{47}{60}$$

The remainder of the bill is:

$$1-\frac{47}{60}=\frac{60}{60}-\frac{47}{60}=\frac{13}{60}$$

4. C: If she has used 1/3 of the paint, she has 2/3 remaining. $2\frac{1}{2}$ gallons are the same as $\frac{5}{2}$ gallons. The calculation is:

$$\frac{2}{3}\times\frac{5}{2}=\frac{5}{3}=1\frac{2}{3}\text{ gallons}$$

5. C: We are trying to find x, the number of red cans. The equation can be set up like this:

$$x+2(10-x)=16$$

The left x is actually multiplied by \$1, the price per red can. Since we know Jessica bought 10 total cans, $10-x$ is the number blue cans that she bought. We multiply the number of blue cans by \$2, the price per blue can.

That should all equal \$16, the total amount of money that Jessica spent. Working that out gives us:

$$x+20-2x=16$$

$$20-x=16$$

$$x=4$$

6. B: Start with the original equation: $x^2 - 2xy + 2y$, then replace each instance of x with a 2, and each instance of y with a 3 to get:

$$2^2 - 2 \times 2 \times 3 + 2 \times 3^2 = 4 - 12 + 18 = 10$$

7. A: To expand a squared binomial, it's necessary to use the *First, Inner, Outer, Last Method*.

$$(2x - 4y)^2$$

$$(2x)(2x) + (2x)(-4y) + (-4y)(2x) + (-4y)(-4y)$$

$$4x^2 - 8xy - 8xy + 16y^2$$

$$4x^2 - 16xy + 16y^2$$

8. D: Factor the numerator into $x^2 - 6x + 9 = (x - 3)^2$, since $-3 - 3 = -6, (-3)(-3) = 9$. Factor the denominator into $x^2 - x - 6 = (x - 3)(x + 2)$, since $-3 + 2 = -1, (-3)(2) = -6$. This means the rational function can be rewritten as:

$$\frac{x^2 - 6x + 9}{x^2 - x - 6} = \frac{(x - 3)^2}{(x - 3)(x + 2)}$$

Using the restriction of x > 3, do not worry about any of these terms being 0, and cancel an $x - 3$ from the numerator and the denominator, leaving $\frac{x-3}{x+2}$.

9. C: The area of the shaded region is the area of the square minus the area of the circle. The area of the circle is πr^2. The side of the square will be $2r$, so the area of the square will be $4r^2$. Therefore, the difference is:

$$4r^2 - \pi r^2 = (4 - \pi)r^2$$

10. B: To simplify the given equation, the first step is to make all exponents positive by moving them to the opposite place in the fraction. This expression becomes:

$$\frac{4b^3 b^2}{a^1 a^4} \times \frac{3a}{b}$$

Then the rules for exponents can be used to simplify. Multiplying the same bases means the exponents can be added. Dividing the same bases means the exponents are subtracted. Thus, after multiplying the exponents in the first fraction the equation becomes:

$$\frac{4b^5}{a^5} \times \frac{3a}{b}$$

Therefore, we can first multiply to get $\frac{12ab^5}{a^5 b}$. Then, dividing yields $12\frac{b^4}{a^4}$.

11. C: If $\log_{10} x = 2$, then $10^2 = x$, which equals 100.

12. B: The determinant of a 2 x 2 matrix is $ad - bc$. The calculation is:

$$-4(-1) - 2(3) = 4 - 6 = -2$$

13. D: If $n = 2^2, n = 4$, and $m = 4^2 = 16$. This means that $m^n = 16^4$. This is the same as 2^{16}.

14. C: To find the y-intercept, substitute zero for x, which gives us:

$$y = 0^{\frac{5}{3}} + (0 - 3)(0 + 1)$$

$$0 + (-3)(1) = -3$$

15. A: The augmented matrix that represents the system of equations has dimensions 4×3 because there are three equations with three unknowns. The coefficients of the variables make up the first three columns, and the last column is made up of the numbers to the right of the equals sign. This system can be solved by reducing the matrix to row-echelon form, where the last column gives the solution for the unknown variables.

16. B: There are two zeros for the given function. They are $x = 0, -2$. The zeros can be found several ways, but this particular equation can be factored into:

$$f(x) = x(x^2 + 4x + 4) = x(x + 2)(x + 2)$$

By setting each factor equal to zero and solving for x, there are two solutions. On a graph, these zeros can be seen where the line crosses the x-axis.

17. B: First, subtract 9 from both sides to isolate the radical. Then, cube each side of the equation to obtain:

$$2x + 11 = 27$$

Subtract 11 from both sides, and then divide by 2. The result is $x = 8$. Plug 8 back into the original equation to check the answer:

$$\sqrt[3]{16 + 11} + 9 = 12$$

$$\sqrt[3]{27} + 9 = 12$$

$$3 + 9 = 12$$

18. D: This system of equations involves one quadratic function and one linear function, as seen from the degree of each equation. One way to solve this is through substitution. Solving for y in the second equation yields $y = x + 2$. Plugging this equation in for the y of the quadratic equation yields $x^2 - 2x + x + 2 = 8$. Simplifying the equation, it becomes $x^2 - x + 2 = 8$. Setting this equal to zero and factoring, it becomes:

$$x^2 - x - 6 = 0 = (x - 3)(x + 2)$$

Solving these two factors for x gives the zeros $x = 3, -2$. To find the y-value for the point, each number can be plugged in to either original equation. Solving each one for y yields the points $(3, 5)$ and $(-2, 0)$.

19. B: Here, f is an exponential function whose base is less than 1. In this function, f is always decreasing. This means that when a is less than b, $f(a) > f(b)$.

20. C: Find $a_2 = 3a_1 - 1 = 3 \times 1 - 1 = 2$. Next, find $a_3 = 3a_2 - 1 = 3 \times 2 - 1 = 5$.

21. C: A full rotation is 360 degrees. Taking the total degrees that the figure skater spins and dividing by 360 yields 6.25. She spins 6 total times and then one quarter of a turn more. This quarter of a turn to her right means she ends up facing East.

22. B: For the side lengths of a triangle, the first rule is that the sum of any two sides must be greater than the third side. Using this rule, Choice *C* can be eliminated because $8 + 4$ is not greater than 13. In order for the lengths to be sides of a right triangle, they must satisfy the equation $a^2 + b^2 = c^2$. For the values in *A*, the equation becomes:

$$7^2 + 8^2 = 12^2$$

Because this does not form a true statement, $113 = 144$, then *A* is not a possible answer. For the values in Choice *D*, $5^2 + 7^2 = 11^2$ is false, so it is also eliminated. The lengths in Choice *B* form a true statement with $3^2 + 4^2 = 5^2$, so the side lengths in Choice *B* contain values that satisfy the Pythagorean Theorem; therefore, they are possible side lengths for a right triangle.

23. D: Denote the width as w and the length as l. Then, $l = 3w + 5$. The perimeter is $2w + 2l = 90$. Substituting the first expression for l into the second equation yields:

$$2(3w + 5) + 2w = 90$$

$$6w + 10 + 2w = 90$$

$$8w = 80$$

$$w = 10$$

Putting this into the first equation, it yields:

$$l = 3(10) + 5 = 35$$

24. A: The total number of coins in the jar is 86, which is the sum of all the coins. The probability of Nina choosing a coin other than a penny or a dime can be found by calculating the total of quarters and nickels. This total is 31. Taking 31 and dividing it by 86 gives the probability of choosing a coin that is not a penny or a dime. The decimal found from the fraction $\frac{31}{86}$ is 0.36.

25. C: Janice will be choosing 4 employees out of a set of 6 applicants, so this will be given by the combination function. The following equation shows the combination function worked out:

$$\binom{6}{4} = \frac{6!}{4!\,(6-4)!} = \frac{6!}{4!\,(2)!} = \frac{6 \times 5 \times 4 \times 3 \times 2 \times 1}{4 \times 3 \times 2 \times 1 \times 2 \times 1} = \frac{6 \times 5}{2} = 15$$

26. B: For the first card drawn, the probability of a king being pulled is $\frac{4}{52}$. Since this card isn't replaced, if a king is drawn first, the probability of a king being drawn second is $\frac{3}{51}$. The probability of a king being drawn in both the first and second draw is the product of the two probabilities:

$$\frac{4}{52} \times \frac{3}{51} = \frac{12}{2,652}$$

This fraction, when divided by $\frac{12}{12}$, equals $\frac{1}{221}$.

27. B: The number of ways to order n objects is given by the product $\binom{n}{1}\binom{n-1}{1} \cdots \binom{1}{1}$.

This is,

$$\binom{3}{1}\binom{2}{1}\binom{1}{1} = \frac{3!}{1!\,2!} \times \frac{2!}{1!\,1!} \times \frac{1!}{1!\,1!}$$

1! is just 1, and the 2! in the numerator and denominator will cancel one another, with a result of:

$$3! = 3 \times 2 \times 1 = 6$$

28. D: Lining up the given scores provides the following list: 60, 75, 80, 85, and one unknown. Because the median needs to be 80, it means 80 must be the middle data point out of these five. Therefore, the unknown data point must be the fourth or fifth data point, meaning it must be greater than or equal to 80. The only answer that fails to meet this condition is 60.

29. A: Let the unknown score be x. The average will be:

$$\frac{5 \times 50 + 4 \times 70 + x}{10} = \frac{530 + x}{10} = 55$$

Multiply both sides by 10 to get $530 + x = 550$, or $x = 20$.

30. A: If each man gains 10 pounds, every original data point will increase by 10 pounds. Therefore, the man with the original median will still have the median value, but that value will increase by 10. The smallest value and largest value will also increase by 10 and, therefore, the difference between the two won't change. The range does not change in value and, thus, remains the same.

Reading Comprehension

1. B: Kate will either buy the CSC or the DSLR, depending on which one is smaller. We know that Kate "has to buy" a camera for her trip and that "it is a requirement that she must pack the lightest equipment possible." Choice *A* is incorrect. In the passage, we don't see Kate's issue of figuring out which equipment to pack, only which camera to buy. Choice *C* is incorrect. We don't have enough information on the price of the camera to make an educated guess. Choice *D* is incorrect; although we know the camera should be able to fit into Kate's pack, we don't know that Kate is worried about this.

2. C: The two major events did not occur at the same. We know that he worried about the kitten in the morning, and that he found out his son got suspended in the evening. Choice *A* and *D* are incorrect, as the passage states he wasn't too concerned about his son getting suspended. Choice *B* is incorrect; we don't have enough information to make an educated guess about why his son getting suspended didn't bother him.

3. A: Hard water has the ability to reduce the life of a dishwasher. Keep in mind that to make an inference means to make an educated guess based on the facts of the passage. The passage says "hard water can reduce the life of any appliances it touches." Since a dishwasher is an appliance, we can infer that hard water has the ability to reduce the life of a dishwasher. Choice *B* is an opinion and not based on fact. Choice *C* attempts to discredit the passage, and Choice *D* might be true, but we have no evidence of these things in the passage.

4. D: The passage essentially states that coaches help kids on sports teams develop their skills and have fun. Choice *A* is an absolute phrase and is not true in every situation. Choice *B* can be implied, but the passage does not mention the stress of the kids. Choice *C* gives advice beyond the statements in the passage.

5. A: The passage says "when warm, humid air near the ground meets cold, dry air from above, a column of the warm air can be drawn up into the clouds." Thus, we can say that tornadoes are formed from a mixture of cold and warm air. Choice *B* is not necessarily the opinion of the passage. Choices *C* and *D* might be true. However, they are not mentioned in the passage.

6. C: As soon as you chew your food, it travels to the esophagus. Choices *A* and *B* might be correct, but there is no evidence mentioned in the passage. Choice *D* is incorrect; food travels to the stomach after the esophagus, not the other way around.

7. B: Jordan has to decide whether personal experience is more important than research experience. Choice *A* is incorrect; while climate change is a hot topic, we don't know from the passage that it's considered more important than food deserts. Choice *C* is incorrect; although this could happen, it's not the *best* inference of the passage. Choice *D* is incorrect because we know that Jordan is very diplomatic and would not choose a topic in an unfair way. This leaves Choice *B*, which is the best choice because Jordan would be considering the evidence each partner has to offer for the best possible presentation.

8. D: At Disney Cruise Lines, hot weather, lines, and crowds will be less extreme than at Disney parks. Choices *A* and *C* are incorrect and state the opposite sentiment of the passage. Choice *B* might be true, but the passage does not state an opinion of this.

9. B: Drowning happens silently and increases as summer approaches. Choice *A* is incorrect, as this expresses the opposite sentiment. Choice *C* is not mentioned in the passage. Choice *D* uses some of the same language of the passage, but the statement itself is incorrect.

10. C: Out of the last 50 years, the ten warmest years have occurred since 2000. The last 50 years is part of the 134 years that the passage mentions, so this is correct. Choice *A* has no evidence in the passage; neither does Choice *B*. Choice *D* is not mentioned in the passage.

11. A: The author's children's books are more popular than her historical fiction novel. This is expressed by the following statements: "it did not sell as many copies as her children's books." Choices *B, C,* and *D* are not sentiments expressed by the passage.

12. C: Choice *A* may be true; however, it isn't supported by the text and therefore, it is not the best answer. Choice *C* is also true and relies on the passage for its information. Choice *B* is incorrect, as there may have been other factors that led Hannah to quit smoking. Choice *D* is not supported by the passage.

13. C: Elephants lose more heat than anteaters. The passage states directly that "an elephant loses a great deal more heat than an anteater" because an elephant is larger. We have no way of knowing if Choices *A* or *B* are true according to the passage. Choice *D* is opposite of the correct answer.

14. D: Apartment B was in front of Apartment C. We know this because it states that Apartment A was in front of Apartment C. We also know that Apartment A was on top of Apartment B, which automatically makes Apartment B in front of Apartment C. The rest of the answer choices are logically incorrect based on the information given.

15. B: People known as "anti-Stratfordians" are people who believe that Shakespeare was not responsible for writing his own plays. Choice *B* is the only answer choice that recognizes this fact in the passage. All the other answer choices are incorrect.

16. A: Oliver prepares Nina a black bean burger with French fries. This meal does not include dairy or meat, so this should be an example of Oliver "consistently preparing Nina food that is suitable to her diet." Choice *B* includes meat. Choices *C* and *D* include dairy, cheese pizza and a milkshake.

17. D: In Samuel's observation, his best students have minimal interference from parents. Choices *A* and *B* are suggestions, and not inferences from the text. Choice *C* is opposite of the correct answer.

18. C: Cynthia doesn't eat meat due to preference as well as necessity. Choices *A* and *B* are incorrect because they don't paint the full picture of Cynthia's situation. Choice *D* is incorrect, as this information is not mentioned in the passage.

19. D: The way Samantha's life is set up currently, she has to decide between working at the restaurant and going to law school. Choice *C* is tempting. However, since Samantha cannot afford college on her own, getting a degree in nutrition is simply not an option in this world. Choices *A* and *B* are opinions, and we cannot make an educated guess with the information provided.

20. B: Barbara returned the $150 bike and bought the $75 bike. Look at the language. In the world of the passage, Barbara "always" went for cheaper machinery but "hardly ever" returned items. Therefore, there was a possibility Barbara would return the bike, but there wasn't a possibility she would keep the more expensive bike. With Choices *A* and *D*, we do not have sufficient information to make an educated guess.

Mechanical Comprehension

1. C: The answer is 1800 seconds:

$$(Desired\ Distance\ \text{in km} \times conversion\ factor\ (\text{m to km}))/current\ velocity\ \text{in m/s}$$

$$\left(45\ \text{km} \times \frac{1000\ \text{m}}{\text{km}}\right)\Big/25\frac{\text{m}}{\text{s}} = 1800\ \text{seconds}$$

2. D: The answer is 3.6 kilometers east of its point of departure. The ship is traveling faster than the current, so it will be east of the starting location. Its net forward velocity is 0.5 m/s which is 1.8 kilometers/hour, or 3.6 kilometers in two hours.

3. B: The answer is 2 m/s²:

$$a = \frac{\Delta v}{\Delta t} = \frac{132\frac{\text{km}}{\text{h}} - 60\frac{\text{km}}{\text{h}}}{10\ \text{s}}$$

$$\frac{72\frac{\text{km}}{\text{h}} \times 1000\frac{\text{m}}{\text{km}} \times \frac{\text{h}}{3600\ \text{s}}}{10\ \text{s}}$$

$$2\ \text{m/s}^2$$

4. C: The answer is 50,000 N. The equation $F = ma$ should be memorized. All of the values are given in the correct units (kilogram-meter-second) so plug them in to the equation:

$$F = 100,000 \text{ kg} \times 0.5 \frac{\text{m}}{\text{s}^2} = 50,000 \text{ N}$$

5. B: The answer is 200 kilograms. This is actually a trick question. The mass of the astronaut is the same everywhere (it is the weight that varies from planet to planet). The astronaut's mass in kilograms is calculated by dividing his weight on Earth by the acceleration of gravity on Earth:

$$\frac{1962 \text{ N}}{9.81 \text{ m/s}^2} = 200 \text{ kg}$$

6. A: The answer is that the upward velocity is at a maximum when it leaves the punter's toe. The acceleration due to gravity reduces the upward velocity every moment thereafter. The speed is the same at points A and E, but the velocity is different. At point E, the velocity has a maximum *negative* value.

7. C: The answer is her angular velocity increases from (a) to (b) as she pulls her arms in close to her body and reduces her moment of inertia.

8. B: The answer is 14 m/s. Remember that the cannonball at rest, y meters off the ground, has a potential energy of $PE = mgy$. As it falls, the potential energy is converted to kinetic energy until (at ground level) the kinetic energy is equal to the total original potential energy:

$$\frac{1}{2}mv^2 = mgy$$

Rearranged to solve for velocity, the equation is:

$$v = \sqrt{2gy}$$

This makes sense because all objects fall at the same rate, so the velocity *must* be independent of the mass; thus, Choice *D* is incorrect. Plugging the values into the equation, the result is 14 m/s. Remember, the way to figure this quickly is to have g equal to 10 m/s² rather than 9.81 m/s².

9. B: The answer is at Point C, the bottom of the arc. The bottom of the arc must have the greatest kinetic energy because it is the point with the least potential energy, since the total energy is equivalent to the sum of both potential and kinetic energy.

10. C: This question isn't difficult, but it must be read carefully:

A is wrong. Even though the total energy is at a maximum at Points A and E, it isn't equal at only those points. The total energy is the same at *all* points. *B* is wrong. The kinetic energy is at a maximum at C, but not the *total* energy. The correct answer is C. The total energy is conserved, so it's the same at *all* points on the arc. *D* is wrong. The motion of a pendulum is independent of the mass. Just like how all objects fall at the same rate, all pendulum bobs swing at the same rate, dependent on the length of the cord.

11. C: The answer is 10 kg × 2 m × 9.81 m/s². This is easy, but it must also be read carefully. Choice *D* is incorrect because it isn't necessary to know the ground elevation. The potential energy is measured *with respect* to the ground and the ground (or datum elevation) can be set to any arbitrary value.

12. B: The answer is 1 centimeter. Remember that the force (F) required to stretch a spring a certain amount (d) is given by the equation $F = kd$. Therefore, $k = F/d = 10$ N/0.5 cm $= 20$ N/cm. Doubling the weight to 20 N gives the deflection:

$$d = \frac{F}{k} = \frac{20 \text{ N}}{20 \frac{\text{N}}{\text{cm}}} = 1 \text{ cm}$$

All of the calculations can be bypassed by remembering that the relation between force and deflection is linear. This means that doubling the force doubles the deflection, as long as the spring isn't loaded past its elastic limit.

13. A: The answer is 981 N. The start-up and sliding friction forces are calculated in the same way: normal force (or weight) times the friction coefficient. The difference between the two coefficients is 0.1, so the difference in forces is:

$$0.1 \times 1000 \text{ kg} \times 9.81 \text{ m/s}^2 = 981 \text{ N}$$

14. C: The answer is 1.6 m². The pressure created by the load is:

$$\frac{50 \text{ N}}{0.1 \text{ m}^2} = 500 \frac{\text{N}}{\text{m}^2}$$

This pressure acts throughout the jack, including the large cylinder. Force is pressure times area, so the area equals pressure divided by force:

$$\frac{800 \text{ N}}{500 \frac{\text{N}}{\text{m}^2}} = 1.6 \text{ m}^2$$

15. D: The answer is $\frac{F}{A} \times (V_2 - V_1)$. Remember that the work for a piston expanding is pressure multiplied by change in volume:

$$W = P \times \Delta V$$

Because pressure is equal to force over an area, $P = \frac{F}{A}$, and change in volume is $V_2 - V_1$, the resulting equation for the work done is:

$$\frac{F}{A} \times (V_2 - V_1)$$

16. C: The answer is 500 J/s. Choice *A* is incorrect because kg $\times$ m/s² is an expression of force, not power. Choice *B* is incorrect because N·m is an expression of work, not power. That leaves Choices *C* and *D*, both of which are expressed in units of power: watts or joules/second. Using an approximate calculation (as suggested):

$$1000 \text{ kg} \times 10 \frac{\text{m}}{\text{s}^2} \times 30 \text{ m} = 300,000 \text{ N} \cdot \text{m}$$

Dividing the total work done by 10 minutes, or 600 seconds, gives the approximate power expended:

$$\frac{300,000 \text{ N} \cdot \text{m}}{600 \text{ s}} = 500 \text{ watts} = 500 \text{ J/s}$$

17. B: The answer is 12,000 Pa. The top of the block is under 12 meters of water:

$$P = Unit\ Weight\ of\ Water\ \times\ Depth\ of\ Block\ Under\ Water$$

$$P = 1000\,\frac{N}{m^3} \times 12\ meters = 12,000\,\frac{N}{m^2} = 12,000\ Pa$$

There are two "red herrings" here: Choice *C* of 14,000 Pa is the pressure acting on the *bottom* of the block (perhaps through the sand on the bottom of the bay). Choice *D* (that it can't be calculated without knowing the top area of the block) is also incorrect. The top area is needed to calculate the total *force* acting on the top of the block, not the pressure.

18. B: The answer is that the water level rises in B but not in A. Why? Because ice is not as dense as water, so a given mass of water has more volume in a solid state than in a liquid state. Thus, it floats. As the mass of ice in Basin A melts, its volume (as a liquid) is reduced. In the end, the water level doesn't change. The ice in Basin B isn't floating. It's perched on high ground in the center of the basin. When it melts, water is added to the basin and the water level rises.

19. B: The answer is 5000 seconds. The volume is $3 \times 25 \times 50 = 3750$ m³. The volume divided by the flow rate gives the time. Since the pump capacity is given in liters per second, it's easier to convert the volume to liters. One thousand liters equals a cubic meter:

$$Time = \frac{Volume}{Flow\ Rate}$$

$$t = \frac{3,750,000\ L}{750\ L/s} = 5000\ s$$

20. A: The answer is 30 m³/s. One of the few equations that must be memorized is $Q = v \times A$. The area of flow is 1 m × 10 m because only half the depth of the channel is full of water.

21. B: The answer is 735.75 N. This is a simple calculation:

$$\frac{9.81\ m}{s^2} \times 150\ kg \times 1.5\ m = 3\ m \times F$$

Rearranging the equation to solve for the unknown force, F, yields:

$$F = \frac{9.81\,\frac{m}{s^2} \times 150\ kg \times 1.5m}{3\ m} = \frac{2207.25\ N \cdot m}{3\ m} = 735.75\ N$$

22. D: The answer is 40 N. Use the equation $F = L \times r/R$. Note that for an axle with a given, set radius, the larger the radius of the wheel, the greater the mechanical advantage:

$$F = 600\ N \times \frac{20\ mm}{300\ mm} = 40\ N$$

23. D: The answer is counterclockwise at 30 RPM. The driver gear is turning clockwise, and the gear meshed with it turns counter to it. Because of the 2:1 gear ratio, every revolution of the driver gear causes half a revolution of the follower.

24. B: The answer is 30 RPM, counterclockwise (B). While meshed gears rotate in different directions, wheels linked by a belt turn in the same direction. This is true unless the belt is twisted, in which case they rotate in opposite directions. So, the twisted link between the upper two wheels causes the right-hand wheel to turn counterclockwise, and the bigger wheel at the bottom also rotates counterclockwise. Since it's twice as large as the upper wheel, it rotates with half the RPMs.

25. C: The answer is $F_1 = 1000$ N; $F_2 = 500$ N. In case (a), the fixed wheels only serve to change direction. They offer no mechanical advantage because the lever arm on each side of the axle is the same. In case (b), the lower moveable block provides a 2:1 mechanical advantage. A quick method for calculating the mechanical advantage is to count the number of lines supporting the moving block (there are two in this question). Note that there are no moving blocks in case (a).

26. C: According to the *ideal gas law* (*PV = nRT*), if volume is constant, the temperature is directly related to the pressure in a system. Therefore, if the pressure increases, the temperature will increase in direct proportion. Choice *A* would not be possible, since the system is closed and a change is occurring, so the temperature will change. Choice *B* incorrectly exhibits an inverse relationship between pressure and temperature, or *P = 1/T*. Choice *D* is incorrect, because the problem addresses all the variables in the *ideal gas law*.

27. A: According to Ohm's Law: *V = IR*, so using the given variables: 3.0 V = I × 6.0 Ω

Solving for I: I = 3.0 V/6.0 Ω = 0.5 A

Choice *B* incorporates a miscalculation in the equation by multiplying 3.0 V by 6.0 Ω, rather than dividing these values. Choices *C* and *D* are labeled with the wrong units; Joules measure energy, not current.

28. C: The weight of an object is equal to the mass of the object multiplied by gravity. According to Newton's Second Law of Motion, *F = m × a*. Weight is the force resulting from a given situation, so the mass of the object needs to be multiplied by the acceleration of gravity on Earth: *W = m × g*. Choice *A* is incorrect because, according to Newton's first law, all objects exert some force on each other, based on their distance from each other and their masses. This is seen in planets, which affect each other's paths and those of their moons. Choice *B* is incorrect because an object in motion or at rest can have inertia; inertia is the resistance of a physical object to change its state of motion. Choice *D* is incorrect because of the reasons given for Choices *A* and *B*.

29. C: In any system, the total mechanical energy is the sum of the potential energy and the kinetic energy. Either value could be zero but it still must be included in the total. Choices *A* and *B* only give the total potential or kinetic energy, respectively. Choice *D* is false because energy is neither created nor destroyed.

30. B: To solve this, the number of moles of NaCl needs to be calculated:

First, to find the mass of NaCl, the mass of each of the molecule's atoms is added together as follows:

$$23.0 \text{ g Na} + 35.5 \text{ g Cl} = 58.5 \text{ g NaCl}$$

Next, the given mass of the substance is multiplied by one mole per total mass of the substance:

$$4.0 \text{ g NaCl} \times \frac{1 \text{ mol NaCl}}{58.5 \text{ g NaCl}} = 0.068 \text{ mol NaCl}$$

Finally, the moles are divided by the number of liters of the solution to find the molarity:

$$\frac{0.068 \text{ mol NaCl}}{0.120 \text{ L}} = 0.57 \text{ M NaCl}$$

Choice *A* incorporates a miscalculation for the molar mass of NaCl, and Choices *C* and *D* both incorporate a miscalculation, so they are incorrect by a factor of 10.

Dear OAR Test Taker,

We would like to start by thanking you for purchasing this study guide for your OAR exam. We hope that we exceeded your expectations.

Our goal in creating this study guide was to cover all of the topics that you will see on the test. We also strove to make our practice questions as similar as possible to what you will encounter on test day. With that being said, if you found something that you feel was not up to your standards, please send us an email and let us know.

We would also like to let you know about other books in our catalog that may interest you.

ASVAB

This can be found on Amazon: amazon.com/dp/1628457759

ASTB

amazon.com/dp/1637750277

AFOQT

amazon.com/dp/1637752938

SIFT

amazon.com/dp/1628458585

We have study guides in a wide variety of fields. If the one you are looking for isn't listed above, then try searching for it on Amazon or send us an email.

Thanks Again and Happy Testing!
Product Development Team
info@studyguideteam.com

FREE Test Taking Tips DVD Offer

To help us better serve you, we have developed a Test Taking Tips DVD that we would like to give you for FREE. **This DVD covers world-class test taking tips that you can use to be even more successful when you are taking your test.**

All that we ask is that you email us your feedback about your study guide. Please let us know what you thought about it – whether that is good, bad or indifferent.

To get your **FREE Test Taking Tips DVD**, email freedvd@studyguideteam.com with "FREE DVD" in the subject line and the following information in the body of the email:

      a. The title of your study guide.

      b. Your product rating on a scale of 1-5, with 5 being the highest rating.

      c. Your feedback about the study guide. What did you think of it?

      d. Your full name and shipping address to send your free DVD.

If you have any questions or concerns, please don't hesitate to contact us at freedvd@studyguideteam.com.

Thanks again!

Made in United States
Troutdale, OR
06/16/2024